The
Dark Energy
Survey

The Story of a
Cosmological Experiment

The
Dark Energy
Survey

The Story of a
Cosmological Experiment

Editors

Ofer Lahav
University College London, UK

Lucy Calder
University College London, UK

Julian Mayers
University of Sussex, UK

Josh Frieman
Fermilab, USA & University of Chicago, USA

World Scientific

NEW JERSEY · LONDON · SINGAPORE · BEIJING · SHANGHAI · HONG KONG · TAIPEI · CHENNAI · TOKYO

Published by

World Scientific Publishing Europe Ltd.

57 Shelton Street, Covent Garden, London WC2H 9HE

Head office: 5 Toh Tuck Link, Singapore 596224

USA office: 27 Warren Street, Suite 401-402, Hackensack, NJ 07601

Library of Congress Cataloging-in-Publication Data
Names: Lahav, O. (Ofer), editor. | Calder, Lucy, editor. | Mayers, Julian, editor. |
    Frieman, Joshua A., 1959– editor.
Title: The Dark Energy Survey : the story of a cosmological experiment / editors, Ofer Lahav,
    University College London, UK, Lucy Calder, University College London, UK, Julian Mayers,
    University of Sussex, UK, Josh Frieman, Fermilab, USA & University of Chicago, USA.
Description: Hackensack, NJ : World Scientific, [2020] |
    Includes bibliographical references and index.
Identifiers: LCCN 2020013218 | ISBN 9781786348357 (hardcover) |
    ISBN 9781786348364 (ebook) | ISBN 9781786348371 (ebook other)
Subjects: LCSH: Dark Energy Survey. | Dark energy (Astronomy)--Research. |
    Cosmology--Research.
Classification: LCC QB791.3 .D36686 2020 | DDC 523.01--dc23
LC record available at https://lccn.loc.gov/2020013218

British Library Cataloguing-in-Publication Data
A catalogue record for this book is available from the British Library.

First published 2021 (hardcover)
Reprinted 2022 (in paperback edition)
ISBN 978-1-80061-166-5 (pbk)

For any available supplementary material, please visit
https://www.worldscientific.com/worldscibooks/10.1142/Q0247#t=suppl

Desk Editors: Britta Ramaraj/Michael Beale/Shi Ying Koe

Typeset by Diacritech Technologies Pvt. Ltd.
Chennai - 600106, India

Contents

Foreword

In the early 1960s, cosmologists were debating whether a Big Bang ever happened. Now we can, with confidence, trace cosmic history back 13.8 billion years, to a moment only a billionth of a second after the Big Bang. Astronomers have pinned down our universe's expansion rate, the mean density of its main constituents and other key numbers, to a precision of 1% or 2%. We're learning how cosmic structures – galaxies and clusters – emerged and evolved from amorphous dense beginnings. When the history of science in recent decades gets written, that amazing progress will be acclaimed as one of its greatest triumphs.

All advances open up new questions that couldn't have previously even been posed. So we are now tackling fundamental questions such as: Why does the universe expand in the way it does? Why does it contain the particular mixture of atoms and radiation that we measure? Answers to such questions lie buried deep in this first billionth of a second. We still have no experimental foothold to give us guidance on the physics of those very early times: everything was then squeezed to super-nuclear densities; particles were moving around with higher thermal energy than experimenters can re-create even in the biggest particle accelerator.

The huge progress already achieved by the global cosmological community is owed to a sustained and cumulative effort, entailing multi-waveband observations using ever-improving instrumentation, on the ground and in space. The cosmic microwave background (CMB), discovered in the mid-1960s, has been mapped in exquisite detail, yielding evidence of the spatial fluctuations 300,000 years after the Big Bang.

The data indicates that cosmic structures are moulded by the gravitational effects of dark matter, which contributes five times as much mass as the stars and gas that we observe. The discovery of dark matter was a surprise – it's thought to consist of a swarm of uncharged heavy particles made in the Big Bang. We still don't have any evidence concerning their properties but there are hopes that clues will emerge from future experiments.

In 1998 there was a bigger and more mysterious surprise. The expansion of the universe was found to be accelerating: some unknown mechanism was 'pushing' galaxies away from one another by increasing the space between them. The exact nature of what is happening is an enigma. Is it Einstein's 'cosmological constant'? Is it a new force that manifests energy latent in 'empty' space? Or are we being fooled because Einstein's equations don't hold on cosmic scales?

Dark energy offers the most spectacular example of how the cosmos provides an arena where we can discover and study fundamental processes that can't be accessed by laboratory measurements. Indeed, most theorists suspect that we won't understand dark energy until we have a theory for the microstructure of space and time. Many theories suggest that space itself isn't infinitely divisible: it has an inherent 'graininess' or texture. But this manifests itself on a scale a billion billion times smaller than an atom and must await a huge leap in our understanding.

Just as the accelerator experiments conducted by particle physicists now involve big teams, there's been a trend for astronomy and cosmology to involve large collaborations – deploying a range of expertise to build expensive instruments, carry out lengthy observing programmes, and analyze large volumes of data.

One such enterprise, the Dark Energy Survey (DES), involved a six-year programme during which 300 million galaxies and thousands of supernovae have been observed – to map out how they are clustered, and to learn how their morphologies have evolved. The prime goal is to infer whether (and by how much) the influence of dark energy has changed during the expansion of the universe. So far, based on the first year of data, the survey has confirmed remarkably accurately the model of cold dark matter and a cosmological constant. The survey has, as a bonus, offered new insights into galaxies, supernovae, stellar evolution, solar system objects and the nature of gravitational wave events.

DES required the multi-year commitment and collaboration of several hundred scientists and technicians around the world, with diverse expertise. Many of them contribute chapters to this book, describing what the survey has revealed and placing it in context.

What makes this book especially fascinating – and a valuable contribution not only to science itself, but to the subject's sociology and history – is that it doesn't just report the results of the project (which will appear in the traditional journals and archival literature). There are chapters that chronicle how the project was conceived, and the years of effort required to forge a collaboration, agree a design, gain the necessary funding, and build the instruments.

It also offers insights into what it's like to spend part of one's scientific career in a collaboration – the frustrations and tensions, and how they are perceived both by senior and by young team members. These people can all now derive satisfaction from having contributed major advances to our knowledge of a key cosmic mystery. It adds to our admiration when we read about the pressures, and the years of prior planning that led to these successes.

Those who edited this book, and all who wrote chapters in it, deserve our thanks for producing an engrossing and enlightening account of how discoveries are made.

Martin J. Rees
Astronomer Royal
February 2019

Preface

The aim of the book is to explain how the Dark Energy Survey (DES) came into being, to document its challenges and achievements and elucidate, for a non-specialist reader, the science involved. Such a book about a large international cosmology project has not been written from the 'inside' before. While the main focus is on the science and technology, we also convey a sense of what it is like to be part of DES from a personal perspective. Scientific collaborations are human endeavours, and we wish to get across some understanding of the kind of every day challenges faced by scientists. For example, what is it like to work nights at the telescope and what kind of problems arise there? What were the challenges involved with designing, building and transporting the camera? What kind of unforeseen problems have come up and how did we overcome them? And in the final part of the book, we reflect on DES from the 'outside' with perspectives from an anthropologist, a philosopher, artists and a poet.

The book is divided into four sections: Building DES, Dark Energy Science, Non-Dark Energy Science and Reflections & Outlook. Listed authors of chapters are usually those who served as working group or project coordinators, as recognition for their contribution to DES over the years (not necessarily for writing the chapter). Parts of some chapters are based on interviews with them, or on their published work.

We also include points of view from many other people who have worked in DES and in doing so we hope to build up a bigger picture of how a huge science project is done: going from an initial idea, to building an instrument, to writing the code to analyze data and finally to publishing the science results. All with the goal of understanding the universe a little better.

Acknowledgements

As well as the named authors, many of whom also checked other chapters, a large number of people have made contributions to individual chapters, or have taken the time to read through them and make constructive comments. The editors would like to thank the following: Rachael Amaro, Eric Baxter, Rutu Das, Richard Ellis, Enrique Fernandez, Agnès Ferté, David Finley, Pablo Fosalba, Enrique Gaztanaga, Tommaso Giannantonio, Satya Gontcho A Gontcho, Davide Gualdi, Ben Henghes, Niall Jeffrey, Elizabeth Johnson, Donnacha Kirk, Pablo Lemos, Maria Marcha, Rebecca Martin, Ramon Miquel, Jeremy Mould, Constantina Nicolaou, Nacho Sevilla Noarbe, Antonella Palmese, John Peacock, Eduardo Rozo, Eli Rykoff, Avila Santiago, Dan Scolnic, Erin Sheldon, Daniel Thomas, Lorne Whiteway and Joe Zuntz.

Funding for the Dark Energy Survey (DES) projects has been provided by the U.S. Department of Energy, the U.S. National Science Foundation, the Ministry of Science and Education of Spain, the Science and Technology Facilities Council of the United Kingdom, the Higher Education Funding Council for England, the National Center for Supercomputing Applications at the University of Illinois at Urbana-Champaign, the Kavli Institute of Cosmological Physics at the University of Chicago, the Center for Cosmology and Astro-Particle Physics at the Ohio State University, the Mitchell Institute for Fundamental Physics and Astronomy at Texas A&M University, Financiadora de Estudos e Projetos, Fundação Carlos Chagas Filho de Amparo à Pesquisa do Estado do Rio de Janeiro, Conselho Nacional de Desenvolvimento Científico e Tecnológico and the Ministério da Ciência, Tecnologia e Inovação, the Deutsche Forschungsgemeinschaft and the Collaborating Institutions in the Dark Energy Survey (DES).

The Collaborating Institutions are Argonne National Laboratory, the University of California at Santa Cruz, the University of Cambridge,

Centro de Investigaciones Energéticas, Medioambientales y Tecnológicas-Madrid, the University of Chicago, University College London, the DES-Brazil Consortium, the University of Edinburgh, the Eidgenössische Technische Hochschule (ETH) Zürich, Fermi National Accelerator Laboratory, the University of Illinois at Urbana-Champaign, the Institut de Ciències de l'Espai (IEEC/CSIC), the Institut de Física d'Altes Energies, Lawrence Berkeley National Laboratory, the Ludwig-Maximilians Universität München and the associated Excellence Cluster Universe, the University of Michigan, the National Optical Astronomy Observatory, the University of Nottingham, The Ohio State University, the University of Pennsylvania, the University of Portsmouth, SLAC National Accelerator Laboratory, Stanford University, the University of Sussex, Texas A&M University and the OzDES Membership Consortium.

Based in part on observations at Cerro Tololo Inter-American Observatory, National Optical Astronomy Observatory, which is operated by the Association of Universities for Research in Astronomy (AURA) under a cooperative agreement with the National Science Foundation.

The DES data management system is supported by the National Science Foundation under Grant Numbers AST-1138766 and AST-1536171. The DES participants from Spanish institutions are partially supported by MINECO under grants AYA2015-71825, ESP2015-66861, FPA2015-68048, SEV-2016-0588, SEV-2016-0597 and MDM-2015-0509, some of which include European Regional Development Fund (ERDF) funds from the European Union. IFAE (the Institute for High Energy Physics) is partially funded by the CERCA (Research Centres of Catalonia) program of the Generalitat de Catalunya. Research leading to these results has received funding from the European Research Council under the European Union's Seventh Framework Program (FP7/2007-2013) including ERC grant agreements 240672, 291329 and 306478. We acknowledge support from the Australian Research Council Centre of Excellence for All-sky Astrophysics (CAASTRO), through project number CE110001020, and the Brazilian Instituto Nacional de Ciência e Tecnologia (INCT) e-Universe (CNPq grant 465376/2014-2).

Introduction to the Dark Energy Survey Project and Science: What Have We Learned So Far?

Josh Frieman & Ofer Lahav

A 'Chapter Minus One'? 'One before zero' suggests 'before we begin'. But it also hints towards the endpoint - the equation of state parameter, w, that the Dark Energy Survey (DES) may confirm as w = −1 (Einstein's cosmological constant) or as something different. In this chapter, we review the long journey since 2013, from the initial idea for a survey, to obtaining funding, building the instrument, observation with the Blanco Telescope, processing the data and extracting the science. The main conclusion from a multi-probe analysis of just the first year of DES data is that dark energy is consistent with a cosmological constant, contributing at present about 70% of the total mass−energy density of the universe. But at this time (mid-2019) the bulk of the data remains to be analyzed, and possibilities for surprises remain. Other exciting results from early DES data include the discoveries of new Milky Way dwarf satellites, of new objects in the outer solar system, and of an optical flash from the first binary neutron star gravitational wave event.

−1.1 What is dark energy?

In the now-standard cosmological model, the universe originated in a hot, dense state 13.8 billion years ago in an event called the Big Bang, in which matter, energy and spacetime itself came into existence. This model is based on Einstein's General Theory of Relativity (1915), the theory of gravity, the dominant force acting on the cosmos. It also rests on the Cosmological Principle, which states that the universe is homogeneous and isotropic: to first approximation, it looks the same at every point and in every direction, when averaged over suitably large scales, and that should hold true for observers in other galaxies as well.

Over the last several decades, a large body of astronomical observations has brought us to a detailed understanding of the evolution of the universe since the Big Bang, the Lambda Cold Dark Matter (ΛCDM) theory. In this theory, the rich hierarchy of observed large-scale structure, from galaxies to clusters to the 'cosmic web', arose from the action of gravity on initially small, quantum mechanical fluctuations in the density of the universe generated during a period of rapid expansion – known as Inflation – a tiny fraction of a second after the Big Bang. Moreover, in this theory an unseen dark energy makes up approximately 70% of the mass–energy content of the universe today. It does not appear to interact directly with matter or radiation, but dark energy is gravitationally repulsive, causing galaxies to move away from each other – and thus the universe to expand – at an accelerating rate. In its simplest form, dark energy is equivalent to Einstein's cosmological constant, Λ. Another 25% of the universe is in another invisible form, dark matter, whose gravity binds galaxies and clusters of galaxies together and which is thought to be composed of yet undiscovered, slowly moving (hence 'cold') elementary particles. These two dark components together make up 95% of the universe, but they have never been observed in the laboratory and their physical natures remain unknown. In this astonishing picture, familiar atomic matter – from which stars and planets are made – accounts for only 5% of the mass in the universe.

Evidence for cosmic acceleration was established in 1998 by two teams of astronomers (Riess et al., 1998; Perlmutter et al., 1999) using type Ia supernovae (SN) as 'standard candles' to chart the history of cosmic expansion. Earlier in the 1990s, galaxy clustering measurements had indicated the possibility of a cosmological constant when combined with the assumption that three-dimensional space is flat (Euclidean), as predicted by Inflation and later confirmed by cosmic microwave background (CMB) anisotropy measurements. In the twenty years since, the evidence for accelerated expansion has been overwhelmingly supported by a host of other cosmological measurements.

Historically, dark energy was introduced into physics by Einstein, who added what he termed a 'cosmological constant', Λ, into his field equations of General Relativity in 1917,

$$R_{\mu\nu} - \frac{1}{2} R g_{\mu\nu} + \Lambda g_{\mu\nu} = \frac{8\pi G}{c^4} T_{\mu\nu}.$$

Here, $R_{\mu\nu}$ and R are the Ricci tensor and scalar curvature of spacetime, respectively; $g_{\mu\nu}$ is the metric tensor; $T_{\mu\nu}$ is the energy-momentum tensor of all matter and radiation; G is Newton's gravitational constant; and c is the speed of light. One way to think about these complicated equations is in the words of John Archibald Wheeler: 'matter tells spacetime how to curve; curved spacetime tells matter how to move.'

Einstein introduced the Λ term in order to obtain static solutions to his equations, since at that time the universe was thought to be eternal and unchanging. However, he abandoned it in 1931, by which time evidence that the universe is expanding had accumulated through the work of Slipher, Lemaitre, Hubble and Humason. Gamow later recalled that, 'When I was discussing cosmological problems with Einstein, he remarked that the introduction of the cosmological term was the biggest blunder he ever made in his life.' Whether or not Einstein really thought this is debatable, and over the course of the twentieth century the cosmological constant periodically came back into vogue. One may argue that Einstein missed a number of opportunities: to predict the expansion/contraction of the universe without Λ; to notice that the static solution he engineered is unstable, so that a universe with Λ would also expand/contract; and to see Λ as a free parameter, which could also lead to an accelerating universe. By the same token, he should post-humously get credit for introducing a term that is now the focus of astronomical projects that collectively cost billions of dollars, with thousands of scientists spending significant parts of their careers trying to understand it.

Einstein introduced Λ on the left hand side of the field equations and considered it a constant of nature associated with the geometry or curvature of spacetime. Alternatively, it can be placed on the right-hand side and considered a contribution to the stress-energy tensor $T_{\mu\nu}$. In this view, if we remove all matter and radiation from the universe, what remains is the stress-energy of empty space, so the 'vacuum' energy density is related to the cosmological constant by $\rho_{vac} = \Lambda c^2/8\pi G$. Unfortunately, the theoretical prediction for the vacuum energy density in quantum field theory is many orders of magnitude larger than the observed value – a figure of 10^{120} is often quoted! This discrepancy is known as the cosmological constant problem (Weinberg, 1989) and has been called 'the worst theoretical prediction in the history of physics' (Hobson et al., 2006). In the absence of a physical explanation, some cosmologists have suggested an 'anthropic' explanation for the small observed value of Λ: if it were too large, it would have prevented gravity from forming large galaxies and life would never have emerged. This reasoning suggests or pre-supposes that there exist a large number of independent universes (the 'multiverse') in which Λ and other cosmological parameters take on all possible values. In this picture, we happen to live in one of the universes with small enough Λ to be habitable. It is the nature of this particular universe that the Dark Energy Survey (DES) was set up to determine.

An intuitive way to picture the effect of dark energy is that it gives rise to a repulsive linear force, uniformly filling all of space and opposing the attractive inverse-squared gravitational force. In the weak-field limit of General Relativity, the equation of motion for a point particle in the gravitational field of a mass M is

$$\frac{d^2r}{dt^2} = -\frac{GM}{r^2} + \frac{c^2}{3}\Lambda r.$$

Two hundred and fifty years before Einstein, a linear force had in fact been discussed by Newton in his *Principia*, in addition to the more famous inverse-square law.

If dark energy is the energy of the vacuum then, unlike matter and radiation, its energy density doesn't dilute as the universe expands. On the other hand, dark energy could be associated with, say, a time-varying scalar field, in which case its energy density would evolve in time. This time evolution is determined by the equation of state parameter of dark energy, defined as the ratio of its pressure p to its energy density ρ,

$$w = p/\rho c^2.$$

The expanding universe can be described by a single function of time, the cosmic scale factor, $a(t)$, which determines how the distances between galaxies change in time. The local conservation of energy–momentum implies that the dark energy density varies with the scale factor a as

$$\rho \propto a^{-3(1+w)}.$$

For vacuum energy, the equation of state parameter is $w = -1$, and the dark energy density does not change with time, consistent with its identification with Einstein's cosmological constant. In other models for dark energy, w may differ from -1 and itself evolve in time, leading to an evolving dark energy density.

In talking about the evidence for dark energy and dark matter, as we do throughout much of this book, we should be careful in stating our assumptions. In fact, what has been shown is that a wealth of cosmological observations are inconsistent with Einstein's equations of General Relativity in the absence of dark energy (or Λ) and dark matter. Introducing dark energy and dark matter resolves the discrepancy, but it is also theoretically possible that the discrepancy could be resolved by replacing General Relativity with an alternative theory of gravity, with no dark components. This logical choice reminds us of two cases in the history of studies of our Solar System. Anomalies in the orbit of Uranus were explained by hypothesizing a previously unseen (aka 'dark') planet, Neptune, within Newtonian gravity, which was subsequently discovered. On the other hand, anomalies in the perihelion of Mercury were successfully explained by invoking a new theory of gravitation beyond Newton, Einstein's General Relativity, instead of by the 'dark' planet Vulcan. History does not provide a definitive guide to choosing between dark stuff or a new theory of gravity in explaining cosmological observations.

So an alternative explanation for the mystery of accelerated expansion may be that General Relativity requires modification at large cosmological scales. Part of the motivation for DES is to combine measurements in a way that could potentially distinguish dark energy from modified gravity as the origin of cosmic acceleration. At present, we do not have a well under-stood theoretical framework that successfully combines General Relativity with quantum mechanics. Ultimately, a full explanation of the accelerated expansion of the universe may require a revolution in our understanding of fundamental physics.

−1.2 The Dark Energy Survey

DES, as its name suggests, has been specially designed to investigate the physical nature of dark energy and the origin of cosmic acceleration, currently one of the biggest mysteries in science. It comprises a wide-field astronomical survey of 5000 square degrees of the southern sky and a twenty-seven square degree time-domain survey to discover and measure supernovae (SN). DES collaboration designed and built the 570 megapixel, three square degree field-of-view the Dark Energy Camera or DECam, which was installed on the 4-m Victor M. Blanco Telescope at the Cerro Tololo Inter-American Observatory (CTIO) in the Chilean Andes. The collaboration involves more than 400 scientists from twenty-six institutions in the United States, the United Kingdom, Spain, Brazil, Germany, Switzerland and Australia. After nearly a decade of planning and preparation, the camera saw first light in September 2012, and we began survey observations on August 31, 2013. The survey ran from August to February for five years, with an additional, final half-season that ended in early January 2019.[1] When fully processed, the survey data will yield a high-resolution map of over 300 million galaxies and a catalogue of thousands of SN.

DES collaboration aims to measure the effects of dark energy on the expansion history of the universe and on the growth of large-scale structure using four complementary probes: galaxy clusters, weak gravitational lensing, type Ia SN and baryon acoustic oscillations (BAO). Specifically, we aim to determine the dark energy equation of state parameter, w, and its time evolution to high precision, along with other key cosmological parameters.

DECam can detect light over the full range of the optical spectrum, from about 4000 to 10,650 Å (broader than the wavelength range our eyes

---

[1] We will refer to DES as a 'six-year' survey throughout except where referring to the original 'five-year' plan.

are sensitive to) but was designed to be especially sensitive to red light, so that it can detect galaxies at great distances. For DES, we recorded DECam images using five optical/near-infrared filters (*grizY*), each sensitive to a different wavelength range of light. Comparing the brightness of a galaxy in the different filters enables us to measure its colour and thereby estimate its redshift, that is, how much its light has been shifted to the red end of the spectrum by the expansion of the universe. While this method allows us to measure the approximate distances to large numbers of galaxies in a short period of time, it is essential to calibrate these 'photometric redshifts' using spectroscopic galaxy samples, for which precise redshifts have been measured. DES survey area (footprint) on the sky was chosen to overlap with previous spectroscopic surveys, including part of 'stripe 82', a strip of sky along the south celestial equator that was observed by the Sloan Digital Sky Survey (SDSS). DES footprint was also designed to overlap the survey area of the South Pole Telescope (SPT), which detected large numbers of galaxy clusters through the Sunyaev-Zel'dovich effect (see Chapter 12), a distortion in the CMB radiation. Finally, the survey area was chosen to avoid the plane of the Milky Way, where extinction by dust obscures the light from external galaxies. The resulting DES footprint is shown in Figure −1.1.

For the wide-field, 5000 square degree survey, individual exposure times for each image were ninety seconds for *griz* and forty-five seconds for *Y*; each part of the footprint was imaged ten times over the course of the survey, with individual exposures added together to make deeper images enabling us to detect fainter galaxies. DES imaged about an eighth of the sky to a depth of approximately twenty-fourth magnitude (about fifteen million times fainter than the dimmest star that can be seen with the naked eye), and the quality of the data is excellent – the typical blurring of the images due to the camera, the telescope optics and the atmosphere is about 0.9 arcseconds. On a typical observing night, the camera took about 200 images; since each DECam digital image is a gigabyte in size, DES typically collected about 200 gigabytes of data on clear nights, a high rate for an astronomy experiment. The data was transferred from the telescope to large computing clusters at the National Center for Supercomputer Applications (NCSA) at the University of Illinois, at Fermilab outside Chicago and at National Optical Astronomy Observatory (NOAO) in

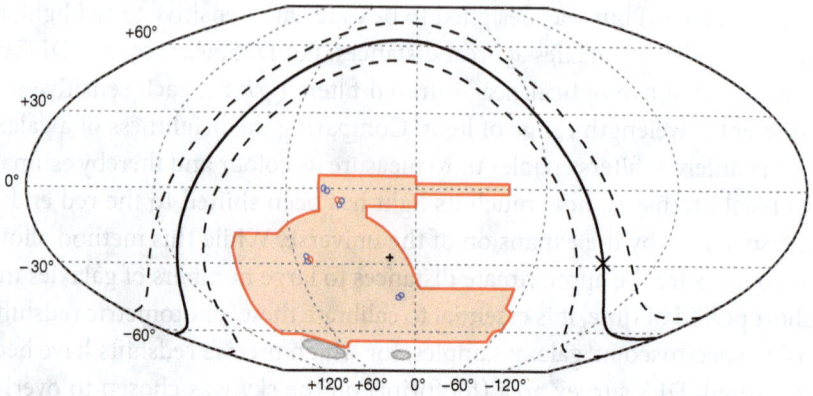

Fig. −1.1 The red shaded area shows the 5000 square degree DES footprint in celestial coordinates (Right Ascension increasing right to left, Declination up-down). The ten small circles indicate the supernova survey fields (eight shallow fields in blue, two deep fields in red). The 'turret of the tank' runs along the south celestial equator at Declination = 0 degrees, and the long base of the footprint between Declination −40 and −60 degrees spans the 2500 square degree region covered by the South Pole Telescope (SPT) Sunyaev-Zel'dovich cluster survey. The band encircling the footprint indicates the plane of the Milky Way Galaxy. The Galactic center (large 'X') and South Galactic pole (small '+') are also marked. The Large and Small Magellanic Clouds, dwarf galaxies that are satellites of the Milky Way, are indicated in grey. Dark Energy Survey Collaboration (2018b).

Tucson, Arizona. The Dark Energy Survey Data Management (DESDM) team led by NCSA has provided the expertise, infrastructure and computational power needed to store the data and process it into useful images and catalogues that can be analyzed by scientists.

In addition to serving the scientific interests of the collaboration, DES data processed by DESDM are being made publicly available on a regular basis and can be downloaded by anyone with an Internet connection. This practice follows on from the success of the SDSS and other surveys, which have become essential resources for the astronomical community. By far the majority of papers written using SDSS data have been authored by people outside the SDSS collaboration. Enabling open access to this data will, we hope, increase the scientific value of the survey far beyond the scope of its original design. It also helps to democratize access to information, enabling students and researchers around the world, including those unable to afford time on large observing facilities, to carry out original research.

Furthermore, DES scientists are working closely with data sets and scientists from other surveys and observatories, including the SPT, the VISTA Hemisphere Survey, the Planck space observatory, the Gemini Observatory, the Atacama Cosmology Telescope and the Anglo-Australian Telescope to enhance the scientific reach of the data. Comparing, or 'cross-correlating' DES data with data from other surveys allows us to obtain new information and make new discoveries.

−1.3 The context of dark energy observations

Following the dramatic conclusions from the SN observations announced in 1998, a number of new surveys were planned to investigate dark energy. Tackling such a fundamental issue requires a combination of strategies involving observations of light over a wide range of wavelengths. Beyond optical and infrared wavebands, key among these are X-rays, radio waves and the CMB, for probing the universe at the earliest times possible. For example, the Planck satellite, launched in 2009 by the European Space Agency (ESA), mapped the fluctuations in temperature of the CMB to an unsurpassed degree of sensitivity (Planck Collaboration et al., 2018).

In 2006, the US Dark Energy Task Force (DETF) report (Albrecht et al., 2006) classified DESs into numbered stages: Stage II projects were on-going at that time; Stage III were near-future, intermediate-scale projects and Stage IV were larger-scale projects in the longer-term future. These projects can be further divided into ground-based and space-based surveys. DES is a quintessential Stage III ground-based survey, which was still in the proposal stage at the time of the DETF report. Stage IV ground-based projects now include the Dark Energy Spectroscopic Instrument (DESI), due to start collecting data at Kitt Peak National Observatory in 2019, and the Large Synoptic Survey Telescope (LSST, recently renamed the 'Vera Rubin Observatory's of Legacy Survey of Space and Time'), under construction on Cerro Pachon in Chile. Space-based missions have the advantage of not having to observe through Earth's atmosphere, which blurs the light and absorbs infared light. ESA is currently building a space mission called Euclid, planned for launch in 2022, which will aim to measure the redshifts and shapes of galaxies up to ten billion years into the past. Even farther ahead, the Wide Field Infrared Survey Telescope (WFIRST, recently renamed the 'Nancy Grace Roman Space Telescope') is a space-based project led by the US National Aeronautics and Space Agency (NASA), which will investigate the expansion history of the universe at near-infrared wavelengths. Furthermore, the detection of gravitational waves by the Laser Interferometer Gravitational-Wave Observatory (LIGO/Virgo) (Abbott et al., 2016) has opened up a completely new way to study the universe. Astronomy is more than ever an exciting area of science to be involved in, and the scientists working in DES are learning invaluable skills to take forward to the next generation of large-scale surveys (see Chapter 27 for an outlook on future projects).

−1.4 Four probes of dark energy

DES was designed to carry out four complementary observational techniques to investigate dark energy. These approaches are type Ia SN, large-scale galaxy clustering including baryon acousitc oscillations (BAO), galaxy clusters and weak gravitational lensing. These were identified by the DETF report as the four most promising and powerful techniques for understanding dark energy; since the techniques are sensitive to cosmological parameters in different ways, we can use them in combination to maximize our knowledge, and in the process control systematic errors. In particular, these techniques are sensitive in different degrees to the expansion history and to the growth of structure in the universe. Since dark energy and modified gravity typically affect expansion history and structure growth in different ways, measuring them both offers the possibility of distinguishing between the two causes of cosmic acceleration.

First, DES discovered and measured several thousand type Ia SN to measure the expansion rate over cosmic time (compared to several tens of SN discovered in the late 1990s to establish the accelerating expansion). Exploding stars of this type are understood to originate in binary systems, in which at least one of the two companion stars is a white dwarf. They are observed to all have nearly the same absolute brightness, or luminosity, when they reach their peak light, so by comparing the apparent brightnesses of different SN we can determine their relative distances. We can also measure the relative size of the universe at the time of the explosion by measuring how much the light wave from the SN has been stretched (redshifted) as it has travelled to Earth over billions of years: measurement of redshift thus tells us how much the universe has expanded since the star exploded. A SN becomes as bright as an entire galaxy of billions of stars and then begins to fade away within a matter of weeks, so DES observed the same ten patches of sky every six to seven nights in order to spot them, measure their evolving brightness, and infer their distances. Using other telescopes to measure the spectra of the SN or of their host galaxies, in particular through the OzDES project on the Anglo-Australian Telescope, we determined accurate redshifts of many ofclumpiness in the mass distribution today DES-SN (about 500 SN spectra and several thousand SN host-galaxy spectra). Other SN redshifts are estimated from the photometric redshifts of their host galaxies.

Second, we are measuring the spatial distribution of galaxies across the cosmos to infer what are called BAO. Until approximately 370,000 years after the Big Bang, the (non-dark) universe was a sea of frequently colliding protons, electrons and photons, forming an opaque, ionized plasma. Gravity pulled the protons and electrons together, while the pressure of the energetic photons of radiation acted to push them apart. This competition between gravity and pressure created a series of sound waves ('acoustic oscillations') in the plasma. As the universe expanded and cooled, the photons lost energy and became less effective at ionizing atoms, until gradually all the protons and electrons combined to form neutral hydrogen. Thereafter, the universe became transparent to photons, which travelled without further interacting with matter. The distance travelled by the sound waves up to that point became imprinted in the spatial distribution of matter, which means that today there is a slight tendency for pairs of galaxies to be separated by about 480 million light years (compared to neighbouring separations). Comparing this scale to how much galaxies *appear* to be separated on the sky, that is, their angular separation allows us to work out how far away they are. Comparing this distance to the redshift of their light, just as we do for SN, tells us about the expansion history of the universe. This helps us distinguish between different theories of dark energy.

The third technique is to count the number of galaxy clusters of a given mass within a given volume of the universe to determine how this quantity has changed over time. Clusters of galaxies form when clumps of dark matter collapse and merge together under gravity: the clumpier the mass distribution at a given time, the more clusters will have formed. The evolution of cosmic clumpiness is in turn determined by the competition between the gravitational attraction of dark matter and the gravitational repulsion of dark energy. Cluster counts thus probe dark energy through both the expansion history (which determines the cosmic volume element) and the growth of structure. The main challenge when interpreting cluster counts is that the mass of a cluster (which is mostly dark matter) is not directly observable, so we have to make inferences using techniques such as the Sunyaev-Zel'dovich effect, X-ray emission and weak gravitational lensing. Powerful computer simulations play a key role in understanding the formation of structure and in assessing the accuracy of these cluster mass estimates.

Weak gravitational lensing, our fourth technique, is also sensitive to both the expansion history and the growth history of density fluctuations. It relies on the phenomenon, predicted by Einstein's General Theory of Relativity, that light rays are bent when they pass through a gravitational field. Since light from a distant galaxy bends as it travels towards us through the gravitational field of intervening dark matter, the image of the distant galaxy is distorted. The apparent distortions of distant galaxies in a small patch of sky will be correlated, since their light has travelled through nearly the same intervening field of dark matter; we can use this correlation to statistically tease out the small lensing distortion. This effect, known as cosmic shear, is a tiny distortion of about 1% on average, and it requires very precise, accurately calibrated measurements of many galaxy shapes to infer. We can use this effect to infer the distribution of the intervening mass and how it evolves. DES has measured the shapes of over 300 million galaxies at different distances, and from these we are constructing a quasi-three-dimensional map of dark matter.

All four techniques are discussed in detail in Part II: Dark Energy Science, in Chapters 9 to 12. Chapter 13 discusses the exciting prospect of combining the probes, which is essential for distinguishing between different explanations of cosmic acceleration. Below, we highlight just a few early DES results.

−1.5 DES joint galaxy clustering and weak lensing: Year 1 cosmology results

The first major DES cosmology results were announced in August 2017, and published the following year by the Dark Energy Survey Collaboration

(2018a). This analysis combined galaxy clustering and weak gravitational lensing data from the first year of DES survey data that covered 1321 square degree, utilizing measurements of three two-point correlation functions (hence referred to as '3 times 2pt'): (1) the cosmic shear correlation function of twenty-six million source galaxy shapes in four redshift bins, (2) the galaxy angular auto-correlation function of 650,000 luminous red galaxy positions in five redshift bins and (3) the galaxy-shear cross-correlation of luminous red galaxy positions and source galaxy shears. The headline results strongly support the ΛCDM cosmological model. Combining DES Year 1 data with Planck CMB measurements, BAO measurements from the SDSS, 6dF and BOSS galaxy redshift surveys, and type Ia SN distances from the Joint Lightcurve Analysis, we find the dark energy equation of state parameter to be $w = -1.00^{+0.05}_{-0.04}$ (68% cluster [CL]), in spectacular agreement with the cosmological constant model, for which $w = -1$.

Comparing DES and Planck measurements yields an important comparison of the early and late universe. Planck measures conditions at the epoch of 'last (photon) scattering' 370,000 years after the Big Bang. Assuming ΛCDM, the Planck measurement predicts the amount of large-scale clumpiness in the mass distribution today. DES measurement of that clumpiness is in reasonably good agreement with that prediction, as shown by the overlap of the blue and green contours in Fig. −1.2. This subject to possible parameter 'tension' due to either systematics or the need for new Physics (both possibilities are currently under further investigation). We can combine DES and Planck results to obtain tighter constraints, shown by the red contours, and find that the matter density is $\Omega_m = 0.298 \pm 0.007$ and the amplitude of mass fluctuations is $S_8 = \sigma_8(\Omega_m/0.3)^{0.5} = 0.802 \pm 0.012$. The future analysis of the full DES data, which covers a larger area of sky and goes deeper than the Year 1 data, will provide stronger constraints on these parameters as well as on the time variation of the equation of state parameter of dark energy, modified gravity and neutrino mass. See more in Chapter 13.

−1.6 DES-SN Year 3 cosmology results

The first DES Supernovae (SN) cosmology results were announced in January 2018 at a meeting of the American Astronomical Society and published a year later by the Dark Energy Survey Collaboration (2019).

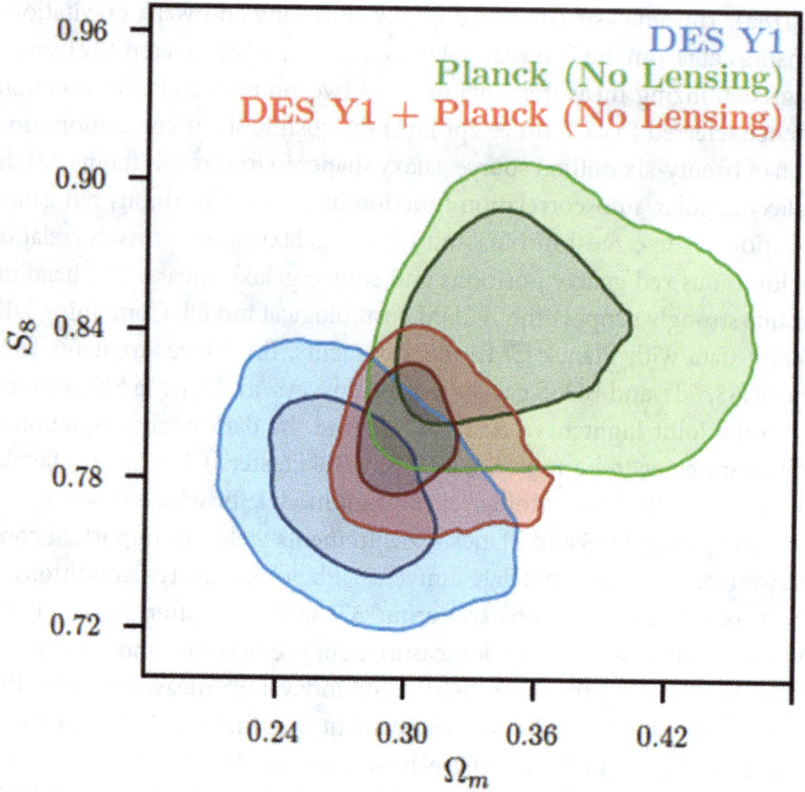

Fig. −1.2 Constraints on cosmological parameters determined by DES Year 1 galaxy clustering and weak lensing (blue), Planck cosmic microwave background (CMB) measurements (green), and the combination of the two (red), assuming the ΛCDM model. Within the measurements' accuracy, the Planck and DES constraints are consistent with each other. Here, Ω_m is the matter density parameter, and S_8 is a parameter related to the amplitude of density fluctuations (see text). For each colour, the contour plots represent 68% and 95% confidence levels. Image from Dark Energy Survey Collaboration (2018a).

The analysis used a sample of 207 spectroscopically confirmed SNe Ia from the first three years of DES-SN, combined with a low-redshift sample of 122 SNe from the literature, thus 329 SNe Ia in total. For the ΛCDM model, combining DES-SN with Planck, we find a matter density $\Omega_m = 0.331 \pm 0.038$; for the wCDM model (with constant w), we find $w = -0.978 \pm 0.059$, and $\Omega_m = 0.321 \pm 0.018$. These results strongly support the ΛCDM paradigm, and they agree with previous constraints using SNe Ia. Future DES-SN analyses will use a much larger sample of several thousand photometrically classified SN from the full survey.

−1.7 Non-dark energy results

Data from the survey have so far been used in a variety of ingenious ways, not all of them anticipated when we began. Early DES discoveries include a distant dwarf planet nicknamed 'DeeDee' and many trans-Neptunian objects in the outer solar system; seventeen dwarf galaxies orbiting the Milky Way; thousands of galaxy clusters; high-redshift superluminous SN and quasi-stellar objects (QSOs); and new insights into galaxy evolution. Furthermore, one of the six minor planets interpreted as indirect evidence for the existence of Planet Nine in the outer solar system was discovered by DES, and DES data is being used to search for this hypothetical ten-Earth mass planet. In August 2017, within hours of the first detection of gravitational waves from the collision of two neutron stars by the Advanced LIGO and Advanced Virgo interferometers, DES team and their collaborators independently discovered the optical counterpart to this event, a kilonova in the galaxy NGC 4993 (catalogued in the New General Catalogue of Nebulae and Clusters of Stars) 130 million light years away (see Chapter 22). Not only was this the first joint detection of gravitational and electromagnetic radiation from a single source (Abbott et al., 2017), it led to a remarkable world-wide observing campaign that captured light from this event across the electromagnetic spectrum over the succeeding weeks. Further non-dark energy results are summarized in the overview paper 'DES: more than Dark Energy' (Dark Energy Survey Collaboration 2016).

Table −1.1 presents an inventory of measured and discovered objects over the full DES. The table is based on the table given in Dark Energy Survey Collaboration (2016).

−1.8 Josh Frieman and Ofer Lahav reflect on DES

Josh on the genesis of DES

In the mid-1990s, I was working in the Theoretical Astrophysics group at Fermilab and teaching at the University of Chicago. My research focus had been the early universe, using the cosmos as a laboratory to constrain fundamental physics, and the theory of large-scale structure and the

Table −1.1 Inventory of measured and discovered objects, over the full Dark Energy Survey (DES).

Objects	Expected from full Dark Energy Survey (DES)
Galaxies with photo-z $> 10\sigma$	300 M
Galaxies with shapes	200 M
Galaxy clusters ($\lambda > 5$)	380 K
Supernovae	1000s identified Photometrically
Superluminous supernovae	15–20
	Confirmed spectroscopically +many more candidates
New Milky Way companions	25
Quasi Stellar Objects at z > 6	375
	Confirmed spectroscopically +many more candidates
Lensed Quasi Stellar Objects	100(i < 21)
Stars ($> 10\sigma$)	100 M
Solar System:	
New Trans-Neptunian Objects	~ 300
New Jupiter Trojans	19
Main Belt Asteroids	~ 200 k
Kuiper Belt Objects	500–1000

analysis of galaxy surveys. In 1995, with several colleagues I proposed a dynamical, particle physics-inspired model for scalar field dark energy (an idea later dubbed 'quintessence' in some quarters). We were motivated by the desire to reconcile the low dynamical estimates for the matter density, $\Omega_m \sim 0.2$, with the prediction from inflation of spatial flatness (which implies $\Omega_{tot} = 1$), then-recent high estimates of the Hubble constant that indicated $H_0 t_0 \simeq 1$ in combination with globular cluster ages, the early-1990s measurements of galaxy angular clustering, plus an inferred upper bound on Λ from strong gravitational lensing statistics (which has since been relaxed). This circumstantial evidence for what my colleague Michael Turner later named dark energy came together several years before the SN discovery of cosmic acceleration announced in early 1998. In fact, in 1997 several of us began organizing a workshop on 'The Missing Energy in the Universe' to be held in May 1998 at Fermilab to discuss this evidence and the theoretical ideas emerging to explain it; in the event, post-SN discovery, interest in the workshop was more intense than we had anticipated.

By the early 2000s, my interests were shifting increasingly to testing models for dark energy and cosmic acceleration with more powerful data. In 2003, I was pursuing several approaches to this. One was developing the concept for what became the SDSS-II Supernova Survey, which discovered 500 type Ia SN at intermediate redshifts in 2005 to 2007. The second was involvement in the SNAP project, led by Lawrence Berkeley National Laboratory, a proposed space-based dark energy mission using SN and weak lensing that later evolved into the Joint Dark Energy Mission (JDEM) and in the Astro 2010 Decadal report morphed again into the WFIRST mission, now slated for launch by NASA in the mid-2020s. The third, which became DES, was a wide-area, optical multi-band survey in the southern hemisphere, originally conceived to provide photometric redshifts and weak lensing mass estimates for the galaxy clusters that the SPT, then under construction, was going to measure for cluster cosmology. Work in 2000 by Zoltan Haiman, Joe Mohr and Gil Holder, then all at Fermilab and Chicago, had pointed out the power of such a cluster census to probe dark energy in a complementary way to SN.

Early in 2003, Joe Mohr, then at the University of Illinois at Urbana-Champaign, who had begun working with John Carlstrom on SPT when he was a post-doc at Chicago, began discussing what became DES concept with me, Jim Annis and Albert Stebbins at Fermilab and with others. Because the expected SPT cluster density on the sky was high (several per square degree) and they are spatially extended, a 'filled' survey covering the 4000 square degree SPT footprint would be efficient. We soon recognized that such a survey would enable a powerful, multi-pronged attack on the nature of dark energy and cosmic acceleration, including but not limited to galaxy clusters. Later that year, we sketched out the case for weak lensing cosmic shear as one of the survey science drivers, relying on earlier work and forecast calculations by Wayne Hu at Chicago.

We originally discussed with UK colleagues the idea of reviving the optimal camera concept for the 4-m VISTA telescope that they were building in Chile as part of their entry to ESO and which at that time was planned to have only a near-infrared, wide-field camera. Fortunately for us, within a few months Joe made contact with Chris Smith at the NOAO, which operates CTIO in Chile, and which was looking to develop a new, wide-field instrument for its 4-m telescope for the astronomical community (see Chapter 1). They were willing to offer a large amount of observing time if a group could design and build that new instrument. This provided us with a fantastic opportunity and was the catalyst for us to form DES collaboration, initially between Fermilab, the University of Chicago and the University of Illinois at Urbana-Champaign. In the early years of the project, my role focused on developing and presenting the science case for DES in myriad proposals and reviews and helping build the collaboration through contacts in the United Kingdom (see below), Spain and the United States. As the collaboration grew, around 2006 Ofer Lahav and I were tasked with organizing the science working groups to prepare for and carry out the analysis of DES data. In early 2010, I became Director of DES when John Peoples, the first Director, stepped down, and in late 2018 I passed the baton to my Chicago colleague Rich Kron, who had served as Deputy Director for those eight years and had previously led the SDSS. It is quite remarkable to consider that the collaboration has grown from a few people discussing the camera and survey concept to an international collective of over 400 scientists from seven countries who have so far written over 200 papers on DES, with the best yet to come.

Ofer on the evolution of DES

In June 2004, I ran into Josh Frieman at 'Neutrino-2004', a conference on neutrino physics in Paris. I was at the neutrino conference to give a talk about the 2dF results and Josh had been in the competing camp to 2dF, the SDSS. Also there was John Peoples, who had been the Director of Fermilab and the Director of the SDSS.

I had been an active participant in the 2dF (two degree-field) Galaxy Redshift Survey, which used the Anglo-Australian Telescope in Australia to map the distribution of a quarter of a million galaxies spectroscopically, but this work was now coming to an end. I moved from Cambridge to University College London (UCL) in February 2004, to head the Astrophysics group as Perren Chair of Astronomy. There was then no cosmology research being done at UCL and I was looking for a suitable project. Two recently appointed UCL lecturers, Sarah Bridle and Jochen Weller, also had an interest in taking cosmology beyond what was known at the time.

Josh, John and I decided to have lunch together in Paris and it turned out to be a serendipitous meeting. The science goals of DES, which were then in a very early stage of development, were well aligned with my scientific interests. Moreover, I had inherited a group that had an established capability of designing and assembling optical systems. The Temple review of the NOAO Proposal (see Chapter 1 for more details) for DES had urged DES collaboration to seek an experienced partner that could procure a state of the art wide-field corrector, since the collaboration did not have that capability. Fermilab was expected to lead the design and fabrication of DECam, the wide-field charge-coupled device (CCD) camera, and this was going to require considerable research and development. John, Josh and I talked extensively about DES and DECam at lunch and we agreed to pursue the idea further. I had then only been at UCL for four months, but I had held initial informal discussions with the instrumentation people, especially Peter Doel and David Brooks, and I thought they had something to offer. DES seemed to be exactly what was needed to animate the newly formed UCL cosmology group.

I began the process of applying to the UK Particle Physics and Astronomy Research Council, PPARC (now the Science and Technology Facilities Council, STFC) for the funding to design and then procure the optical corrector. I had never applied for grants of this magnitude before, but I found it an interesting challenge and learned that you have to start by building up small sums. I managed to procure 11,000 pounds from the UCL Provost Fund to get things going. It became clear that I could not apply for money just for UCL, so I helped to put together a consortium of several universities in the United Kingdom. I started off simply by calling friends. I told them about the survey and asked if they would at least like to have their name on the proposal and could give me some advice about how to proceed. We had to propose what is called an SOI, a Statement of Interest (of only two pages), and I was given some good advice about that. Then the question was how much money should we ask for? The whole project was estimated to be forty million dollars, so do we ask for one million dollars? How do we convince them? We were in stiff competition with another group, who also wanted funding for a similar goal on a different telescope. This was not something I had experienced before: to take a fairly esoteric concept, bordering on philosophy at that point, dark energy; then combine it with instrumentation, theoretical modelling, computing and also with politics and fundraising. How could we convince committees that this is one of the most pressing questions in science?

I found the whole process fascinating. For the first year or two, from 2004 to 2006, it was all about lobbying: going around the United Kingdom giving talks about DES and raising our profile. Eventually, DES-UK secured 1.76 million pounds from PPARC in 2006 and this was very important in helping to convince the United States funding agencies that DES was serious. We also brought together some very good people in the United Kingdom, and resources from their universities. Eventually DES:UK Consortium included, in addition to UCL, scientists from Cambridge, Edinburgh, Manchester, Nottingham, Sussex and Portsmouth.

What was pleasing about it was that PPARC didn't just write a cheque and send it to the United States. The optical corrector was assembled in UCL, in the basement of the Physics and Astronomy Department.[2] This was very special for all of us in the group. Our involvement with DES also served as a way of obtaining grant money for post-docs to work on the theoretical aspects. It turned out to be a very good move for us, strategically. I was the Head of Astrophysics for seven years so I had to look after the whole group, which involved a lot of other things, but DES was always very central.

My other DES chapter was to co-chair DES Science Committee from its inception in 2006 to 2016, alongside Josh Frieman and then Gary Bernstein, setting up a dozen Working Groups, many analysis challenges, and coordinating some of the work and papers on the Science Verification

---

[2]The lenses were cast in New York and polished in France before being transported to London for alignment in cells cast in the north of England. See Chapter 2 for more details.

data. Josh has counted that he and I exchanged 3000 emails over those ten years! I was succeeded as the Science Committee co-chair by Scott Dodelson, and Gary was succeeded Elisabeth Krause. They have led the impressive Year 1 analyses described throughout the book.

References

Abbott, B. P., Abbott, R., Abbott, T. D., Abernathy, M. R., Acernese, F., et al. (2016). The advanced LIGO detectors in the era of first discoveries, *Physical Review Letters* **116**, 13, p. 131103, doi:10.1103/PhysRevLett.116.131103.

Abbott, B. P., Abbott, R., Abbott, T. D., Acernese, F., Ackley, K., et al. (2017). Multimessenger observations of a binary neutron star merger, *The Astrophysical Journal Letters* **484**, L12, doi:10.3847/2041-8213/aa91c9.

Albrecht, A., Bernstein, G., Cahn, R., Freedman, W. L., Hewitt, J., et al. (2006). Report of the Dark Energy Task Force, *ArXiv Astrophysics e-prints*.astro-ph/0609591.

Dark Energy Survey Collaboration, (2016). The Dark Energy Survey: More than dark energy - an overview, *Monthly Notices of the Royal Astronomical Society* **460**, pp. 1270–1299, doi:10.1093/mnras/stw641.

Dark Energy Survey Collaboration, (2018a). Dark Energy Survey year 1 results: Cosmological constraints from galaxy clustering and weak lensing, *Physical Review D* **98**, 4, p. 043526, doi:10.1103/PhysRevD.98.043526.

Dark Energy Survey Collaboration, (2018b). The Dark Energy Survey: Data release 1, *The Astrophysical Journal Supplement Series* **239**, 2, p. 18, doi:10.3847/1538-4365/aae9f0.

Dark Energy Survey Collaboration, (2019). First cosmology results using Type Ia supernovae from the dark energy survey: Constraints on cosmological parameters, *The Astrophysical Journal* **872**, 2, p. L30, doi:10.3847/2041-8213/ab04fa.

Hobson, M. P., Efstathiou, G. P., and Lasenby, A. N. (2006). General Relativity, doi:10.2277/0521829518.

Perlmutter, S., Aldering, G., Goldhaber, G., Knop, R. A., Nugent, P., et al. (1999). Measurements of Ω and Λ from 42 High-Redshift Supernovae, *The Astrophysical Journal* **517**, pp. 565–586, doi:10.1086/307221.

Planck Collaboration, (2018). Planck 2018 results. VI. Cosmological parameters, *ArXiv e-prints*. arXiv:1807.06209.

Riess, A. G., Filippenko, A. V., Challis, P., Clocchiatti, A., Diercks, A., et al. (1998). Observational evidence from supernovae for an accelerating universe and a cosmological constant, *The Astronomical Journal* **116**, pp. 1009–1038, doi:10.1086/300499.

Weinberg, S. (1989). The cosmological constant problem, *Reviews of Modern Physics* **61**, pp. 1–23 doi:10.1103/RevModPhys.61.1.

Building the Dark Energy Survey

We describe how the Dark Energy Survey (DES) came to life, from the early days when it started as an idea, through the struggle for funding, the renovation of the Blanco Telescope, the building of the camera, to the ingenious solutions for running the survey and dealing with the unprecedented amount of new data.

Early Days of the Dark Energy Survey

Josh Frieman, John Peoples, Chris Smith & Alistair Walker

The Dark Energy Survey (DES) may not have happened at all if not for a unique combination of factors. One might say that all the stars were aligned. This chapter records the story of these different circumstances and how they brought together a diverse set of people and organizations from around the world to work towards a common goal: to increase our knowledge of the universe and understand the nature of dark energy. The path was not straightforward, but the time was right for DES, and with enough belief and hard work it gradually started to become a reality. In 2008 the US Department of Energy authorized full funding for the Dark Energy Camera Project and the National Science Foundation agreed to support the Data Management System; after that there was no going back.

1.1 Planning a future for the Blanco Telescope

In 2001, the US National Research Council issued 'Astronomy and Astrophysics in the New Millennium', the Decadal Report by a panel of experts in physics and astronomy. Following input from the astronomy community and a number of sub-panels, the Decadal committee makes recommendations to the US funding agencies concerning the space and ground-based astronomy projects to be prioritized for the following decade. Topping the list of major initiatives was the James Webb Space Telescope, now due to be launched at the end of March 2021. Next in line was the Giant Segmented

Mirror Telescope (GSMT), the next generation of very large (20 to 40 m aperture) ground-based telescopes, a concept that the National Optical Astronomy Observatory (NOAO) had been deeply involved with for the preceding two to three years. In order to participate fully in such a project, Jeremy Mould, the new NOAO director, instructed the scientific staffs of his two observatories, Cerro Tololo Inter-American Observatory (CTIO) near La Serena, Chile, and Kitt Peak National Observatory (KPNO) near Tucson, Arizona, to re-prioritize activities so as to cut some 25% (two million dollars) each from their yearly budgets. At CTIO this meant that, among other cuts, there would be no funds either for new instruments for its telescopes, or for the associated technical people, in the foreseeable future. Not being able to have state of the art instrumentation on the flagship 4-m Victor M. Blanco and Nicholas U. Mayall telescopes would mean that cutting-edge science would not get done, and would likely lead to a death spiral ending in irrelevance and closure of these facilities.

Paradoxically, this rather bleak view of CTIO's future contrasted with the recommendation of the 'Program Review Panel', a committee that reports directly to the National Science Foundation (NSF) each year on NOAO's proposed yearly work plan, and also with the report of the 'Observatories Visiting Committee', which was appointed by the Association of Universities for Research In Astronomy (AURA). AURA operates NOAO through a cooperative research agreement with the NSF. NOAO operates Cerro Tololo and Kitt Peak observatories for the astronomical community, with funding from the NSF Astronomy Division. The Observatories Visiting Committee visited CTIO in December 2002. In their report to AURA they said that there had been more formal recognition at all levels that 4-m class telescopes still have a very important role to fill in the future of NOAO 'public' astronomy. How then to reconcile the reality with a need to build new instrumentation? At CTIO, this meant developing an instrument for the Blanco 4-m telescope, but what new instrument would be best?

During 2002, the CTIO science staff and the NOAO director discussed whether or not they should upgrade the prime focus imaging capability of the Blanco, which at that time was the Mosaic imager, and the conclusion was, 'Drastically!' Given the wide-field capability of this telescope, it was an attractive path, and now with encouragement from these high-level committees there was a need for action. The question, of course, was how to pay for it, but NOAO Director Jeremy Mould addressed this problem in his January 2003 address at the American Astronomical Society (AAS) meeting, introducing the concept of 'instrument partners'. CTIO had a telescope and a highly developed infrastructure for supporting it and could offer time on that telescope, nominally valued at 10,000 dollars per night. Accordingly, then-CTIO Deputy Director Alistair Walker called for a meeting of the CTIO scientific staff, on February 6, 2003, to discuss these issues. Those present included Chris Smith and Tim Abbott. Specifically, the question they asked themselves was, 'Would we give up 30% of Blanco time over five years to someone who would help us to build a 16K × 16K pixel charge-coupled device (CCD) camera to be installed at the prime focus of the telescope?' Such a large focal plane would also require a new optical corrector with six times the field of view of the Mosaic imager, which had been in use on the Blanco Telescope since 1999.

With strong support from the CTIO scientific staff, the following activities were initiated on two fronts: (1) selling the instrument part-ner concept to various oversight committees, and (2) exploring optical corrector designs to see where the focal plane size limits were. CTIO made a call for optical design studies in May 2003, awarding a contract to Prime Optics, but they were disappointed that the preferred design only covered 1.65 degrees diameter (angular diameter of the sky as seen from Earth). For comparison, the diameter of the full moon is 0.5 degrees. So they issued a second contract to Y. Terevich (Sternberg), who produced an all-spherical design that could give a 2.4-degree diameter field with a 1.1-m diameter front element, the largest lens in the corrector. It was also clear that downsizing a little to about a 2-degree diameter field was possible with a 0.9-m diameter front element, which was not beyond the state of the art, together with one or two aspherical surfaces on the smaller elements. To reduce complexity, they also decided that the optical corrector should not include an atmospheric dispersion compensator,[1] which is usually made from two prismatic elements, each a doublet (two lenses paired together).

At the same time, CTIO started looking for instrument partners. They realized that it would be difficult to follow the (usual) route of university

---

[1] The atmosphere acts like a weak prism, causing stars to appear as short 'rainbow' streaks. The effect is small, except when the telescope points to low altitudes.

partners, because NSF funds could not be used to pay for the instrument, and universities often rely on NSF grants to support their instrument building projects.

1.2 An instrument partner for the Blanco

Independently of the CTIO staff discussions about a future instrument for the Blanco, members of the South Pole Telescope (SPT) Consortium, who were building a 10-m telescope at the South Pole to study the cosmic microwave background (CMB), began discussing the need for a multi-band optical survey to cover the same part of the southern sky that the SPT planned to observe. The consortium expected to discover about 25,000 clusters of galaxies over 4000 square degrees through the Sunyaev-Zel'dovich (SZ) effect. Confirming these clusters and measuring their redshifts photometrically with optical follow-up was essential to exploiting them as a cosmological probe.[2]

The SZ effect is a distortion of the apparent brightness of the CMB (radiation that has been travelling through the expanding universe since 380,000 years after the Big Bang) due to high-energy electrons in galaxy clusters scattering the low-energy CMB photons. High-energy electrons in galaxy clusters are known to be among the most active sources of X-rays in the universe. The SZ effect is measured as a decrease in the intensity of the CMB at long wavelengths in the direction of a cluster; this decrement provides an estimate of the mass of the cluster. However, since it arises from scattering, the SZ effect is insensitive to the redshift of the cluster. The redshift is the stretching of the wavelength of light from a galaxy due to the expansion of the universe and is directly correlated to its distance. If the

---

[2]At that time, the modelling of how many clusters there should be was at a relatively early stage. In addition, the amplitude of density fluctuations in the universe (denoted by σ_8) was estimated to be higher than we now know it to be. The higher the value of σ_8, the more clusters there will be, because clusters form at peaks of the density field. In the event, with lower σ_8, SPT found fewer clusters than originally expected. Furthermore, because SPT started its survey in 2007 and DES in 2013, a number of SPT clusters were already observed by other optical telescopes before DES. Nevertheless, the combination of DES data with SPT data forms a key part of the science program of both.

redshifts of clusters of galaxies could be measured independently, the SPT team realized they could make a census of clusters by mass and volume and thereby probe dark energy. A wide-field (4000 square degrees), multi-band, optical survey of the SPT galaxy clusters could provide the photometric redshifts for the SPT clusters.

John Carlstrom at the University of Chicago was the principal investigator of the SPT project, and he was very interested in knowing whether such a wide field survey was possible. He encouraged Joe Mohr (then at the University of Illinois), a member of the SPT Consortium who had worked with John at Chicago, to reach out to CTIO. Joe wrote a 'breakthrough letter' to Chris Smith on July 8, 2003, putting forward his interest in building a large optical camera for the Blanco Telescope as a joint project with NOAO. This email had the subject title, 'large solid angle camera for CTIO 4 m'. After some initial family greetings, Joe began, 'I'm writing to find out some details about a camera upgrade for the CTIO 4 m'. He described the SPT and its capabilities and noted that the SPT Consortium was exploring the possibility of building an optical camera for VISTA, a 4-m telescope under construction on the European Southern Observatory (ESO) site at Cerro Paranal, Chile. The email closed with, 'It would be wonderful if we could join forces on a single camera on a US telescope ... please let me know what you think might be possible'.

Chris Smith replied the same day: 'Not only is this possible, it is currently under pretty serious consideration!' Within the next forty-eight hours, Joe Mohr, Tim Abbott, Alistair Walker and Chris Smith had exchanged ten emails on the subject. On July 10, 2003, Alistair sent an email to Jeremy Mould that started, 'We have a big fish enthusiastically nibbling and may want to move fast...' By the very next day Jeremy Mould had permission from Craig Foltz (NSF) to begin the engagement! Alistair then advised the group that we had to move through formal channels: 'We have to do all this by the rules, which involves the Announcement of Opportunity (AO) and so on, as was done for the Mayall'. In an email to Joe Mohr on July 17, he outlined the process for this AO: an open call for research proposals based on the Blanco 4-m Telescope. A competition would promote excellence and ensure fairness by giving the entire scientific community a chance to respond.

1.2.1 Think what we could do with a camera like that!

Things moved quite fast once this initial connection was made. At that time, Fermilab was active in R&D for the SuperNova Acceleration Probe (SNAP) and there were a number of Fermilab high energy physicists interested in survey cosmology. Josh Frieman, Jim Annis and Albert Stebbins (Fermilab) had been discussing the wide-field camera concept for a southern hemisphere telescope with Joe Mohr for several months. Jim had earlier joined the Sloan Digital Sky Survey (SDSS) while he was still a postdoc and had helped it grow into a success; like Joshua Albert was a cosmology theorist. Joshua invited Joe to present the camera concept to a group of scientists who might be interested in such a project. A telecon was arranged for July 23, 2003; the participants were Robert Brunner, Joe Mohr, Jon Thaler (all from the University of Illinois), Jim Annis, Peter Limon, Huan Lin, Vic Scarpine, Albert Stebbins, Douglas Tucker, William Wester (all Fermilab), Josh Frieman and Rich Kron (Fermilab and the University of Chicago), John Carlstrom (University of Chicago) and Connie Rockosi (University of Washington).

During the meeting, we considered both the technical requirements of such a survey and the possible science that could be obtained. Joe discussed the possible refurbishment of the Blanco Telescope. In addition to a new camera, this would require a new optical corrector to ensure an accurate wide field of view for the camera. These additions would enable a multi-band survey of the SPT area. He noted that the CTIO staff had begun studying the feasibility of putting a new wide-field camera on the Blanco several years earlier. CTIO did not have the funds to acquire the instrument, so they were planning to issue an AO that would offer the successful team 30% of five years of observing with the Blanco in exchange for delivering the instrument to CTIO. In the past, such AOs had yielded a single proposal submitted jointly by an outside group in partnership with NOAO. The remainder of the observing time would be allocated to the astronomy community. The CTIO staff included Alistair Walker, Chris Smith, Tim Abbott, Nick Suntzeff and the then-current CTIO Director Malcolm Smith. They wanted to make the Blanco a state-of-the-art facility in order to secure a long-term future for the telescope.

The telecon participants also realized that such a survey would enable other probes of cosmic acceleration beyond the cluster counting method, making it much more powerful. The minutes of the July 23 telecon were prepared by Joshua. The key conclusions show that the science goals we considered possible with such a survey included (1) photometric redshifts to about redshift one ($z \sim 1$) for the approximately 25,000 clusters forecast to be discovered by the 10-m SPT via the SZ effect; (2) a larger, optically selected sample of clusters as a function of richness; (3) weak lensing mass estimates of these clusters; (4) weak lensing cosmic shear measurement with photo-z tomography; (5) measurement of galaxy clustering on large scales to $z \sim 1$ and (6) galaxy–galaxy lensing, and many other extragalactic spin-offs. These measurements could be used to determine the properties of dark energy in ways that were complementary to type Ia supernovae.

Driver (1) set the minimum technical requirements for the survey, and the participants reasoned that satisfying them should make the other goals possible to an extent determined by details of the exposure times, atmospheric conditions at the site, image quality of the camera–telescope system and total observing time available. We envisaged a survey in about three optical passbands to obtain the photometric redshifts (estimated accuracy of delta $z \sim 0.02$ for clusters), with an exposure time of the order of one hour on a 4-m telescope in each passband, to reach a limiting magnitude of 24 to 24.5. For comparison, Sirius, the brightest star in the night sky, has an apparent magnitude of −1.5. Since the magnitude scale is logarithmic, an object of apparent magnitude 24.5 appears billions of times fainter than Sirius. The 4-m Blanco Telescope equipped with the existing Mosaic II camera, which had a field of view of about half a degree, would require several decades of observing time to deliver a survey that could achieve these science goals, which was not practical. A new wide-field corrector and a 2-degree field of view CCD camera mounted on the Blanco would enable such a survey to be carried out in about 500 nights, a much more reasonable plan.

The telecon participants also reviewed other paths that could lead to a wide-field survey of the southern sky and provide the redshifts for the SPT clusters. One possibility that had been under discussion since January 2003 was to team up with a UK team building VISTA, a 4-m telescope equipped with a 2-degree field of view near-infrared (NIR) camera, which was being built at the ESO on Cerro Paranal in Chile. The VISTA team had funding for the NIR camera and the telescope, but not for an optical camera. The cost estimate for the optical camera was eleven million dollars. While the British PI of VISTA, Jim Emerson, had been interested, the transatlantic politics would certainly be a challenge in terms of both securing US funding for an ESO telescope and securing agreement from ESO to carry out the survey we wanted.

Another possibility was to move the SDSS camera on the 2.5-m telescope at Apache Point in New Mexico to the Dupont 2.5-m telescope at Las Campanas Observatory in Chile, run by Carnegie Observatories. Advantages included the fact that this telescope was well-matched to the SDSS telescope and could therefore make efficient use of its gappy focal plane through drift scanning; the camera already existed; and the SDSS was expected to finish its imaging phase sometime between 2005 and 2008. The Las Campanas site has excellent seeing, and there was potential interest at Carnegie in a wide-field instrument/survey for that telescope. Disadvantages included the smaller telescope aperture, which would increase the required survey time by a significant factor, and the political reality that the SDSS camera was owned by the Astrophysical Research Consortium.

These other paths seemed less promising than putting a new wide-field camera on the Blanco, and the telecon concluded with a discussion of possible technical contributions that the various institutions might make. This was extremely preliminary and would ultimately depend to large extent on who would eventually lead the project. The CTIO team had suggested an aggressive timeline for development that involved an alert to the community through the NOAO Newsletter on September 1, 2003, and an AO on October 1. Proposals would be selected in April 2004, and with luck the project might see first light in 2007.

1.3 The Dark Energy Survey collaboration comes together, 2003 to 2004

In mid-September 2003, John Peoples, who had recently joined the discussions, had a chance encounter with John Carlstrom as they were on their way to Topics in Astroparticle and Underground Physics (TAUP) 2003, an astrophysics conference in Seattle. Carlstrom had attended the July telecon, and he encouraged Peoples to get Fermilab more involved in the project.

When Peoples returned to Fermilab, he described this conversation to the Fermilab group. They wanted to talk with the scientists at CTIO to learn what needed to be done to turn it into a real project, and Peoples contacted Jeremy Mould to ask him to suggest people to contact at CTIO. Mould suggested Alistair Walker, then-director of CTIO, and Tim Abbott.

Peoples sent a letter to Alistair Walker and Tim Abbott on October 8, stating that Fermilab scientists were very interested in the science that could be done with a wide-field camera for the Blanco Telescope. He wrote, 'We would like to explore whether we could make a major contribution to such an instrument in return for the observing time that would allow us to create a uniform optical, near-infrared survey and thus advance our scientific interests'. He proposed that contact should be made between their two groups in the near future and asked whether the CTIO staff thought Fermilab would be a good partner in the project. In response, Alistair wrote that 'Fermilab is indeed a desirable partner for NOAO'. He brought John up to date on the progress that CTIO had made towards issuing an AO to solicit proposals to acquire the instrument and proposed a teleconference call on October 22, just before the NOAO user's meeting. The idea of including a supernova survey was also raised at this time, although it had been talked about earlier.

1.3.1 Preliminary allocation of responsibilities

Joe Mohr, Jon Thaler, Josh Frieman, Jim Annis, Brenna Flaugher, John Peoples and other scientists from Fermilab and the University of Illinois began holding conference calls to discuss the wide-field camera and to prepare for the teleconference call with Alistair Walker (CTIO), Jeremy Mould and David Sprayberry (both from NOAO) on October 22. The conference calls allowed the participants to make a preliminary assignment of responsibilities. Fermilab wanted to lead the instrument construction; Chicago was interested in developing the optical corrector; the University of Illinois at Urbana–Champaign (UIUC) High Energy Physics group wanted to lead the development of the data acquisition (DAQ) system and the UIUC Astronomy Department, with support from the National Center for Supercomputing Applications (NCSA), wanted to lead the data processing and the creation and distribution of the archive of data products. The remaining responsibilities would be assigned later.

During the October 22 meeting, the participants from Fermilab and UIUC learned that NOAO would be responsible for the telescope modifications and that the consortium that won the bid to build the camera would be responsible for everything else. The successful consortium would have to obtain all of the funds to build the instrument, the data processing system and the archive. Although some of the participants had heard this before from Mould and Walker, the others needed to hear it a few more times before it would sink in.

Brenna Flaugher joined the discussions and brought her knowledge of building silicon detectors to the planning. As a Fermilab high energy physicist, she managed the Silicon Detector Facility (SiDet), which was building silicon vertex detectors for CDF and D-Zero, the flagship collider detectors at the Fermilab Tevatron Collider, then the highest energy particle accelerator in the world. In early September, Mike Witherell, the Fermilab director, cancelled the upgrade of the silicon vertex detectors for these experiments when the Department of Energy (DOE) informed him that it could no longer fund them. This was a devastating blow, and the collaborations resisted the decision. The engineers and technicians were stunned and didn't have a new project to turn to. They were relieved when Brenna quickly engaged them in developing an engineering design for the wide-field camera. This unexpected gift of technical support didn't last very long, but it lasted long enough to evaluate some of the issues that the 'Dark Energy Camera' (DECam) posed. Their work led to a reference design of DECam that would become a critical part of the proposals that the DECam team would submit to Fermilab and later to NOAO.

1.3.2 Visit to KPNO

On November 17 and 18, 2003, Jim Annis, Juan Estrada, Brenna Flaugher, Huan Lin, Peter Limon and John Peoples, all scientists from Fermilab, travelled to Tucson for a meeting with the NOAO staff. They were joined by Greg Derylo, Del Alspach, Jim Fast and Guilherme Cardoso, experienced Fermilab engineers who had been working on the silicon detectors for CDF and D-Zero. Jon Thaler and Joe Mohr from the University of Illinois also joined the visiting team as did Todd Moore and Mike Haney, engineers in Jon's High Energy Physics group. On the first day, they broke into smaller groups and discussed DECam and its challenges with the staff at NOAO in person and the staff at CTIO by video, including Alistair Walker, Chris Smith, Tim Abbott and Nick Suntzeff, and on the technical side Gustavo Rahmer and Ricardo Schmidt. On the second day, they travelled to Kitt Peak to see the 4-m Mayall Telescope, the twin of the Blanco at CTIO. This gave the visitors a much better understanding of the massive size and weight of the prime focus cage and the telescope. The 'instrument partner' would need to provide a new prime focus cage that contained the complete survey instrument, its utilities and its connections to the telescope.

1.3.3 Announcement of Opportunity

Alistair recalls that at CTIO they were of the opinion that a very large CCD imager would be the instrument of choice, and it would be straightforward to write a good science case for it, but the AO for a Blanco instrument partnership did not specify the type of instrument. It did say that a proposed instrument was expected to exploit the Blanco's wide-field capability. The AO appeared in the NOAO Newsletter of December 1, 2003 (Issue 76), under the name of Alistair Walker, with a link to full details on the CTIO website. It stated that proposers would need to submit 'a science plan, a technical plan and a management plan'. The due date for Letters of Intent was March 15, 2004, with full proposals by August 15, 2004, and an announcement of the selected instrument by NOAO by October 15, 2004.

1.3.4 DES workshop at Fermilab

More teleconferences and meetings followed, culminating with a workshop on December 4 and 5, 2003, at Fermilab. Participants included scientists and engineers from the Lawrence Berkeley National Laboratory (LBNL), Fermilab, the University of Chicago and UIUC. The dates of the meeting were chosen so that Alistair Walker, Chris Smith and David Sprayberry could attend and give the participants a clearer understanding of the AO and what CTIO could provide. The goals of the workshop were to define the primary science goals of the Dark Energy Survey (DES); to agree on the initial assignment of responsibilities for constructing the DECam and for developing the data management system for DES and to explore how to respond to the NOAO AO.

Each institution described their capabilities and the contributions that they could make to the project. Mike Levi and Saul Perlmutter from LBNL described their interest in designing and procuring the CCDs for the camera. Their engineers had been developing relatively thick CCDs, made from ultra-pure silicon, which were exceptionally sensitive to red and infrared light and could therefore detect distant, high-redshift galaxies. Brenna Flaugher described the status of the design work on the DECam that was already underway and proposed that Fermilab design and construct DECam. The SiDet staff gave the participants a tour of the facility and showed that it was well suited to constructing and testing the camera, including the prime focus cage. Joe Mohr and his colleagues described the facilities at NCSA and said that they were suitable for developing the data processing system and the archive.

Chris Smith presented a set of slides arguing for a supernova component to be added to the DES observing plan, based on work on the ongoing ESSENCE survey using the Mosaic II camera on the Blanco. There was still some debate at this early stage of the project about whether or not to include supernovae in the mix of cosmological probes. In the summer of that year, Gajus Miknaitis and Chris Stubbs (both then at the University of Washington), and in parallel a small group led by Josh Frieman at Fermilab, were exploring and simulating what later became the SDSS-II Supernova Survey. Gajus carried out the first cosmological supernova simulations for DES in a midnight rush to get them ready for the December meeting, using John Tonry's code. They were actually in the middle of running the code at the time of the meeting.[3] Levi and Perlmutter also described their interest in adding a supernova search to the science goals.

By the end of the workshop, the participants had agreed to explore the formation of a collaboration to respond to the AO. Each group would have to negotiate the different and complicated processes required by the government funding agencies before they could secure the funds to carry out the project. The Fermilab participants also needed to request

---

[3]Chris Smith wrote to Gajus immediately after his talk to tell him, 'People were impressed that, (a) we could get that many supernovae in so little time and (b) that we constrain w pretty well. However it was clear that they wanted to focus on their core science (dark energy studies from galaxy clusters) so were a bit skeptical about including it in the core of the proposal.'

permission from Michael Witherell, the Fermilab director, for the incipient collaboration to submit a Letter of Intent to the CTIO director in response to the AO. John Peoples prepared an Expression of Interest Letter to Mike Witherell that set out their requests. He asked for the engineering support from Fermilab that would be needed to prepare a credible proposal to NOAO and also requested funds and engineering support to start an R&D program with LBNL on the development of an appropriate red-sensitive CCD.

Mike Witherell agreed that John Peoples could write the Letter of Intent on behalf of the collaboration. John sent the Letter of Intent to Alistair Walker, stating that DES intended to submit a proposal to build the survey instrument, DECam, in exchange for 525 nights of observing time with DECam over a five-year period. The letter of intent was acknowledged by Alistair on March 12, 2004.

1.3.5 Presenting DES to the Program Advisory Committee

In response to DES Expression of Interest, Mike Witherell asked the collaboration to prepare a detailed proposal containing their requests and to submit it to the April 2004 Program Advisory Committee (PAC) meeting. The PAC advises the Fermilab director on its program and in particular makes recommendations on proposed new initiatives. Josh Frieman prepared the science case for the proposal, emphasizing the value of using complementary approaches to measure dark energy in a wide-field survey. Joe Mohr prepared the case for using clusters as one of the complementary approaches. Brenna Flaugher contributed a reference design of DECam, a wide-field CCD camera with a 2.2-degree diameter field of view, a 60-CCD focal plane with an angular area of 2.8 square degrees, and four filters with g, r, i and z bandpasses. Jim Annis contributed a strategy to observe a 5000-square degree area of the southern sky that overlapped the SPT SZ survey. Everyone from the emerging collaboration made a contribution to this proposal, and it brought the group closer together. Jim Annis and Brenna Flaugher submitted the proposal, and they presented DES to the April PAC meeting.

The PAC noted that the science was very important and recommended that the director support the preparation of an even more complete proposal to submit to NOAO, which would satisfy the AO. The PAC also wanted to know what the project would cost and how the collaboration intended to pay for it. Director Witherell agreed to support the preparation of the complete proposal for NOAO. He also said that the NOAO proposal would be reviewed by a team of physicists and engineers, chaired by Ed Temple of Fermilab, and the results of this review would be presented to the Fermilab PAC at its June 2004 meeting. All of this had to be done before he would allow the Fermilab scientists to submit the proposal to NOAO. This was the standard procedure when requesting support for an experiment that would use the facilities at Fermilab. It was the first step along a new path through the forest. While the PAC proposal was incomplete, it prepared the DES collaboration with a template for the NOAO proposal.

1.4 The NOAO proposal for DES

There were just two months to prepare the DES proposal for the Temple review, and it was a major effort. The main objectives of the review were to determine whether the science was sufficiently compelling for Fermilab to use its resources to manage the construction and commissioning of DECam and to establish whether Fermilab was prepared to support a request for DOE funding when the time came. The proposal had to contain features that were not part of a typical high energy physics proposal to Fermilab. DECam would be a community instrument, and astronomers who were not connected with the DES collaboration could request to use it during the time that it was not being used by DES. In addition, the DECam data obtained by the DES collaboration would be made public one year after it was obtained. For these reasons, the proposal contained a section on 'Community Service'. Director Witherell and the PAC also wanted to know about other projects that had the same science goals as DES and the commitments that DES collaborators had made to other projects. For this reason, the proposal contained a section on the 'Relationship of DES to other Astrophysics Projects'.

The primary science goal of DES was to measure w, the dark energy equation of state parameter, to a precision of order 5% using four complementary techniques:

(1) Galaxy Clusters,
(2) Weak Gravitational Lensing,
(3) Galaxy Angular Power Spectra, and
(4) Supernova Light Curves.

These four techniques were described in detail in the proposal. The table of contents of the NOAO proposal shows that the scope of the project was already well developed:

The primary science requirements were to observe a 5000-square degree area to the required depth in each filter within 525 nights of observing time over a five-year period. These requirements, derived from the science goals, in turn determined the observing strategy and the design characteristics of the camera, including the imager and the corrector. DECam needed a field of view of 2.2 degrees to meet the primary science requirements.

Mike Gladders (Carnegie, and later University of Chicago) and Steve Kent substantially refined the design of the optical corrector that CTIO had examined much earlier. In particular, they determined that glass blanks of up to 1.2 m in diameter could be procured from a qualified vendor and established the feasibility of obtaining the large optical components for the corrector. The Survey Instrument section also describes the program that the camera builders planned to follow to acquire and package red-sensitive CCDs for the camera. The considerations that had to be given to the weather, survey strategy and operational efficiency, among others, were discussed in enough detail in the Survey Planning section to make the request for 525 nights over five years plausible.

The camera team led by Brenna Flaugher developed the reference design of the survey instrument, which included the camera, the cooling system, the wide-field corrector, the data acquisition and the mechanical and electrical interfaces to the Blanco Telescope and its infrastructure. The reference design was not the final design, since a number of technical choices were still being evaluated. Nevertheless, it allowed the collaboration to develop a thorough enumeration of the tasks that had to be executed in order to build DECam.

The Data Management section described the preliminary plans to develop the astronomy pipelines that would remove the DECam instrument signature from the camera images and process the DECam data at NCSA when it arrived from Chile. It described the Science Data Products that needed to be created in order to carry out the science analyses; the System and Software Architecture needed to create and support the proposed Science Data Products and the Public Data Release for the public archive of the survey data that would be maintained by NOAO and NCSA. It also described the procedure for the nightly transfer of data from the instrument on the Blanco to the data archives at NCSA and NOAO in Tucson. This section demonstrated that the overall plan to process and archive the data products was feasible with then-current technology.

1.4.1 Temple review

On June 7 and 8, 2004, there was an extensive review of the NOAO proposal, chaired by Ed Temple, who had developed the cost and schedule process for the DOE during the 1980s. The reviewers included scientists from Fermilab and other institutions. The proposal, along with the recommendations of the Temple review panel, was presented to the June meeting of the Fermilab PAC by Hugh Montgomery, the Fermilab Associate Director for Research. While the reviewers found that the project was very well motivated and the reference design was quite complete, they had several concerns. In particular, they noted: the need for a more experienced team of optical designers to manage the procurement of the large optics, given that there were very few vendors that could meet the requirements; that the CCD fabrication and testing plan assumed an optimistic yield of 25% and they recommended the project adopt a yield of 10% for planning purposes; and that the data management plan was at an early stage. The results were conveyed to the participants at the review closeout on June 8; none of the panel's concerns came as a surprise, given the short amount of time that the collaboration had been working on the proposal.

On the basis of the positive PAC recommendations, Director Witherell granted the DES 'stage 1 approval' and allowed the collaboration to submit the proposal to NOAO in order to fulfill the terms of the NOAO AO. Stage 1 approval meant that DES would be included in the Fermilab Program, but it did not provide the funding needed to carry out the project. That would require the approval of the DOE Office of High Energy Physics. Witherell did agree to support the continuation of the design effort and the early phase of the CCD development at LBNL and the CCD testing effort at Fermilab.

1.4.2 Blanco Instrumentation Review Panel review

On July 15, 2004, the DES collaboration submitted the DES proposal to NOAO. NOAO Director Jeremy Mould formed the Blanco Instrumentation Review Panel (BIRP) to review the DES proposal. The review took place in Tucson on August 12 and 13; it was quite extensive and made a number of recommendations. The BIRP concluded that the survey instrument, pipelines and archive (the whole package) were worth the requested observing time (30% of the nights over five years). It therefore recommended that the DES collaboration form a partnership with NOAO. Mould accepted the recommendation and asked Peoples to prepare a Memorandum of Agreement to clarify the responsibilities of the collaboration and CTIO. We had finally reached agreement in a very broad sense.

1.4.3 Memorandum of Understanding

The preparation of the Memorandum of Agreement was not so simple. First, it had to be a Memorandum of Understanding (MOU) to satisfy the DOE. Moreover, the NSF Astronomy Division and the DOE Office of High Energy Physics, which would be funding the project, had never worked with each other in this way before. While DOE and NSF would not be signing the MOU, they were very concerned with its content and with the legal status of the parties that would sign it. In addition, DOE did not want Fermilab to proceed too far until it was clear that DOE and NSF would be authorized to fund the construction and operations phases of DES. These hurdles were difficult to get over, and it became an enormously complicated proposal. Ultimately, it took until May 2008 to sign the appropriate document. DOE and NSF had different procedures to follow before they could authorize their respective programs to request funding. In the end, DECam had to be submitted as a Major Item of Equipment within the DOE Congressional Budget Request, and a great deal had to be done to reach that stage.

In our favour, interest in probing cosmic acceleration and dark energy had been growing substantially, and the DOE and NSF wanted to participate in funding experiments that could explain the phenomenon. In January 2005, Jeremy Mould visited Mike Witherell at Fermilab, and this helped push the project forward. At that point, the DOE hadn't agreed it would commit to supporting DES, but the DOE Office of High Energy Physics wanted to be helpful. There were a lot of hoops to jump through, but Kathy Turner, in charge of the cosmic program at DOE, helped DES to jump though them, while Nigel Sharp did the same at NSF.

1.5 Presenting DES to the scientific advisory committees

Once NOAO had accepted the proposal from DES, the next step for the DES collaboration was to build support for the project in the scientific communities of Particle Physics (High Energy Physics) and Astronomy. The collaboration had to persuade various advisory committees that DES would be an important next step in understanding the nature of dark energy and that the camera would be a valuable resource for the astronomy community. The first opportunity that arose was to present DES to the Science Advisory Group on Experimental Non-Accelerator Physics (SAGENAP), a panel of scientists that had been charged to advise the High Energy Physics Advisory Panel (HEPAP) on how to proceed with proposals that did not require accelerators.

HEPAP advises the DOE Office of High Energy Physics and the NSF Division of Particle Physics on the priority to be given to accelerator facilities and experiments. SAGENAP was a subpanel of HEPAP that was traditionally charged with recommending priorities for experiments to explore cosmic rays and solar neutrino experiments. HEPAP asked SAGE-NAP to review the proposals for dark energy and dark matter experiments, using telescopes or detectors sensitive to the dark matter, and to make recommendations on how these experiments might be incorporated into the programs of the agencies. Brenna Flaugher and Joe Mohr presented DES to SAGENAP on April 14, 2004. SNAP, LSST and other dark energy teams also made presentations that day. The presentations went well, but DOE and NSF were not prepared to fund DES. The path to approval through the agency advisory committees didn't exist for dark energy. It would have to be created.

HEPAP ran another important panel called P5, the Particle Physics Project Prioritization Panel. P5 was charged with recommending a complete and detailed program for US particle physics for the next decade, given budget constraints provided by the agencies. It was asked to recommend when to terminate operations of the US accelerators for particle physics. It had to establish the priority for US participation in the European Organization for Nuclear Research (CERN) Large Hadron Collider, R&D for the International Linear Collider and plans of the DOE laboratories engaged in particle physics, as well as all experiments that were seeking funding in the next ten years. P5 had to address everything that the Office of High Energy Physics was expected to support, and now the dark energy experiments were included in their mandate. The P5 panel met for two years from 2003 to 2005. Among its recommendations for small experiments, P5 recommended that the DES collaboration be funded to build DECam beginning in the fiscal year (FY) 2007. To our great relief, HEPAP accepted the recommendations of P5.

Normally, the NSF Astronomy Division plans its programs based on the National Research Council Astronomy & Astrophysics Decadal Survey. This report identifies key questions facing astronomy and astrophysics and makes recommendations for the coming decade. However, at the time of the 2001 Decadal Survey, dark energy was still a relatively new subject and ideas for new projects were still germinating. However, by 2003 to 2005, NSF Astronomy needed to respond to the dark energy proposals it had received and had to work out a strategy to deal with this emerging area of research. Eventually, the directors of the DOE Office of High Energy Physics and the NSF Division of Astronomy developed a plan to review DES. This last step took until October 2006.

The Astronomy and Astrophysics Advisory Committee (AAAC) was also advising the National Aeronautics and Space Administration (NASA), NSF Astronomy and DOE High Energy Physics on the programs in astronomy and astrophysics that used their facilities. The AAAC recommended to its sponsoring agencies that it would create a task force to study how the agencies could respond to the growing interest in dark energy. The resulting Dark Energy Task Force (DETF) decided to organize the proposals that had been submitted in stages. Stage I were just completed programs/experiments; Stage II were experiments/programs that were funded and underway; Stage III were experiments/programs that were ready for construction funding; and Stage IV were future experiments that were ready for development. Stage IV consisted of large programs, and Stage III consisted of medium-sized projects. The DETF was not asked to set priorities for individual experiments. Josh Frieman and Joe Mohr were invited to give presentations on DES science to the DETF, and the presentations went well. DES was an exemplar of a Stage III project, while the Large Synoptic Survey Telescope (LSST) and the Joint Dark Energy Mission (JDEM; formerly SNAP) were Stage IV projects. DETF recommended that the agencies invest in both Stage III and Stage IV projects. This helped a lot, since DES now had a defined place in relation to other surveys.

1.6 Building the DES collaboration

The preparation of the NOAO proposal for DES and the reviews of the proposal persuaded most of the DES collaborators that DES would be a significant undertaking that required more collaborators and additional funding. The collaboration had grown to thirty-seven members by the time of the BIRP review in 2004. The collaboration recognized that they needed to set up a formal management structure and formed a management committee to coordinate the search for additional collaborators and additional funding. Each institution selected one member to represent their institution on the management committee. The chair was John Peoples. Brenna Flaugher, the DECam project manager, was also a manager on the committee. The initial focus was to seek institutions that could contribute funding and expertise to the optical system and the electronics.

1.6.1 The United Kingdom joins the DES collaboration

As described in Chapter -1.8, Josh Frieman and John Peoples met Ofer Lahav from University College London (UCL) in June 2004, at Neutrino-2004, a conference on neutrino physics in Paris. Their discussions eventually led to UCL joining the DES collaboration in the spring of 2005.

Along with UCL, the Universities of Cambridge and Edinburgh officially joined DES at around the same time, followed by the University of Portsmouth. Eventually, the UK DES collaboration comprised six universities, including also Nottingham and Sussex. Scientists from the University of Manchester joined under a separate arrangement when Sarah Bridle moved there from UCL in 2013.

DES-UK's successful application for a grant from the UK's Particle Physics and Astronomy Research Council in 2006 was an important step towards the collaboration winning approval from the US funding agencies.

1.6.2 Spain joins the DES collaboration

Also attending the Neutrino-2004 conference, where Joshua met Ofer, were Enrique Fernandez, director of the High Energy Physics Institute of Barcelona (IFAE), and Ramon Miquel, then at LBNL. Ramon had worked for years on particle physics experiments and was also involved in SNAP. He was considering joining IFAE. IFAE was participating in high-energy gamma ray astronomy and had excellent capabilities in building electronics. During the conference, they discussed DES with John Peoples and expressed an interest in joining the collaboration and in making a significant contribution to DECam.

In the autumn of 2004, Josh Frieman discussed DES with Enrique Gaztanaga, a cosmologist from the Institute of Space Sciences (IEEC/CSIC) in Barcelona, who had held a Fulbright Fellowship at Fermilab in 1990 to 1993. Joshua and Enrique were meeting to continue their collaboration on galaxy clustering analysis of wide-field galaxy surveys (the Automated Plate Measurement Galaxy Survey, the Center for Astrophysics Redshift Survey, the Infrared Astronomical Satellite and the Sloan Digital Sky Survey). This collaboration included studies of perturbation theory, galaxy biasing (how light traces the mass) and one of the first attempts to measure cosmic shear (weak gravitational lensing distorting the shapes of galaxies) on large scales using the wide-field camera on the Isaac Newton Telescope in La Palma. The group at the IEEC/CSIC also included Francisco Castander, who had worked with Joshua on the SDSS project in Chicago. Together with Pablo Fosalba, the group had experience in the cosmological analysis of galaxy surveys and were interested in joining DES to help build DECam (e.g. CCD testing and camera guiding), developing simulations and coming up with new science ideas to exploit with DES.

Fernandez and Gaztanaga's institutions were next to each other in the Campus of the Universitat Autonoma de Barcelona and they decided to join forces to participate in DES, to bring expertise in both hardware and galaxy surveys. Remarkably, it was through Joshua and John that the two Barcelona institutes learned about each other's interest in joining DES. They attended the inauguration of the Fermilab Center for Particle Astrophysics in December 2004, where they met with John Peoples, Mike Witherell and Hugh Montgomery. They also met with Michael Levi to discuss the status of the DES CCDs and with Ofer Lahav to discuss the United Kingdom's interest in joining DES.

In January 2005, Brenna Flaugher and John Peoples visited Barcelona to explore the contributions that the two teams (IFAE and IEEC/CSIC) could make to DECam. The teams agreed that they would contribute to designing and building the front-end electronics, helping with CCD testing and the camera guider system. They would submit a joint proposal to the Spanish Funding Agency for Science, for the funding to build their contributions, in exchange for membership in the DES collaboration. The group was successful, although it took some time. Just as in the United States, the funding agencies were not initially prepared to receive a proposal from high energy physicists who wanted to work in astronomy.

DES-Barcelona was formally admitted to DES on February 6, 2005. They were joined in September 2005 by another Spanish group from CIEMAT (in Madrid) led by Eusebio Sánchez, which had collaborated previously with IFAE in high energy physics projects. The CIEMAT team was also joined by Juan Garcia-Bellido, a cosmologist at UAM (also in Madrid). In November 2005, a proposal from the newly formed DES-Spain Consortium was submitted to DES and approved, including also Ramon Miquel from IFAE.

Together, the three Spanish groups were awarded funding in different programs from 2005 to the end of the DES construction. This funding included a total of about one million Euros for DES construction. The institutions also provided laboratory and human resources (e.g. engineers) to support the Spanish contribution to building DES.

1.6.3 The wider DES collaboration

The efforts to bring DES to the attention of the astronomy and particle physics communities were successful and attracted several groups that wanted to work on DES, including those from the University of Michigan, the University of Pennsylvania and the Ohio State University; they each joined the DES collaboration and agreed to provide a mixture of cash and in-kind effort to build specific parts of the instrument.

By the time of first light in 2012, there were a total of thirteen US institutions in the DES collaboration. The full list included Fermilab, the University of Chicago, NOAO, the Ohio State University, Texas A&M University, UIUC/NCSA, LBNL, the University of Michigan, the University of Pennsylvania, Argonne National Laboratory and the Santa Cruz-SLAC-Stanford DES Consortium.

The DES-Brazil Consortium negotiated during 2006 and officially joined DES in January 2007.

The final European institutions to join were the Ludwig Maximilian University of Munich (LMU), the Munich Institute for Astro and Particle Physics (MIAPP) and the Swiss Federal Institute of Technology in Zurich (ETH Zurich). Emmanuel Bertin in France also formally joined the DESDM team in 2007, as an external collaborator.

The OzDES Consortium in Australia began working with DES through a broad scope external collaboration agreement (with the supernova working group) and a normal external collaboration agreement (with the QSO working group), set in place in 2013. They became full members of DES in 2016.

1.7 The DOE–NSF proposal for DES

Once the directors of the DOE Office of High Energy Physics (Robin Staffin) and the NSF Division of Astronomy (Wayne Van Citters) agreed on how to review DES, they signed a formal document called a Record of Agreement. They then sent a short email to John Peoples, Hugh Montgomery, Thom Dunning (then director of NCSA) and Jeremy Mould on October 18, 2006. The email read as follows:

Gentlemen: NSF and DOE recently met to discuss what process could be put in place were we to jointly consider a DES proposal as a possible new initiative. For DES to be considered adequately, we concluded that we need to get an official proposal that describes the overall experiment end-to-end, including the R&D and fabrication of the camera, Data Management System and telescope upgrades, as well as a description of the commissioning and operations phases. It should also include the following:

- *Information on the science case and how it will relate to and satisfy the recommendations put forth in the recent Dark Energy Task Force report*
- *Scope, technical, cost, schedule and management aspects*
- *Costs and plans for the commissioning and operations phases*
- *Information on all anticipated funding requests to NSF and DOE to support the R&D, fabrication, integration, commissioning, operations and science activities*
- *Expected contributions from NOAO/CTIO operations, NOAO/DPP and any other laboratory contributions, including programmatic deliverables and estimated equivalent cost*
- *Expected contributions from university and foreign partners*

For such a proposal to be considered in a timely way, it needs to be submitted to DOE and NSF by mid-December in the 'NSF-style' proposal format. A waiver on the page limit could be granted. If received, DOE and NSF plan to jointly review its scientific merit and technical feasibility. Please contact Nigel Sharp (NSF) and Kathy Turner (DOE), if you have any questions.

Best Regards,
Wayne Van Citters and Robin Staffin

This meant there was finally a path forward to obtain the approval (or rejection) of DES for funding by both agencies.

1.8 Funding DES

The process of approval for funding by the DOE Office of High Energy Physics involved DES passing a series of 'Critical Decision' (CD) reviews, the requirements of which became increasingly detailed and specific as the project progressed. See Appendix A for more details. After passing CD-0, in November 2005, Fermilab was authorized to spend some R&D money on DECam, and this amount increased when CD-1 was approved in October 2007.

The following appeared in the FY 2008 DOE Congressional Budget Request:

> Fabrication of the DES project. The FY 2008 request includes a new major item of equipment, the DES project. This effort will provide the next step in determining the nature of dark energy, which is causing the universe to expand at an accelerating rate, by measuring the distances to approximately 300 million galaxies. DES employs several methods to measure the effects of dark energy on the distribution of these galaxies and other astrophysical objects. The DES scientific collaboration will begin fabrication of a new CCD camera and associated lenses, electronics and Data Management System to be installed on the Blanco Telescope at the CTIO in Chile. The project is planned as a partnership between the DOE and the NSF, which operates the telescope. The scientific collaboration is led by Fermilab and includes participants from laboratories and universities in the U.S., England and Spain. Funding for fabrication in FY 2008 is contingent on successful scientific and technical readiness reviews by the interested funding agencies.

The inclusion of DES funding in President G. W. Bush's budget request was significant. Following it, John Peoples wrote a comment in the agenda for the Management Committee: 'The language in the Congressional Budget Request puts us (DES) on a very good road to the promised land; but we aren't there. The last sentence in the paragraph above is very important.

It clearly states we are not approved until we successfully pass the review process of DOE and NSF and we do not get the money until Congress appropriates the money.'

DES had now passed through five years of meetings, presentations, proposals and reviews, so John Peoples was right to feel cautious. Nevertheless, the DOE authorized full funding for the DECam Project (for the DES Experiment) at Fermilab when it approved the 'Critical Decision-3b' in May 20, 2008. The total amount granted was 35.1 million dollars. This was fantastic news and it meant the project could finally go ahead, but the funding applied only to DECam, for R&D and construction. Funding for everything else had to be found elsewhere.

The glass for the optical corrector lenses was purchased by UCL; this cost was reimbursed by the Universities of Chicago, Ohio State, Pennsylvania and Portsmouth, using funds from their endowments. UCL financed the glass polishing, the installation of the finished lenses in the Fermilab-provided barrel and the related testing through a PPARC (later STFC) grant of 1.76 million dollars. The Spanish groups funded the production of the CCD electronics boards, which were fabricated in Spain. The University of Michigan provided the filter changer. Ohio State, the University of Illinois and DES-Spain did the data acquisition software, the online programming, the guider software and the cost of assembling the data for transmission to La Serena. Some of this was reimbursed by DOE through Fermilab, but much of it was an in-kind contribution. There were also more in-kind contributions for DECam.

The cost of upgrading the Blanco Telescope for DECam was the NOAO contribution and was not part of the 35.1 million dollars. This funding came out of the NOAO budget from the NSF. NOAO also paid for the installation of DECam. When CTIO and NOAO labour is included, the cost to NOAO was several million dollars. This was a major effort and was treated as a separate project.

The cost of developing the software to process the data, the equipment to process the data at NCSA, the storage and distribution of the data and the catalogues and science products was funded by a series of grants from the NSF and is not included in the DECam cost. Some of this work was part of the agreement with CTIO to get 525 nights for DES observations. For example, there was a requirement that the DES collaboration provide a

Community Pipeline, the software that NOAO uses to process DECam data for community users of the camera. DES paid for this work with money from the universities in the collaboration. This was also treated as a separate project.

Taken together, the cost of the three projects that produced a working DECam on the upgraded Blanco Telescope, along with the data link from CTIO to NCSA and the software and hardware systems needed to process the data and distribute it to the DES collaboration, probably exceeded fifty million dollars. This fifty million dollars just made it possible to start the survey observations and process and distribute the data, it did not include the operating funds. As this chapter should make clear, a major astronomy survey is no small undertaking! The journey from idea to reality for DES required the dedicated effort of hundreds of people for over a decade. Hundreds more became involved once DECam started taking data. Eventually all the data and catalogues will be available as a permanent resource for anyone with an internet connection to analyze. This book is a record of this remarkable project.

CHAPTER 2

The Dark Energy Camera

David Brooks, H. Thomas Diehl, Peter Doel &
Brenna Flaugher

This chapter focuses on the Dark Energy Camera (DECam) [Flaugher et al. (2015)] and its five giant lenses: the optical corrector. We describe how they work and the story behind their construction, from planning to installation on the Blanco Telescope. DECam is the result of nearly a decade of collective effort by hundreds of people who used their ingenuity to push forward the limits of technology at the time.

2.1 DECam *(Brenna Flaugher, Fermilab)*

The design of the Dark Energy Camera (DECam) was driven by the need to survey a 5000 square degree area during a total of 525 nights, with excellent image quality, high sensitivity in the near infrared and low readout noise. To meet these requirements, within tight budget constraints and with a limited amount of time for construction, we designed a camera with a three square degree field of view, a five-lens optical corrector and a large focal plane with seventy-four fully-depleted, 250-micron thick charged-coupled devices (CCDs). DECam is conceptually similar to a familiar single-lens reflex (SLR) hand-held camera, albeit much bigger and heavier and much more sensitive. The optical corrector, filters and shutter would be the biggest ever made for optical astronomy up to that time. The camera and optical corrector together would weigh about 4650 kg. This weight was due to the large size and the need to use a stiff steel barrel structure to keep the corrector well-aligned in the different orientations the camera would have while on the telescope. We acquired an original design study for the telescope and verified that the Serrurier truss could take the load. This giant camera was to be mounted at the prime focus of the stately Blanco Telescope, which would swing it around 50 feet above ground, at the top of Cerro Tololo mountain in Chile.

2.1.1 Rising from the ashes: 2003 to 2008

DECam and my involvement in it may not have come about at all if not for a very particular set of circumstances. I was the project manager for one of the projects to upgrade the reach of the Fermilab proton–anti proton collider that involved the construction of a new silicon vertex detector for the Collider Detector at Fermilab (CDF) detector. In September 2003, Fermilab Director Mike Witherell cancelled the majority of these Run IIb detector upgrade projects, and so I began looking around for a new challenge. My husband, H. Thomas Diehl, had been working on R&D for the SuperNova Acceleration Probe (SNAP) satellite project for cosmology. Tom knew that colleagues at Fermilab were also discussing a different, cheaper, ground-based project to investigate dark energy and suggested I talk with them, with the idea of becoming Project Manager for DECam. I was instantly hooked. As a graduate student and postdoc on the Fermilab collider program, I had studied the clusters of particles produced as a result of collisions of quarks and gluons inside the protons and antiprotons in the Fermilab beams. The issues associated with doing science with clusters of particles turned out to be similar to the issues of doing science with galaxy clusters. To me, the idea was both familiar and an opportunity to use what I had learned as a particle physicist in a new and challenging way. I saw the opportunity to establish a new project for Fermilab that would re-energize the engineering and technical team from the cancelled Run IIb project.

The skills and equipment used to construct the silicon vertex detectors for the Fermilab collider program seemed to be ideally suited to those required to build a large CCD camera. Fermilab had also been involved in the Sloan Digital Sky Survey (SDSS), in particular with the data handling and the mechanical infrastructure. In November 2003, we took the team of Run IIb engineers to Kitt Peak, near Tucson, where the Mayall telescope (twin to the Blanco) is located. We toured the telescope and discussed ideas with the technical people as well as with the leaders of the National Optical Astronomy Observatory (NOAO). The Fermilab engineers were excited and confident that they had the skills to pull it off. In December 2003, Alistair Walker and a few others came to Fermilab for the first official meeting between the Cerro Tololo Inter-American Observatory (CTIO) and Fermilab.

DECam's CCDs had to be sensitive to the long-wavelength, redshifted light from galaxies seven billion light years away, halfway back to the Big Bang. This distance is about twice as far as previous wide-area surveys such as SDSS had covered. Lawrence Berkeley National Laboratory (LBNL) had developed CCDs for the SNAP project that met these specifications and were ideal for the Dark Energy Survey (DES) science goals. LBNL was one of the early institutions to join DES collaboration. At the time, they had limited experience as a CCD production facility, but they were enthusiastic about this opportunity to demonstrate their CCDs in such a large camera.

Building in part on the experience and success of SDSS, DES planned to depend on photometric redshifts ('photo-zs'), rather than spectroscopic redshifts, which would take far too long to obtain. SDSS and other surveys had shown that photometric redshifts, which are obtained in DES using five filters (g, r, i, z and Y-band) spanning the wavelength range from 400 nm to 1065 nm, can be used to approximate the true redshift of galaxies in the field of view. This approximate redshift is precise enough for the measurements DES needed to make to constrain dark energy.

DECam filters are housed in a filter changer located behind the shutter, between the third and fourth lens of the optical corrector. The corrector is supported in a steel barrel and mounted in the Blanco prime focus cage, with a hexapod that provides focus, lateral positioning and the capability to tip and tilt the camera. The CCDs are cooled with a closed-loop liquid nitrogen system and housed in a vacuum vessel connected directly to the optical corrector. The CCD electronics are housed in thermally controlled crates mounted to the vacuum vessel. See Figure 2.1 for a schematic of DECam in its new prime focus cage.

The Temple review in June 2004 was one major step along the path to approval for the project. At that point, we had already made a preliminary design of the camera and had a good understanding of the equipment that would be needed to test the CCDs. The science was very compelling, and the proposal was given a generally positive review by the committee.

Fig. 2.1 Engineering drawing of the Dark Energy Camera and prime focus cage. The primary mirror is out of the picture, to the left side of the camera. Starting from the right side of the diagram, the major components are the electronics crates (red), the imager vessel (green), the barrel that supports the optical elements elements (blue), the filter changer (grey) with the sides removed so that the filters (green) can be seen in the out position and the arms of the hexapod (white). The leading surface of the first corrector element (C1) can be seen almost protruding from the left side of the barrel. The camera is attached to the cage at the heavy-duty 'hexapod ring'. The cage is attached to the telescope (not shown) by four 'fin' structures, also shown.

However, they expressed some concerns about our experience in the procurement of optics for the corrector. The chance meeting between Ofer Lahav (University College London [UCL]), John Peoples and Josh Frieman (Fermilab) in Paris later that same month occurred at exactly the right moment. It brought in expertise in optics from UCL and opened the way to collaborations with other European universities. A UK Consortium of researchers (see Chapter 1) applied for funding from the Particle Physics and Astronomy Research Council (PPARC) in the United Kingdom and ultimately they were successful. This meant they could purchase the glass for the optical corrector in 2007. This was very important to us, because it was something real. It's a big step in a project to go from ideas and pieces of paper to something tangible.

Given that some of the Fermilab participants came from cancelled projects, the question of whether DES would truly happen was in every-one's minds during the first few years of the project. DES had many strong qualities and did not feel like a project that would be cancelled. For example, it had a strong, unique science program; it was comparatively inexpensive and could be built on a relatively short timescale (compared to so-called Stage Four dark energy projects like the Large Synoptic Survey Telescope [LSST]); John Peoples was a former director of Fermilab with stature and he knew his way through the Department of Energy (DOE) system; and we had strong collaborators and money from UK and US universities. Nevertheless, we ended up having to go through almost four years of reviews before the project was officially approved for construction. There were reviews of each individual camera component, such as the optics and the CCD readout. There were progress reviews. There were reviews of whether or not the Project Team (I was Project Manager, but there were another half-dozen 'Level Two' managers, and even more 'Level Three' supervisors) understood their roles in the Project Management as 'Cost Account Managers' working within an 'Earned Value Management System'. The Project Execution Plan, complete with risk management and change control, was all described in monthly and quarterly reports discussed in Project Planning Meetings. For each of these reviews, there were three levels: a practice review, a Temple review (internal to the lab) and a review by the DOE. There were also Integration, Commissioning and Operational reviews.

Eventually, we became comfortable with the review process. The reviews were valuable both because we needed to prepare thoroughly, with documented evidence, and because our reviewers, who were experienced in making astronomical instruments, often provided us with valuable suggestions and advice. We learned what to expect from the reviewers and could anticipate most of their questions. We were trained to conform to a uniform standard in our review presentations, with a common basic appearance in our PowerPoint slides and the presentation of a common set of information in a prescribed order. Towards the end of the process, our reviewers were impressed that all the Level Two Project Managers even dressed in 'DES colours', wearing black trousers (or skirts) with dark red shirts for each review.

During this time, we were able to set up a CCD detector research and development program, using contributions from the Universities of Chicago and Illinois. We developed a CCD test lab that led to greater understanding of the yield of the CCDs (the number of CCDs that met our requirements compared to the number fabricated) from LBNL. Uncertainty in the yield was one of the big questions, and determining it minimized a lot of the fabrication risks later. Fermilab had experience with handling and packaging silicon electronics for the silicon vertex detector, but had never worked with optical CCDs. We built a prototype imager in 2006 and worked on CCD readout electronics and the focal plane cooling mechanism. The CCDs must be held at a stable operating temperature of −100 degrees Celsius for maximum efficiency, so they are housed within a vacuum vessel and cooled by pumping liquid nitrogen around it in a closed loop.

We passed the important DOE CD2 review in January 2008 and that June they finally said, 'OK, go ahead and build it'.

2.1.2 Staying afloat: 2008 to 2010

Once funding for DECam was officially confirmed in the 2008 Congressional Budget appropriation, we could start buying parts for the camera. There's a standard process for procurement, but you also talk to all the people who've ever done anything like this before and then visit the vendors to try to get them excited about the project. For nearly the entire project (2004–2012), I spent a large percentage of my time travelling around the world, initially to select vendors and then to stay up to date on progress at the ones that we selected.

DECam shutter, which weighs 35 kg, was designed and fabricated by a group led by Klaus Reif at Bonn University and the Hoher List Observatory. It was the largest shutter ever built, and the precision measurements required by the survey placed stringent demands on the accuracy and uniformity of the exposure times. DECam hexapod and its controls were designed and built by ADS International in Valmadrera, Italy. The hexapod is a device consisting of six adjustable legs (actuators), attached between two flat rings (flanges). The lower flange is bolted in a fixed position to the prime focus cage (see Figure 2.1), while the upper flange is attached to the optical corrector barrel and can move it around. This system allows for adjustments to the focus and to the alignment between the corrector and the primary mirror. ADS International had not been asked to make a hexapod this large before and this, together with the required accuracy, drove them to develop an innovative design.

DECam filter changer mechanism was designed and built by DES group at the University of Michigan, using machined components fabricated at Leonard Machine Tool Systems in Michigan. The mechanism provides positions for eight filters and at present they are (*u, g, r, i, z, Y, VR* and *N964*).

While every component of DECam required innovation and thus risk, early funding and effort on the CCDs and the lenses were successful at mitigating those risks. We knew the filters would also be a challenge, but they could be added late in the project without causing delays, and we didn't have funds for them early on. Each filter is 13 mm thick, 620 mm in diameter and has a mass of about 9.95 kg; and each has to filter light uniformly over the entire surface to enable the precise photometric redshifts critical to DES science. Nobody had built filters of this size with our uniformity requirements before. We approached the vendors who had built smaller filters and asked them to put in bids. In the end, we chose a Japanese company called Asahi Spectra, because they had an excellent reputation and they needed to develop similar filters for a Japanese telescope project on roughly the same timescale as DES. Due to the way the funding works in Japan, and the synergy with the development for the Japanese telescope, they covered the cost of all the machinery and R&D and only charged DES the cost of fabricating the actual filters. We knew it was going to take time for them to build up all the required infrastructure, so we made a case to start early. The DOE helped us do that by moving money into the project in 2009, two years earlier than scheduled, so we made the contract with Asahi and sure enough it took them over a year to build and commission the huge coating chamber as well as the custom cleaning, polishing and testing equipment. Just when they had figured out how to fabricate filters that met our specifications, in March 2011, the Tohoku magnitude 9.0 earthquake struck Japan, causing a tsunami that initiated the Fukushima Daiichi nuclear disaster, followed by a major aftershock in Fukushima prefecture in April.

Asahi was based about 50 miles from the centre of the second earthquake, which caused disruptions to the power supply for months. They needed power for seven consecutive days to make a filter, so they were unable to complete our order. This was the very last piece of the project, and Fermilab, DOE and NOAO were very concerned about the progress. We even investigated switching to other vendors, but the added cost and development time was beyond what the project could bear. In the nick of time, Japan got back to reliable power and, remarkably, within three months Asahi finished the job. During this process, when they were very late, their spokesman would end every email with, 'We will keep trying'. I was afraid that they would just say, 'Forget it, this has cost us too much and we're going to quit', but apparently they were thinking the same thing about us! On our second to last visit, the spokesman picked us up at the station, we chatted for a bit and then he said, 'I'm really glad you're here; I was afraid you would give up on us'. I said, 'Really? I'm really glad you're here because we're counting on you and I was afraid you would give up on us too!'

In the end, the transmission of the filters turned out to be excellent, exceeding DECam transmission requirement of greater than 85% by a substantial amount, with excellent uniformity over each entire surface. Asahi Spectra went on to build filters for Hyper Suprime-Cam on the Subaru telescope.

2.1.3 Wrapping it up: 2010 to 2012

Meanwhile, at Fermilab, we built a full-scale reproduction of the telescope top-rings called the 'telescope simulator'. It had pitch and roll capability, allowing us to position the camera in any orientation that it might meet on the Blanco. Constructing it was a lot of effort, but it was essential for testing and debugging all the systems prior to shipping to Chile, and including it in our plans was critical for getting us through the early DOE technical reviews.

There was also strong, steady progress on the testing of the focal plane detectors. DES technical requirements demanded CCDs with low dark current, low noise, and high quantum efficiency, with high sensitivity to both optical and near infrared light. To achieve this, the CCDs were made of silicon about ten times thicker than conventional CCDs. The LBNL CCD group supplied CCDs to Fermilab as bare rectangular, 2 cm by

4 cm, 250 micron-thick 'chips' of silicon. Assembling and testing each CCD was a long process. We needed sixty-two 2048 × 4096 pixel CCDs for imaging and twelve smaller format 2048 × 2048 pixel CCDs for guiding and focus/alignment. The yield improved dramatically towards the end of the project, but we had to make the decision to order the last two batches before we knew how good they would be. In the end, 124 of the focal plane CCDs passed all the post-production tests and we chose the best of these. The seventy-four CCDs are arranged in a hexagonal pattern on the focal plane, where images are recorded (see Figure 2.2), and the field of view is so wide that about ten copies of the full moon would fit into a single image. In total, the imager has 570 megapixels.

The design and development of the front-end electronics was a joint effort among multiple institutions in the United States and Spain, including CIEMAT in Madrid, IFAE and IEEC/CSIC in Barcelona, Fermilab, and the University of Illinois. The imager vacuum and focal plane cooling systems were operational at Fermilab from June 2010. Our aggressive construction schedule was designed so that by December 2010 most of DECam systems were complete in time to be brought together at Fermilab for full system tests (apart from the optics) on the telescope simulator. Mock observing and extensive testing in a full-scale test dewar at Fermilab led to the discovery that the CCDs could be damaged if exposed to excessive light

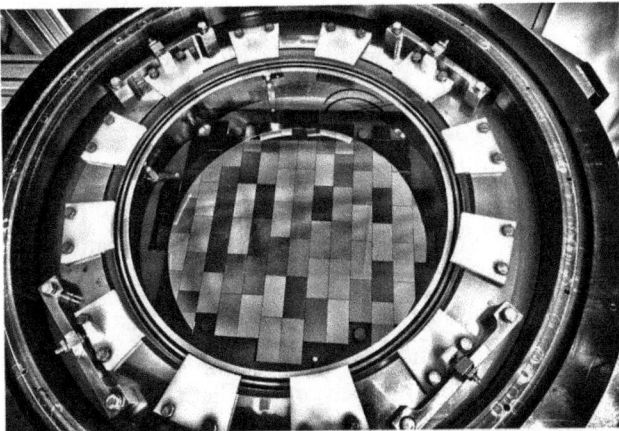

Fig. 2.2 The Dark Energy Camera (DECam) focal plane, with sixty-two 2k × 4k and twelve 2k × 2k charge-coupled devices (CCDs).

while they were turned on. This was contrary to the long experience of the astronomy community. Extensive study eventually revealed that the effect was real, and probably also affected other types CCDs to lesser or greater extent as well as ours: excessive uniform light exposure to the CCDs reduced the full well depth (which determines the maximum charge/light that can be collected by a single pixel). We also found that the sensitivity of the individual CCDs varied, such that some would become nearly unusable while others were not damaged. This discovery allowed us to develop procedures that turned the CCDs off whenever excessive light was a potential hazard and to install sensors on the focal plane that would turn off the CCD power if excessive light was detected.

DECam and its supporting hardware were shipped to Chile over the course of the next few months, with large, heavy parts travelling separately by boat and the more delicate pieces by air. Testing of the imager was completed by July 2011, and it left Fermilab four months later, arriving at CTIO in December. Within a week of being unpacked, it was put under vacuum, cooled down and, to everyone's relief, working beautifully. As the components arrived at CTIO, their performance was checked using DECam mountain-top instrument control and data acquisition system, named Survey Image System Process Integration (SISPI). SISPI was developed by Ohio State University with contributions from Illinois, Fermilab, Barcelona, CTIO and SLAC . The r, i, z and Y filters arrived from Japan in the summer of 2011, and the optical corrector arrived soon after the imager. Continuing with the theme of natural disasters conspiring to delay the filters, the very last component, the g filter, got diverted during its trip from Japan to Chile due to a volcano eruption in South America and finally reached the mountain in April 2012.

All our project milestones were met on schedule and DECam project was completed in June 2012 at 1.2 million dollars under budget. This money could then be used to help support DES operations. In total, 118 people in DES collaboration worked on DECam, plus many engineers, technicians, and computer professionals at DES institutions in addition to all the commercial vendors. Of course, the camera and its support systems still needed to be installed on the Blanco Telescope at CTIO and commissioned (i.e. readied for scientific observations).

2.2 The Optical Corrector *(Peter Doel, UCL)*

When Ofer Lahav returned from Paris to UCL in 2004 and told us about DES project, we were very excited about the opportunity to be involved. The Optical Science Laboratory (OSL) at UCL had previous experience in producing astronomical instruments, such as the UCL Echelle Spectrograph for the Anglo Australian Telescope, but DECam corrector would be a considerable step up in terms of the sizes of the lenses involved. A UK Consortium of researchers from UCL and the Universities of Cambridge, Edinburgh and Portsmouth (PI: Lahav) put in an application for funding to what was then PPARC, later the Science and Technology Facilities Council (STFC). We started by submitting a Statement of Interest (SoI) to the PPARC Science Board in 2005. This was recommended for review by the STFC Project Review Panel (PRP). Ofer Lahav and I appeared in front of this panel during the National Astronomy Meeting in Birmingham in 2005. We were competing against a proposed project to build a visible wide field survey camera for the Visible and Infrared Survey Telescope for Astronomy (VISTA) telescope, but we were asking for a much smaller portion of funds. John Peoples came over especially from the United States to join the meeting. After many approval steps, DES-UK group was successful and we were awarded a grant of 1.76 million pounds in 2006. This, along with a £250,000 contribution from the University of Portsmouth, was the first serious financial support for DES. These funds helped to build momentum for the bid to the DOE, the National Science Foundation (NSF), and other funding agencies, because they could see that the United Kingdom considered it a worthwhile venture.

DECam optical corrector allows the Blanco Telescope to view a large section (several square degrees) of the sky, so we can take images containing a hundred thousand galaxies in one go. Cassegrain telescopes like the Blanco have a primary mirror that collects and focuses light towards a secondary mirror, which then reflects the light back down through a hole in the centre of the primary to a focal plane. This gives excellent image quality over a field of view of a few tens of arcminutes, but at larger off-axis angles the image quality rapidly degrades. The solution for DES was to remove the secondary mirror and place a camera at the focus of the primary mirror. However, without additional optics, although the image quality in the centre of the field of view would be good, away from the centre the image quality would still deteriorate very rapidly. What the optical corrector does is improve the off-axis performance across the primary focal plane, so you can image a wide field of view with good image quality.

In designing the corrector, there are a number of issues to take into account. To save on cost, it is better to use as few lenses as possible, with as few aspherical surfaces as possible, because they're harder to make. This must be balanced against the fact that providing good correction across a very wide field is easier with more lenses and more aspheric surfaces. Early optical designs were developed by Mike Gladders (U. Chicago) and Steve Kent (Fermilab), who investigated various configurations and different combinations of glass types. The final optical design was done by Rebecca Bernstein, then at the University of Michigan, and in this design the corrector consisted of five fused-silica lenses with two of the lenses having an aspheric surface. We used it to produce technical drawings for optical fabrication to send to the polishers. The baseline optical design is shown in Figure 2.3. Fused silica was chosen because it has a high transmission and can be produced with excellent homogeneity across the full wavelength range required for DES and into the u-band (this is of interest to the astronomy user community, so CTIO later contributed a u-band filter for DECam).

There were several steps in the fabrication of the corrector. The production of the lenses is a lengthy task, so it was important that this was started as early as possible. The early availability of the funds from Portsmouth allowed the order for the glass to be placed in 2007. Corning Inc., based in Canton in upstate New York, were selected as the fused-silica supplier. First the disc shaped lens blanks were cut out of large blocks, called 'boules', of fused silica, which are produced by a flame deposition technique. Corning then milled the blanks for lenses C2 to C5 to approximately the correct lens shape. C1 was designed with a very deep, curved shape. Instead of being ground, the flat piece of fused silica was heated almost to melting point and 'slumped' over a former (a mould) to get close to the required lens shape before a final milling process.

The shaped lens blanks were then shipped to Thales SESO, in Aix-en-Provence, France. SESO did the fine grinding and polishing so that the fused-silica pieces attained the exact shape we needed for the camera. This

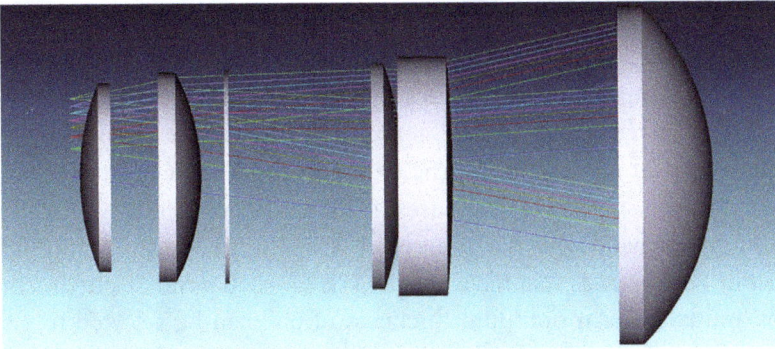

Fig. 2.3 The baseline optical design for the Dark Energy Camera (DECam). The elements, from right to left, are C1, C2, C3, a filter (one of four positions is shown), C4, C5 (Dewar window) and the focal plane array. The primary mirror is to the right, approximately 8.9 m from the vertex of C1. The total length of the camera from C1 to the focal plane is approximately 1.9 m.

is painstaking and difficult work, because a single mistake of over polishing means the working surface needs to be ground off and the polishing restarted. There were some hiccups in the process, and in particular C5, the smallest lens, had to be ground ~3 mm thinner than planned in order to remove an annulus that was accidentally ground into the edge. This was a concern, because C5 also acts as the vacuum window to the instrument Dewar, so it has to take a lot of pressure. We worked out that it is equivalent to about thirty people standing on it. Though we were confident that our calculations showed the lens was sufficiently strong, it was still a relief when it didn't crack when we did the first vacuum tests at UCL. To mitigate against light loss and scattered light, an anti-reflective multi-layer coating was applied to all the lenses except the biggest one, C1. C1 was left uncoated because of the difficulty of guaranteeing sufficient uniformity of the coating. SESO coated the C2, C3 and C5 lenses, while C4 was coated by Reynard Corporation in the United States.

The barrel, comprising a larger 'body', a smaller 'cone' and a new prime focus cage, was assembled at Fermilab out of steel, which is a strong, relatively cheap material. The barrel had very stringent requirements on its flexure, while keeping within a strict weight allowance. It also had to incorporate inspection hatches and the filter changer mechanism. The choice of steel for the barrel material posed a problem. The coefficient of thermal expansion (CTE) of steel, that is, the amount it expands with increasing temperature, is about 12 ppm/ °C, which is very high compared to that of fused silica, at 0.57 ppm/ °C. This is a problem, since the expansion and contraction of the steel barrel with temperature changes could impose stresses in the lenses if they were coupled to it directly. A solution that avoids this is to use Room Temperature Vulcanization (RTV) silicone rubber pads that have a high CTE of ~200 ppm/ °C, between the lens and barrel. An optimum thickness of pad can be selected to give a stress-free solution with temperature change. However, for a simple steel-RTV-fused-silica system the required thickness of the pads (about 5 mm) would mean that the lens would not be held stable enough to meet its imaging quality goals. To avoid this, a design was developed, based on work done on the MMT Telescope,[1] whereby each lens is first mounted via RTV pads into a lens cell (essentially a mounting ring) made of a nickel-iron alloy that has a CTE of approximately 3 ppm/ °C. The required RTV pad thickness for a stress-free solution is much smaller and thus the lens is held more stable. The mismatch in CTE between the lens cell and barrel is accommodated for by the use of thin flexures machined into the cell.

The cells were cast in Sunderland by Jennings Foundry Co. Ltd and milled in Abingdon, near Oxford, by Ashby Engineering Ltd. They were then sent to Fermilab, which aligned them into the barrel using a coordinate measuring machine, fixed them into position with fiducial pins, then unpinned them and sent everything back to UCL. At UCL the lenses were integrated with the lens cells and barrel.

The alignment and assembly of the lenses in the cells and barrel posed a considerable challenge. In order to meet the imaging performance requirements, each lens had to be positioned in centring and tilt to within a tolerance of approximately ±0.05 mm. To compare, an A4 sheet of paper is about 0.1 mm thick. The smallest lens weighs about 24 kg, and the largest is 173 kg, and we had to align them to a precision of half the thickness of a piece of paper! Each lens and each cell was supported separately on an XYZ motion and tip/tilt system placed on a precision rotary table (see Figure 2.4). The lens and cell could be accurately positioned with these

---

[1] The MMT Observatory in Arizona is known for its innovation in telescope design.

Fig. 2.4 Alignment of a lens in a cell, in the lab at University College London (UCL).

systems to micrometre precision by measuring the runout with the use of contact dial gauges as the parts were rotated. When the lens and cells were in the required position, they were gradually mated together. The lens was fixed in place in the cell by gluing RTV pads, that are mounted on radial cell inserts, on to the lens. The lens and cell could then be installed in the barrel in their correct positions using the Fermilab fiducial pins. Each lens position in the barrel was then checked using a laser pencil beam aligned along the axis of the rotary table. By measuring the reflected beam and transmitted beam positions as you rotate the lenses, the lens tilts and decentres can be determined. The lens spacing was controlled by the metal spacers placed between the cell and barrel. These had been milled to required thicknesses to place the lenses within 50 μm of their nominal positions.

The optical corrector was crated up in two sections and sent to CTIO in Chile in late 2011, travelling by truck and by air. We (David Brooks, Peter Doel and Michelle Antonik) flew in afterwards to assemble it at the telescope. We repeated the laser pencil beam test to check that the position of the lenses had not changed.

Due to the size and complexity of the optics and project timescales, a full aperture test of the optics was not planned. (A full aperture test involves replicating the light beam as it would be from the primary mirror and looking at the quality of the images formed.) Instead, we relied on measuring and aligning every single component very accurately and hoped we hadn't forgotten anything. The first time true images were produced by the optics was when the shutter of the camera was opened on the sky for the very first time. We were confident in our alignment technique, but it was still a relief to see that those first pictures were beautifully sharp and in focus!

Reference

Flaugher, B., H. Thomas Diehl, H. T., Honscheid, K., Abbott, T. M. C., Alvarez, O., et al. (2015). The Dark Energy Camera, *The Astronomical Journal* **150**, 150, doi:10.1088/0004-6256/150/5/150.

CHAPTER 3

Installation and First Light

Tim Abbott, H. Thomas Diehl & Alistair Walker

Once the Dark Energy Camera (DECam) was built and delivered to Chile, it needed to be installed on the telescope – a major undertaking in itself – take its first images, and undergo a series of commissioning tests to ensure that it would meet the stringent requirements demanded by the Dark Energy Survey (DES) to reach its science goals. These steps required a sustained, coordinated effort by many groups of people from disparate science and engineering traditions. This chapter is written from the perspective of three of those groups: Cerro Tololo Inter-American Observatory (CTIO), Fermilab and the survey commissioning team. No one narrative tells the whole story, and there is an overlap between them, but in combining all three a richer and fuller picture of the process is painted.

3.1 Installation and First Light *(Tim Abbott, Cerro Tololo Inter-American Observatory)*

This section is dedicated to Oscar Miguel Saá Martinez, Dario Guajardo Balcazar and Rolando del Carmen Puño Ramirez, invaluable colleagues sadly missed.[1]

Preparations at Cerro Tololo Inter-American Observatory (CTIO) for the Dark Energy Survey (DES) began a number of years before any

---

[1]Author's note: the work described here involved many people from many places over more than a decade and it would be impossible to give proper credit to everyone in the space available. I have therefore elected not to name anyone besides these three gentlemen and trust that those who were there know who they are and that they have our gratitude for a job very well done.

hardware from the project was delivered to Chile. The Blanco 4-m Telescope (see Figure 3.1) achieved its own first light in 1974. Nearly identical to the Mayall Telescope at Kitt Peak, it rides on an equatorial mount with its major rotational axis (right ascension) pointed at the south pole on the sky, instead of the altitude-azimuth arrangement common in modern telescopes, and it was designed in an era where computer-aided design tools were completely unknown. However, many of the original documents are still available and in occasional use, and it is a fascinating exercise to explore some of these hand-drawn diagrams and schematics with their accompanying mathematical, as opposed to digital, analyses. The Blanco is built like the proverbial battleship, its total structure weighing about the same as the newer Gemini 8-m telescope on nearby Cerro Pachon. Nevertheless, everyone who works on this telescope comes away impressed by what was achieved with the tools of the time, by the elegance applied to solving fundamental problems, and by its resilience to the ravages of use. It is fair to say that it was one of the best of its era, and the telescope has a long and distinguished history at the front line of astronomy research, including key contributions to the discovery of cosmic acceleration.

However, even before DES was conceived, it was recognized that the facility had suffered from a lack of attention as CTIO directed the majority of its resources to the development and deployment of the more modern Southern Astrophysical Research (SOAR) telescope next door to Gemini. As SOAR neared completion, CTIO moved to rectify this imbalance and as DES gathered steam, this developed urgency. There was an explicit feeling in some sectors of the community that the telescope was not, and could not be, opto-mechanically stable enough to make the precise observations required. Some of the difficulty was due to irreducible flexure in the telescope. Even steel bends significantly as the telescope moves around the sky. Such flexure is, however, repeatable and predictable and could be addressed by incorporating a 'hexapod' in the design of the Dark Energy Camera (DECam) to move it precisely into position on the optical axis of the primary mirror as the telescope moved. To function properly, the movements of the mirror would have to be predictable at the 10-micron level, but they were not. Indeed, if a suitably strong giantess could have picked up the 100-tonne telescope and moved it around, she would have felt the mirror sliding around inside. This movement was not repeatable, could

Fig. 3.1 The V. M. Blanco 4-m telescope at Cerro Tololo Inter-American Observatory (CTIO) in Chile, with star trails behind. Photo credit: Reider Hahn (FNAL).

not be resolved with the hexapod and its cause needed to be identified and resolved.

The primary mirror is supported around its edge by a system of twenty-four gravity-driven levers – the radial supports – which compensate for the changing direction of gravity at different telescope orientations. These supports prevent the mirror from bending and thus maintain its optical figure. These levers were known to detach from the mirror at the rate of about one per year, and more frequently in the late 1990s and early 2000s. Careful measurement and detailed computer and cerebral analysis of the design revealed a fundamental flaw in the attachment assemblies that could not have been anticipated by the original designers. A full set of new mechanisms was designed and installed in 2009. Since then, no further radial supports have failed, and the telescope behaviour is now predictable at the level required by DECam. The success of this refurbishment was necessary for DES to be able to make precise weak lensing measurements.

Several other tasks also preceded the arrival of DECam at CTIO. The telescope control system (TCS) is at the heart of any major telescope, providing a means of precisely pointing the telescope at any object in the sky and holding it on that position as Earth turns. Blanco's TCS was long overdue for an overhaul; a number of components were so obsolete that spares could only be obtained by scouring the Internet for used parts. Fortunately, a CTIO team had just completed installing a modern TCS on the SOAR telescope. Despite that telescope's altitude-azimuth design, the control problem is fundamentally the same for an equatorial mount telescope, and the port of the SOAR TCS to Blanco proceeded smoothly, being functionally completed in time for the installation of DECam. Considerable further tuning would need to be carried out during commissioning, once the instrument was installed. Upgrades to the design now proven on the Blanco are being retrofitted to SOAR, and the basic TCS is being developed further for the Large Synoptic Survey Telescope (LSST).

A class 100 cleanroom[2] was designed and built in a spare room in the telescope building to provide a pristine space in which DECam imager and other equally sensitive instruments could be opened up and worked upon. Other telescope and dome infrastructure were expanded, particularly in the area of thermal control and cooling. This latter work has continued to be refined even after DECam was installed, as we learn better how to maintain thermal equilibrium between the telescope, the building and the mountain atmosphere. This has provided the best possible environment for DECam's observations within the limitations of the site and building. In late 2011 and again in 2018, the primary mirror was re-coated with fresh aluminium to maximize the observational efficiency of the telescope. As with all major telescopes, this is done on a near regular schedule as the old coating degrades.

Meanwhile, apart from building the instrument itself, DECam team were also preparing for installation by building a 'telescope simulator' at Fermilab. This consisted of a near copy of the telescope top rings mounted in a race that was in turn mounted on a coplanar axle. The final result

---

[2] A room in which no more than hundred particles of size 0.5 μm or larger are permitted per cubic foot of air. For comparison, the air outside in an urban environment typically contains millions of particles of this size or larger.

was an instrument mount that could be rotated into any orientation so that DECam could be tested for the effects of gravity in any orientation of the camera. In practice, it was used only at a small number of key orientations sufficient to demonstrate adequate function, but the simulator was also useful as a platform to practise installing the 4350-kg instrument (not including the support cage) on the telescope itself. Initial tests failed, requiring adjusting the existing fittings to get the instrument into position in the top ring. However, subsequently, with CTIO engineers in attendance, DECam was mounted on the telescope simulator at Fermilab using fittings identical to the over-constrained mounts on the telescope itself. It was tested and declared ready for shipment.

As with any major project, DECam development was not without its hiccups. Quite late in the fabrication phase, it was realized that a misunderstanding over the telescope weight allowance had resulted in a design that was significantly heavier than anticipated. This was despite an extensive and active oversight program of as many as four project reviews a year. Dozens of managers, scientists and engineers failed to realize that Fermilab was building an instrument that would weigh as much as the existing prime focus assembly plus the weight of the inner top ring (i.e. an additional 2500 kg of instrument would need to be carried by the Blanco Telescope).[3] To lead the engineering team responsible for the installation, CTIO had recently hired a new Chilean mechanical engineer whose fresh eyes quickly spotted the discrepancy. After we all calmed down, 1960s' telescope design came to our rescue again. Analysis showed that a total of about 5600 kg of counterweight would need to be installed at the back end of the telescope, the mechanism could easily bear it, and the resultant flexure would be well within the hexapod's range to correct. More modern telescopes do not have such heavy duty designs and can tolerate only small extra masses – which would have forced us back to the drawing board and added years to the project.

Such heavy duty construction is not entirely advantageous of course, or modern telescopes would be built to the same rules. Our plans for the

---

[3] All told, the telescope now weighs 10% more than before DECam, and much of the extra counterweight is mounted on the outside of the Cassegrain cage, making the Serrurier truss uniquely unbalanced.

Blanco and DECam took us well beyond what the original designers of the Blanco had intended or foreseen, and after forty years of use there were a number of unknowns to face during the installation. In particular, the top end had not been adjusted for decades, some linkages were known to be loose and some were sufficiently corroded that we were not sure how easily they could be disconnected. We would also be making significant changes to the way the telescope is used. For example, the 1.3-m, 1000-kg f/8 secondary mirror would be installed in front of DECam rather than behind it as with previous prime focus imagers. Also, it would be completely removed from the telescope when not needed instead of simply rotated out of the way.

For the installation, we would be in large part disassembling and reassembling the telescope. With the telescope locked and braced in position pointing directly up (this being an earthquake zone, proper telescope bracing was all the more critical), we would remove the Cassegrain cage and move it outdoors for reinforcement to carry the additional counterweight. The precious and irreplaceable primary mirror would be removed and placed in a reinforced protective housing, built to withstand the impact of one of the largest bolts being accidentally dropped onto it from the top of the telescope. On the dome floor, we would have to move the bulky aluminizing chamber to one side to provide better access to the telescope underside; a boom lift would be brought in to allow access to the telescope top end; a custom-built, two-tier work platform would be bolted onto the top of the telescope and DECam itself would be raised up through the telescope into position at the top end. All of this work would then be reversed to leave the telescope usable again.

Essentially, we would be converting an operating facility into a construction site and back again over a few months. The import of this realization lies in the differing work patterns of the two circumstances. An operating telescope is normally a quiet and peaceful place in which the goal is consistent, high-efficiency observation from night to night, with minimal disturbance beyond what is necessary to maintain this steady state. Only when something needs changing or goes wrong does a disciplined and enthusiastic response become apparent. Like long-distance driving, there are long periods of relative quiet as the scenery slides by and short periods of intense activity when negotiating a complex junction. Conversely, the installation of DECam would require the coordinated activity of many staff, from several institutes, for several months, all working in a confined space around heavy and sensitive machinery in various states of functionality and disassembly. Careful planning and tracking of work was necessary, effective communications vital, and a new safety culture would be required. All this was laid out, reviewed and distributed before we began.

Nevertheless, no matter how careful the preparations, accidents happen. The very first day of the installation taught us an expensive, upsetting but ultimately salutary lesson. The f/8 secondary mirror was prematurely removed from the telescope and installed incorrectly on its handling cart. Flipping over, it fell face first onto a carpet-covered concrete floor, crushing the toe of one worker and badly bruising the leg of another. Work effectively stopped while we considered the causes and consequences of the incident. Fortunately, the injuries sustained by the workers were not severe, although the crushed toe took some time to heal; the f/8 mirror was repaired and continues in use to this day as if it never underwent such an unfortunate experience. Safety procedures were reviewed again and reinforced, and everyone learned substantially better discipline for the rest of the installation. In the end, we were delayed by only two weeks.

It is noteworthy that the two major components of DECam were not introduced to one another until they both arrived at Cerro Tololo: the imager, from Fermilab, containing the focal plane mosaic of CCDs and the corrector, from University College London, containing the lenses that would focus light from the primary mirror onto the imager. Additionally, there was no way to make a full pupil test of the assembled instrument without the telescope. Such decisions are trade-offs between time, cost and risk in all large projects. The Hubble Space Telescope lost a similar gamble in its first configuration and we learned from that to make extensive and redundant metrology of all critical interfaces wherever possible. In the event, the imager and corrector were successfully mated in the temporary cleanroom on the Blanco building's ground floor before the installation began, but we would not be certain that the optical prescription was correct until observing the sky through the telescope.

The installation crew was made up of people from several institutes and a wide range of nationalities. Many old hands from Cerro Tololo made key contributions – many of them learned their unique skills through long experience at the observatory and some first arrived decades ago, without the kind of formal education assumed in today's market. Four crew from Kitt Peak National Observatory outside Tucson, very familiar with the Blanco's twin there, the Mayall Telescope, joined us for several weeks, making up an additional shift for the most intense stages. A number of scientists, engineers and technicians participated from Fermilab. From time to time, further workers from the University College London, the University of Michigan, the Argonne National Laboratory, SLAC and others joined us as their skills were required. Since a lot of work would take place at significant height, CTIO hired a specialist mountain rescue team to be in attendance – fortunately their services were not needed. In the aftermath of the f/8 secondary incident, Fermilab provided a safety expert in addition to CTIO's own to ensure full oversight. Managing such a broad range of people and disciplines, not to mention the expectations of the two major funding agencies – the Department of Energy (DOE) and National Science Foundation (NSF) – required considerable diplomacy, not always successfully applied or accepted. We adopted a practice of team meetings at the beginning and end of each day with notes distributed throughout the project. The whole exercise was led on a daily basis by a capataz, or foreman, who was responsible for monitoring and directing all work on site through the team leaders for each work package (group of related tasks). Given that the whole process would take about seven months, it was decided that the site would be vacated every weekend, apart from a caretaker staff, to ensure that everyone was adequately rested.

For the installation, the telescope was fixed in its vertical position. This meant that all forces on the telescope would be symmetrical, and the prime focus cages could be hoisted from a point along their long axes, making them more stable to handle. Nevertheless, matters were complicated by the fact that the primary mirror had to be removed and the prime focus cages had to be rotated repeatedly about their short axes to get them under the lower telescope structure. The alternative would have been to install DECam with the telescope pointed near to the horizon, but this would have required an even more elaborate choreography in the long run, and tweaking alignment to equalize the forces in the spider structure supporting the cage within the telescope would have been more difficult. The choice was not controversial.

Before the old prime focus cage was removed, the position and alignment of its optical axis were located within the telescope tube using removable cross-wires in order to establish a fiducial by which the new cage would eventually be aligned. Thus it was possible to retain, in large part, the alignment of the prime focus with the primary mirror, which had been carefully established over a number of years of prior observations. Before installing the corrector in the telescope, the new empty cage was hoisted into place in order to test its fit and make sure no modifications were required. This dry run was recommended by a review panel before starting and, while the exercise was well motivated, no problems were found.

Eventually, DECam cage and corrector were installed and alignment checked against the fiducial (displacements were at about the 1 mm level). Work platforms were removed, the primary mirror rescued from its protective shell and reinserted into the telescope, and the now-reinforced Cassegrain cage brought back inside and reinstalled. It was now necessary to balance the telescope again before it could be moved. While the drives deliver considerable torque, they are quite low power (about one-third horsepower, with about 50 foot pounds of torque) and exquisite balance is necessary to allow the telescope to move from position to position on the sky. Given that the top end now weighed three tonnes more than before, and six tonnes of lead had been added to the back, this was not trivial to achieve. That done, we released the telescope restraints and began to move it gingerly around the sky. Were our calculations correct and would the drive be able to handle it? Yes! In the event, better than anticipated. Some tuning of the control system was inevitable, but we were shortly able to move the telescope over the western horizon and begin 'dressing' it.

DECam is cooled using a closed-cycle liquid/gaseous nitrogen system. Liquid nitrogen is pumped from a reservoir mounted on a platform in the dome through hoses down through the floor, up the telescope superstructure, around three rotational axes and on to the imager at the top of the telescope – and back. En route, some of the liquid nitrogen boils off and becomes gas, but a near-liquid state is maintained by keeping the whole system at a high pressure (100 PSI or around 7 atmospheres). To reduce thermal losses, the nitrogen hoses are enclosed in an evacuated jacket. The hoses are rigid where they can be, but some stretches must be flexible.

Managing hoses like these is not easy – just installing them took several weeks.

On top of this, numerous control and power lines were installed along similar routes as well as the bulky 'light pipe' fibre optic for the calibration system. Finally, we installed the comparatively trivial and apparently flimsy fibre optic line that would carry all the data from the instrument as it is collected.

Once 'dressed' and rebalanced, the telescope was ready to receive DECam CCD imager, as shown in Figure 3.2. At the heart of DECam, this delicate component was reassembled and checked out in the new cleanroom close by the telescope. On the scale of the project, wheeling it out and plugging it in behind the corrector was a quick and simple procedure, but nevertheless an event laden with tension and care. Once connected up to all its subsystems, the imager was turned on and checked out satisfactorily. On the night of September 11, 2012, a variety of nervous managers, scientists and engineers gathered in the newly modernized

Fig. 3.2 The Dark Energy Camera mounted at the prime focus of the Blanco Telescope. Photo credit: Reider Hahn (FNAL).

control room. The dome was opened, the telescope pointed high in the sky and the first image of the night sky taken with DECam. Initially slightly out of focus, it was moved to correct and stars and galaxies swam into view. Observatory rules prohibit alcoholic beverages and so a number of bottles of rather foul, children's 'champagne' were cracked and the new instrument toasted.

3.2 The path to First Light (H. Thomas Diehl, Fermilab)

DECam installation was a more than two-year effort by several teams of technicians, engineers and scientists. It required many trips to the telescope by members of the collaboration and many of us spent a significant part of 2011 and 2012 away from our homes. Most visiting team members lived in the mountaintop dormitory or in one of the two small mountaintop houses that the camera project rented. We had minimal personal items, but such as we had with us made our 'second home' on the mountaintop more comfortable. When we weren't working, entertainment was provided by the Internet. On some of the weekends, we were sent off the mountain and stayed in one of the hotels in the nearest city, La Serena. We took advantage of these weekends by visiting some of the sights in the Coquimbo region such as Isla Damas with its Reserva Nacional Pingüino de Humboldt and the Parque Nacional Bosque Fray Jorge. On weekends when we stayed on the mountain we liked to go for Sunday dinner to the 'Halley' restaurant[4] in the nearby town of Vicuña. A few elected to live in La Serena and had to make the ninety-minute trip twice a day.

3.2.1 Preparations off the Blanco Telescope

CTIO planned for the Blanco Telescope to continue to be able to use instruments at the Cassegrain focus after DECam's installation. Prior to DECam installation, the inner ring at the top of the telescope could be flipped over, orienting either the old prime focus instrument to receive light from the primary mirror or the f/8 mirror to re-reflect that light back down through the hole in the centre of the primary mirror to the Cassegrain instruments. However, the large vacuum-jacketed lines that

---

[4]Named after the comet discoverer.

would carry liquid nitrogen (LN2) to cool DECam would limit the rotation of the inner ring to 120 degrees, which was not enough for this trick to work for f/8 observing. DECam project provided the solution. A new f/8 handling platform was designed and built by engineers and technicians at Argonne National Laboratory. The downward end of the cage could be temporarily rotated 120 degrees towards this platform, which had movable fixtures that could be used to attach or remove the secondary mirror from the end of the cage. It was part of the equipment that arrived by boat in October 2010 and was the first piece of equipment to be unpacked. Its reconstruction and alignment, on the floor at the north end of the dome, were performed by Ken Schultz and Cheryl Jackson, two technicians from Fermilab, during January 2011.

Operating DECam would require many new racks of computers and disc drives and there was insufficient space in the old computer room on the ground floor of the dome. Plans were drawn up to relocate this room, taking over much of the space occupied by the old control room, where the observers and telescope operators performed their nightly work. A comfortable new control room was constructed by CTIO and outfitted with arrays of modern flat screen displays. CTIO finished the remodelling work in April 2011.

Meanwhile, critical camera components, including the filter changer, hexapod and shutter, arrived at CTIO. These were unpacked and taken up to the coudé room. Each piece of equipment was assembled and tested in a stand-alone configuration to verify it had not been damaged during shipping. In addition, each was operated through the same Survey Image System Process Integration (SISPI) interface, which would ultimately control the components when they were assembled together to make the camera. Establishing computer control of these elements prior to their installation ensured that no surprises would arise later on.

DECam imager, with all its CCDs, was also retested at CTIO before it was attached to the barrel and optics. In order to do that, we needed to install the utilities required to run the instrument: the LN2 system and chilled water system. Sensibly, we decided to install them into their final locations. The biggest of these components was the LN2 storage tank, which also contained a small pump used to push the LN2 to the camera, as well as equipment to reliquify nitrogen that had boiled to a gaseous state. Along with the 9-foot tall tank was a valve box that could switch the direction of the LN2 flow down to the coudé room for testing, or up the telescope once we got into observations as well as the vacuum-jacketed lines themselves. A team of five from Fermilab – Andrew Lathrop, Rolando Flores, H. Thomas Diehl, Project Manager Brenna Flaugher and Engineer Herman Cease – arrived at CTIO in early July 2011 to help the CTIO team with the job.

Each workday began with an 8:30 a.m. 'Breakfast Meeting', run by the CTIO 'Daily Foreman'. The agenda was to review the tasks to be done during that particular workday and discuss how we would coordinate our activities so as to eliminate potential safety hazards that could arise during work. After that, we would reconvene in the dome at the site of the particular day's work.

'Somebody get me a drill!' Brenna commanded. It was the morning of July 13, 2011, and one of the ten people standing around looking down at the dome floor ran off to fetch one. 'Herman, tap on the floor from below so I know where to knock the first hole, and can somebody else get me a saw'?

Soon Brenna was knocking holes in the wooden floor as we all looked on in amazement. Immediately after that, a 2-foot-wide penetration was cut for the run of the LN2 lines to the coudé room and to the telescope itself. That's how we got started. Over the next three weeks, the Fermilab and CTIO teams worked together to install these important components that would cool the camera and its electronics. Before the end of the trip, we had filled the nitrogen tank and pumped LN2 through the lines that circulated through the coudé room. When anybody went into the dome after that, the sight of the equipment reminded them that DECam was on its way.

Just before dinner one Friday during this same trip, Hernan Tirado called us to the control room. While we had been working, a strong blizzard had struck the mountain. Visibility was poor, and the roads were too slippery for the economy cars we had driven to the telescope. Luckily, Hernan's four-wheel drive pickup truck had chains on the tyres, so he was able to cart us off to dinner and respective accommodation. The wind howled most of the night. When we woke up, we found that our corner of the mountain was buried under 16 inches of heavy snow. The cacti were encrusted with ice. The Andes surrounding us had changed from desert taupe to a spectacular frosty white. By Monday, the road up and down the mountain was cleared. The snow on Tololo was mostly melted within a week.

Work on the interior of the camera required a very clean environment with temperature and humidity controls, so CTIO constructed a cleanroom in the coudé room. This room was first used for preparation of a Cassegrain infrared imager called 'NEWFIRM'. In mid-October, the environmental controls were working and the cleanroom was outfitted with the equipment needed for work on the interior of the imager. Assembly of the optics, which was to occur in early 2012, required a cleanish area, allowing access to a crane that would be dedicated to this process. A new crane was installed in the ground floor garage of the telescope building and a simple clean hut was constructed underneath it.

DECam imager arrived at CTIO in late November 2011. It was unpacked and moved into the cleanroom. On December 1, we removed the stainless steel cover exposing the interior of the Dewar to find that the delicate CCDs and the electronics connected to them were undamaged. The detailed engineering planning that went into packing and shipping the Dewar and its contents had paid off. It was what we expected, but nevertheless seeing the focal plane had survived relieved an unspoken tension. After a day of inspection and tidying up, the Dewar was closed up again, with the steel flange replaced by a 2-inch-thick flat window with a protective cover. The Dewar was then rolled out of the cleanroom and the three readout crates and the heater crate were attached to their final locations. The camera was hooked up to the electrical system and the cooling system we had installed in August. On December 5, we cooled it down and Marcelle Soares-Santos read out the first exposure. The long cosmic ray muon trail across the focal plane and subsequent exposures indicated that all the detectors were working. Following this initial test, a Fermilab team including Juan Estrada, Marcelle, postdoc Jiangang Hao, electrical engineer Terri Shaw and technicians Lee Scott, Kevin Kuk and Steve Chappa spent long weeks at CTIO verifying the imager was working perfectly.

The barrel, optics and alignment and testing equipment from University College London (UCL) arrived by ship in mid-December. Optical assembly commenced in January 2012. First, the alignment of the individual optical components with their barrel sections was checked. Then the two barrel segments were attached using an alignment key to ensure they were in the planned positions. The cage was left without the hexapod and barrel attached for a time, because a 'practice installation' on the telescope was to be carried out by CTIO using the empty cage. Once that was completed, the opto-mechanical team returned to Chile. DECam cage was stood on end and the hexapod centred within it and secured. Then the barrel assembly was lowered onto the hexapod, aligned and bolted in place. This assembly was stowed safely in the ground floor garage near the access for the in-dome crane. At this point, all the camera components were ready for final assembly and DECam was at last ready to be installed on the telescope.

Up until this time (early 2012), the installation work was carried out in daytime while visiting observers used the telescope at night. We always looked for, and sometimes saw, the 'green flash' as Sun set over the ocean. At night, the installation team stayed out of the dome and away from the working astronomers. On clear nights, we invariably gathered on the mountaintop to gaze at the southern sky and identify interesting Milky Way objects, such as the globular clusters 47 Tucanae and Omega Centauri, which could be seen with the naked eye. The Milky Way is a spectacular sight in the southern hemisphere as the centre of the galaxy is at times nearly overhead. Arturo Gomez, one of the CTIO support staff, sometimes set up a small telescope aimed at an interesting object. He also kept us abreast of interesting night-time events, such as a flyover of the International Space Station (ISS). On the occasion of the US Space Shuttle's last flight, on July 21, 2011, we were able to see the spectacular sight of the Shuttle Atlantis in orbit just ahead of the ISS.

3.2.2 Preparations on the Blanco Telescope

DECam imager installation was almost the final task of a seven-month period during which the old prime focus cage was removed and the new one installed, containing the barrel, the supporting hexapod and all the optics carefully aligned in their mounts. All utilities (e.g. electrical, vacuum, cryogenic, cooling water) were installed via moveable cable wraps and fixed segments, running from various off-telescope locations, up the telescope structure, then across to the prime focus cage. In parallel during this installation period, tests of DECam imager in the Blanco Instrument Laboratory reconfirmed that the camera met the performance specifications as measured at Fermilab in advance of installation. During the week of August 20, 2012, the telescope was moved to the northeast and the filter changer and shutter were inserted into their slot in the barrel, which was horizontal at that telescope location. The calibration system electronics and lamps were mounted on the telescope top ring.

Finally, on August 27, 2012, the imager itself was installed on the telescope. This was an exciting event. The imager, with electronics crates attached, mounted on its large wheeled stand, was rolled out of the coudé room onto the main floor of the telescope (something old). A specially designed lifting fixture was attached to the overhead dome crane hook and to the imager (something new). The imager was uncoupled from the stand and in a few minutes it was soaring 50 feet above the floor, which was fairly terrifying to watch. It was oriented over the northwest platform, which had previously been used to access the instruments in the prime focus cage and also occasionally to remove/install the f/8 mirror from the top of the cage using a mobile fixture. An updated imager-handling fixture was in place on the northwest platform, ready to receive the imager. The history of that fixture is notable, for parts of it had originated from the f/8 mirror handling fixture on the Mayall Telescope's southeast platform at Kitt Peak. While visiting Kitt Peak in 2009, Brenna saw it rusting under some bushes, identified it and realized she could use it. She had it sent to Fermilab, sandblasted and repainted, and it was used as the base for the imager handler (something borrowed). Now, in Chile, it received and supported the imager when that was unhooked from the crane. The next step was to use the fixture to line up the imager with the flange on the end of the barrel, to close the gap, and to marry the imager to the optics for the first time. It took two attempts because there was an unwanted conduction path between the barrel and the imager that had to be identified and removed so that the imager and electronics would be electrically isolated from the rest of telescope. The main body of the Blanco Telescope provided the blue. See Figure 3.2.

Then it was, 'all hands on deck' for the next two weeks as each team of subsystem experts needed to hook up and energize their components, and complete the final testing that would verify the parts' functionality. The 'top end' components that were last to be installed included the power distribution chassis, vacuum instrumentation, camera control computers, all the cabling and utilities and finally the cage-balancing weights. The telescope was decoupled from its parking station at the northwest platform on September 3. After it was balanced around the RA and DEC axes, DECam's 'first flight' took place. It was inspiring to see the Blanco swing the heavy camera around the dome and park it at the zenith. The camera didn't look so big up there. Within a few days, three of the filters were installed into the filter changer. On Saturday September 8, over four hours, the camera was cooled to operating temperature. Integrated system tests were carried out over the next two hours, including CCD readout noise studies, cosmic ray studies and the use of star trails, to check that the that the camera was well aligned with the cardinal directions and that the hexapod range of motion contained the focus. The Data Transport System (DTS) (responsible for efficiently moving DECam images over the microwave link from Tololo to La Serena and then to Tucson and National Center for Supercomputing Applications [NCSA]) was exercised. As this intense procedure drew to an end, members of the Fermilab team began returning home.

3.3 First Light! *(Alistair Walker, CTIO)*

Finally, on September 11, 2012, the Commissioning Scientist, Alistair Walker, reviewed the progress made since installation of the imager and gave the go-ahead for first light to take place that evening. Among those present were Tim Abbott, Alistair Walker, Chris Smith, Mauricio Rojas, David Rojas, Manuel Martinez, Darren DePoy, Brenna Flaugher, Marcelle Soares-Santos, Steve Kent, Theresa Shaw, Herman Cease and H. Thomas Diehl. Many people attended by Internet communication, including Nicole van der Bliek, David James, Klaus Honscheid, Josh Frieman, Adam Sypniewski and Juan Estrada. With both DES and CTIO commissioning members present, the telescope was pointed near the zenith and the telescope tracking turned on. Marcelle was at DECam controls.

A series of eight r-band exposures each a few seconds long were taken, adjusting the focus by 250 microns between each, over a range 1750 to 3500 microns. Careful metrology by Roberto Tighe and Patricio Schurter of the distance between the telescope primary mirror and DECam had indicated that the focus should be approximately at 3000 microns, that is, a 3-mm displacement from the midpoint of the hexapod focus range (approximately ± 20 mm). However, the temperature was a little warmer than when these measurements were made and the change in focus with temperature is approximately –120 microns/°C. The initial coarse focus scan indicated we were close to focus, and a subsequent fine scan decided on a best focus position of 2620 microns. This was the first observing milestone, finding that the focus was well centred within the range of adjustment. While maintaining that focus, some longer exposures were taken with three filters installed (g, r, Y), of fields containing the galactic globular cluster 47 Tucanae, the Small Magellanic Cloud and the Fornax Cluster of galaxies (see Figure 3.3). It was immediately obvious that the image quality (1.2 arcsecond full-width half-maximum [FWHM][5]) was good, with tight circular star profiles and, even more importantly, it was constant over the whole focal plane area. This was great to confirm, as image quality of the mounted optical corrector and filters had not previously been tested as a unit. The facility differential image motion

---

[5]In order to describe the apparent size of a celestial object, the entire sky can be imagined as a celestial sphere. A complete circle around Earth is divided into 360 degrees and there are 60 arcminutes in a degree and 60 arcseconds (arcsecs) in an arcminute. The angular diameter of the full Moon is about half a degree, that is, it subtends an angle of 0.5 degrees. FWHM or 'full-width half-maximum' is used to describe a measurement of the width of an object, when that object does not have sharp edges. A point-like source, such as a star, becomes a fuzzy blob when observed by a telescope looking through the atmosphere. The image has a profile close to a Gaussian curve. The approximate size of the star's image as seen by the eye can be represented by the width across the profile when it drops to half of its peak, or maximum, value. The FWHM can be used to compare the quality of images obtained under different observing conditions. In general, it is measured for a selection of stars in a single image and the 'seeing' or image quality is reported as the mean value. If the image quality is one arcsecond FWHM, objects one arcsecond apart on the celestial sphere can be seen as distinct.

Fig. 3.3 NGC 1365, also known as the Great Barred Spiral Galaxy, is a barred spiral galaxy about fifty-six million light years away in the constellation Fornax. It fits almost perfectly into just one of the sixty-two charge-coupled devices (CCDs) on the Dark Energy Camera (DECam) focal plane.

monitor (DIMM), which would have allowed an independent measure of the atmospheric seeing, was not working, but the 0.9- and 1.3-m telescopes were getting similar image quality to that being achieved at the Blanco with DECam.

All was not perfect, however: the telescope tracking and pointing were poor, there were issues with the primary mirror supports control computer and DECam and the telescope were not communicating reliably. But these were all things that could be tackled in the coming days. The critical initial tests of the optical performance resulted in the same focus across the whole focal plane. This meant that the optical corrector was functioning as designed, and the charge-coupled devices (CCDs) and other hardware such as the hexapod, the shutter and the filter changer were performing normally. It was a huge encouragement to the team and we toasted the new instrument with some fairly awful, non-alcoholic 'Kiddie Champagne'.

A message from DES Director Josh Frieman to DES collaboration a few hours later read:

'We are very happy to announce that DECam achieved its first light images (through the *g, r* and *Y* filters) on the Blanco Telescope tonight (around 2 a.m. September 12 UT). Although there are many tests and adjustments to be done during commissioning, so far the instrument appears to be working very well and has already delivered round 1.2 arcsecond images in its first couple of hours of operation (on a night when the seeing at nearby telescopes was reported to be comparable to this). This is a tremendous achievement by DECam team and by DES collaboration, and our congratulations go out to all those who have put in years of effort in the design, construction, installation, and now commissioning of what will be the world's most powerful survey imaging instrument. On the current schedule, we expect commissioning to run through October, and science verification will occupy the bulk of November. Pending success in those endeavors, DES could start in December'.

While the final statement proved optimistic (DECam 'season' is nominally August through February), DES began its first full season of operation on August 31, 2013.

CHAPTER 4

Commissioning the Dark Energy Camera on the Blanco Telescope

Aaron Roodman & Alistair Walker

For the Dark Energy Camera (DECam), we defined the commissioning phase as starting at the time of installation of the imager on the telescope, together with all ancillary equipment needed to support its operation. At the formal completion of commissioning, that is, when the instrument and telescope are basically functional, activities move to a 'science verification' (SV) phase, where scientists attempt to do a variety of astronomy projects (in this case including a proto-Dark Energy Survey [DES]) designed to test the whole instrument-telescope-facility system, in order to flush out problems and identify inefficiencies. In this chapter, we describe this phase in more detail, including the use of 'donuts' to focus DECam.

4.1 Commissioning

The activities carried out during commissioning are systems tests to evaluate the performance and to identify causes where specifications are not met. It should be emphasized that by the very nature of commissioning it is not possible to predict what problems will occur or be identified, and whether they should be attended to immediately, or later. The ability to respond to the unexpected with fast decisions is important, as is being able to adjust the schedule in near real time depending on weather and other circumstances, while still trying to keep within the bounds of the time allocated. Usually, commissioning a new astronomical instrument is scheduled in blocks of time, with the blocks separated by weeks or months during which other instruments are scheduled. In this case, DECam and the prime focus were the only option for the Blanco Telescope at this time, so such a strategy was not chosen. The f/8 secondary mirror that reflects the light from the telescope primary mirror back to instruments mounted below the primary in the Cassegrain cage was still under repair, and in any case the system for installing the f/8 mirror in front of the DECam corrector needed to be completed and tested. This arrangement was very different than that in place before DECam, where the f/8 mirror was permanently mounted at the top end of the prime focus cage and then the whole cage flipped by 180 degrees to bring the mirror into the correct position. Such a flip was not possible with DECam. Therefore DECam was tested as thoroughly as possible prior to installation and then commissioning was essentially a single block of time. As described below, this was a successful strategy, although it did mean several telescope issues were not fixed by the time of transition to science verification (SV), and some SV programs were impacted.

Given the dual purpose of DECam – to conduct the six-year the Dark Energy Survey (DES) using 30% of the Blanco Telescope time, and to function as a facility-class instrument for use by the general community – the requirements encompassed more than just the DECam instrument; they included requirements for the telescope, the specifications for a deliverable reduction pipeline for the community, and requirements that covered the use of the telescope in f/8 mode with other instruments. One important interface that was tested before and during commissioning was that between the data acquisition system (Survey Image System Process Integration [SISPI] and the Data Transport System [DTS]). The latter is made up of National Optical Astronomy Observatory (NOAO) hardware and software operated by the NOAO Science Data Management (SDM) group, and is responsible for efficiently moving the DECam images over the microwave link from Tololo to La Serena, and then on to the United States (Tucson and National Center for Supercomputing Applications [NCSA]).

Commissioning activities were derived from two defining documents that were written early in the project: the DECam Technical Specifications and Requirements, and the Community Needs for the Dark Energy

Camera and the Data Management System. A summary spreadsheet was prepared, The Compliance Matrix, that listed all the requirements and stated at what stage of the project they should be checked (e.g. installation, commissioning, SV) and in many cases, re-checked. A commissioning plan was then prepared, writing the requirements into a schedule. The first version was written in September 2009 and then regularly revised over the subsequent three years. The plan covered both commissioning and SV, and by the end of SV, the great majority of the requirements should have been tested.

4.2 The days and nights following First Light

The days and nights following first light were extremely busy, systematically exploring the performance of the telescope and instrument, and evaluating whether the facility supplies and systems were performing reliably and to specifications. Work was under the leadership of the commissioning scientist, Alistair Walker, with an at-telescope team of typically four or five DES and Cerro Tololo Inter-American Observatory (CTIO) scientists, with more engineers and scientists participating remotely when needed. There was good news and not so good news. As already mentioned, the delivered image quality appeared to be very good, and after a few days the hexapod system used to align DECam and the primary mirror was under control and able to put the optics into good alignment with the telescope pointing near the zenith. On an excellent quality night, September 14, with the atmosphere stable and DECam well aligned, Steve Kent (Fermilab) was able to report:

'I've just done an analysis of the focus sweep frame, and the shape of the FWHM versus focus step is basically invariant over the entire focal plane. This says that the focal plane departure from the surface of best focus (FWHM-based) is no more than about 25 microns, and likely less'.

This is hugely important – those few words mean that the performance of the optics, filters and focal plane all met or surpassed specifications, on a night where DECam was producing 0.75 arcsecond images! Given that the optics and instrument could not be tested together prior to commissioning, this was an enormously satisfying result. Furthermore, we found that the basic instrument systems, such as filter selection, shutter operation; and the telescope support systems, such as compressed air, chilled glycol/water, power, all functioned as expected from off-telescope tests. In addition, the complex liquid nitrogen cooling system appeared to be working as designed, after tuning by its builder, Herman Cease (Fermi National Accelerator Laboratory [FNAL]). Big kudos to the CTIO-FNAL team that installed all the complex flexible and fixed tubing all the way up the telescope to the prime focus cage.

In contrast, we were finding that the telescope was not performing on the sky so well, with pointing poor and tracking drift not as good as it should be either. Some of this was due to software bugs in the relatively new telescope control system (TCS), and these were attended to by the CTIO TCS team led by German Schumacher. Some of the problems arose because the characteristics of the telescope were rather different with the extra weight of DECam; and there were various random faults that needed to be diagnosed and then fixed. Our servo specialist, Michael Warner, was indispensable in this effort. A parallel effort involved testing the operating software (SISPI), swatting the bugs and making sure that the data from the charge-coupled devices (CCDs) and the telemetry were correct; Klaus Honscheid, Marco Bonati, Ann Elliot and Kevin Reil worked tirelessly in this area. Examples ranged from the simple, replace the wired keyboards on the instrument computer with wireless ones to avoid static discharges crashing the program, to the more complex, locating and fixing subtle sources of noise on the astronomical images. Relocating the ion pump controller removed these and afterwards the images were impressively clean. Klaus, based at Ohio State University, later displayed a plot showing the time he had spent on Skype talking to the mountain-top. It started at twelve hours a day or more, but (thankfully) rapidly descended to two or three hours a day.

One unexpected discovery was that the filter change mechanism had some light-emitting diodes (LEDs) inside and this light could reach the CCDs. So we scheduled a shut down for a week in early October to take the filter changer off the telescope and zap the LEDs. Experts from the University of Michigan came to lead this effort, which we felt was just since they had designed-in the LEDs in the first place! We also discovered that the hexapod x and y axes were 15 degrees from the expected orientation. No one owned up to this 'interface issue', but it required just a simple software fix.

Over the next few weeks a lot of all-sky calibration data was taken. The aims were to test all six filters; to quantify the flexure around the sky; to build in corrections to the optical alignment control; to improve the telescope pointing model and to test systems such as the guider. All this took many hours of observation and an equal number or more hours of analysis. An early success came when Aaron Roodman and Kevin Reil brought the telescope focus under the control of the DECam wavefront sensor – see below – and this subsequently proved to be very reliable and stable. By late October 2012, tests became dual purpose: recording engineering data for use in building instrument subsystem control systems and monitoring performance, while at the same time taking data similar to that which would be taken by an astronomer during the project. Also during this period the DECam Calibration System (DECal) was commissioned by the Texas A&M University building team led by Jen Marshall and Darren Depoy, with CTIO astronomer David James.

4.3 Donuts at DECam

Just like a personal camera, DECam needs to be focused to attain the best possible images. Ironically, it's harder to focus a telescope's camera than to focus an iPhone's camera. A cell phone camera has an auto-focus: it varies the focus and, by analyzing the images, can set the camera to the best focus. To do the same kind of thing, DECam would have to repeatedly collect extra images, but it takes around twenty seconds to collect the enormous images from DECam. We would much prefer to spend that time taking images to study dark energy, rather than in focusing the camera.

DECam has eight CCDs on the edge of the focal plane whose job is to focus the camera. These special CCDs are out-of-focus; they are placed above and below the focus plane by 1.5 mm. How, you may ask, can these image sensors placed out-of-focus help you focus the camera? The answer is donuts! The out-of-focus CCDs see stars not as points of light, but as the annuli you can see in Figure 4.1. The key point is that the bigger the donut, the further the camera is from focus. And since we have CCDs placed both above and below the focal plane by the same distance, we know we are focused when the donuts are the same size in both sets of sensors. We collect data from these focus CCDs for every image taken with DECam, so

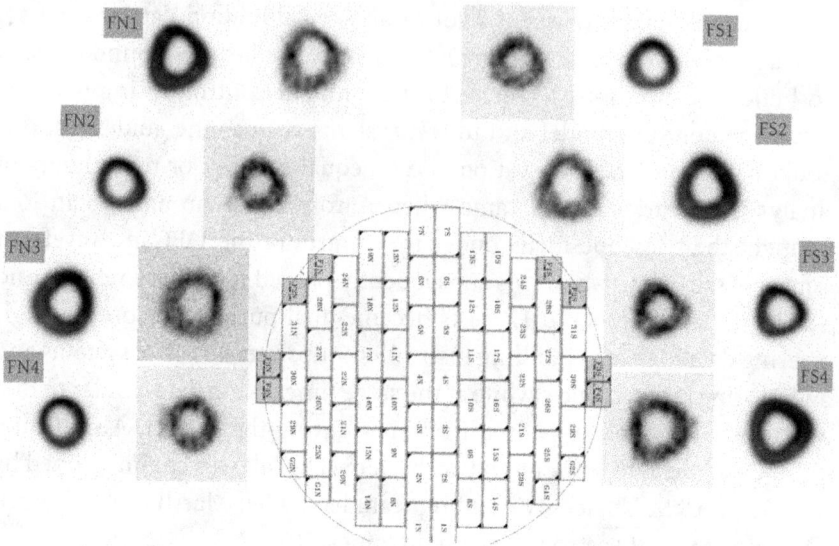

Fig. 4.1 The Dark Energy Camera (DECam) focal plane, with the eight out-of-focus sensors highlighted. Examples of out-of-focus stars, or donuts, from each are shown (interior), along with the corresponding fitted model (exterior). These donuts come from an image taken during DECam commissioning; the camera was roughly 200 m out of focus at the time. Credit: A. Roodman.

no extra images are needed and we don't waste any precious telescope time. Better still, we check and then adjust the focus on every single DECam image. Telescope cameras without this feature need to refocus every so often and astronomers constantly worry that they are slightly out of focus. With donuts there is no wasted time and no anxiety.

The auto-focus, or Active Optics, system for DECam was developed by Aaron Roodman and Kevin Reil, from Stanford Linear Accelerator Center (SLAC), with lots of help from across DECam. During the Commissioning period in the autumn of 2012 both spent about a month on Cerro Tololo, as well as continued work back in California. Of course, commissioning is exciting because usually multiple things don't work quite as planned. The first unplanned circumstance, observed in some of the very first images from DECam, was that two of the eight out-of-focus sensors had a problem with their maximum signal, or 'full-well'. Fortunately, after some study, we determined that this would not cause a significant problem, since only the very brightest donuts would be affected. The second unplanned circumstance was more serious. It became apparent from the beginning of commissioning that the telescope pointing was not yet as accurate as planned. This had consequences for the Active Optics, and the DECam guider as well, since both were planning to use a catalogue to select stars given the direction of each image. The telescope pointing was off enough that the catalogue would not work and we needed a plan B, fast.

Instead of using a catalogue to pick out the stars we expected, we had to look over each CCD and find good donuts from scratch. This kind of image processing is routine, but the Active Optics system also needed to complete its work in a fixed amount of time, which was only about seven seconds between exposures, and standard image processing could use up most of that time. So plan B was to hack a very fast, very simple-minded, piece of code to locate good out-of-focus stars. Aaron says, "This would normally have taken me a week or two to write and test, but in the pressure of commissioning, I wrote the algorithm during one night's shift. And since it was a very fast piece of code, I called it 'damn-fast donut finder', or 'dfdFinder'." The code worked well enough that we kept using it even when the telescope pointing was fixed.

At this point in commissioning, we had the basic components of the Active Optics working and were ready to test it out in normal observing conditions. A single setting in SISPI controlled whether the auto focus, or the auto-focus and auto-alignment, were to be used. One night during commissioning, we asked Klaus to flip the switch and turn the auto-focus system on. What was supposed to happen was that, after the image was taken, donuts would be located and analyzed, and the Active Optics algorithm would determine how much the focus should change for the next image. This calculation would be sent to the DECam hexapod, and the hexapod would move by the desired amount. A typical motion might be only 25 microns – less than the thickness of a piece of paper. Everything worked perfectly, except for one detail. Inside the code was a calibration constant, to convert from the donut measurements to the change in focus (the constant is 172) and we just divide what we measure by this number to calculate the hexapod motion. But somehow, we got our signals crossed and multiplied by 172! So this first time, the software told the hexapod to move by something like an enormous 10 mm. We watched the hexapod move with mounting alarm, fervently hoping it would not hit the hardware limit switches and wishing there was an emergency 'off' button. There was nothing we could do besides wait for it to finish, and when it did we breathed a real sigh of relief. Embarrassment averted!

One more thing we observed during commissioning is that the donuts themselves had an unexpected feature. The donuts are the image of the optical pupil, or the shadow that DECam makes against the Blanco primary mirror. We knew that this shadow would have a complicated shape, because the filter changer and shutter stick out from DECam's cylinder, and also because the field of view is so wide that we see DECam's shadow from an angle. However, we observed that the donuts had a strange additional feature, one that looks a little like jelly oozing out the side of an actual donut. And it turned out that this 'ooze' affected the donuts in a noticeable way. After some thinking we found a way to account for the ooze but, interestingly, we still don't have a proper explanation for this effect.

Commissioning the Active Optics system didn't end with the end of the commissioning period, and we kept working on the focus and alignment of DECam throughout the SV period and beyond.

4.4 The transition to SV

Finally, at the end of October 2012, the schedule read, 'Transition to Science Verification'. We selected a number of experienced astronomers, including people from DES team, to carry out short projects to test the whole DECam system and give a fresh evaluation of whether DECam was ready for normal operations. DES program consisted primarily of a 100-square degree survey to the full six-year depth, to allow detailed analysis of the data quality and the testing of analysis software. The community programs were chosen to range over a variety of science, from solar system to distant galaxies, testing DECam and the telescope in diverse ways.

DES SV was only scheduled to last about a month, to be followed by a transition to scheduled programs. However, due to concerns about image quality this phase was extended to the end of February while DES collaboration worked with CTIO to try to find out what was going on. The ups and downs of the SV period are described in Chapter 5.

Community programs, on the other hand, began as scheduled on December 1, 2012, and it was entirely fitting that the very first program was led by Jeremy Mould, who had been NOAO Director when DES began back in 2003. Other December programs included Scott Sheppard (the Carnegie Institute for Science) searching for solar system planets beyond Neptune, and David Nidever (then University of Virginia, now NOAO) with the 'DECam Magellanic Clouds Survey Pilot Program'. Both of these large programs continued in subsequent years and have been enormously successful.

In March 2017, we reached a milestone: 500,000 DECam images taken since first light in September 2012, corresponding to 0.5 petabytes (500 million million bytes) of raw data, and close to 200 refereed papers in the literature with the total increasing every month. Looking back on the whole installation-to-commissioning process, what is the view from almost five years down the road? Naturally, we held a 'Lessons Learned' meeting, where Large Synoptic Survey Telescope (LSST) folks were an interested audience. With my (Alistair) perspective as CTIO Director, DECam Commissioning Scientist and DECam Instrument Scientist, I think we demonstrated that a Department of Energy (DOE) Lab and an National Science Foundation (NSF) facility can build a complex telescope-plus-instrument system, successfully commission it, and then use it over the subsequent five years at near-100% cadence. Observers love it, the data is great, but (of course) behind the scenes it is a beast that needs a lot of care and feeding. So far, however, it's been very successful and we are now planning for the next ten years, including a major role as part of the support system when the LSST begins its survey in 2023.

CHAPTER 5

The Dark Energy Survey Early Observations and Science Verification

Gary Bernstein & Klaus Honscheid

After the Dark Energy Camera (DECam) was completed and installed on the camera, there was a commissioning period to establish the basic functionality of the instrument. This was followed by a science verification (SV) phase, with the purpose of providing an end-to-end test of all the systems needed to conduct the survey, including the camera, telescope, operations and data-handling systems. The initial allocation for SV was only twenty-three nights, from November 1 to 24, 2012, split in half between Dark Energy Survey (DES) activities and observations selected from a call for proposals from the astronomical community. After the completion of SV, DES would officially begin, and we would be 'on the clock', using up our 525-night allocation of Blanco time. It was critical, therefore, that by the end of SV we could be confident we were taking images that met the standards desired for the survey, and that we could run the survey efficiently. Telescope time is a precious resource, and we could not afford to waste it. Continuing improvements to camera and telescope systems were another key component of the SV program. SV would also be our first chance to take data in the patterns of the full survey, allowing us also to test DES Data Management (DESDM) and community pipelines (see Chapter 6). Ideally, these test runs during SV would also jump-start DES science programs. The capabilities of DECam were such that just ten nights of good data could produce a survey competitive with the best previous weak lensing surveys. In this chapter, we describe in more detail what SV in a large astronomy survey involves, the particular challenges faced in DES and how these issues were overcome. We review the state of the instrument and survey readiness at the end of SV and close with a brief summary of the surprising science analyses enabled by the SV data set.

5.1 Getting in gear

By the time of the Dark Energy Survey (DES) collaboration meeting in Munich in May 2012, it was clear that the collaboration did not have a handle on what activities needed to be done during science verification (SV) to demonstrate readiness for the survey, never mind a plan for who would do them and how to prepare. The director created an SV task force at this time; Gary Bernstein and Klaus Honscheid were appointed co-chairs. The task force was charged to verify that all systems were functioning at the required level and that image data was being produced with sufficient quality and efficiency to be included in DES. An additional goal was to exercise DES Data Management (DESDM) processing to determine whether quantities derived from image data were meeting DES requirements.

Perhaps a greater challenge was to convince the bulk of DES collaboration to start thinking of DES as a 'real' project! Of course, many DES members had been working hard for years building the detectors, electronics, optics and other systems. But the majority of DES members were astrophysicists with little to no involvement in the construction phase. They had helped scope the survey, forecast its performance, justify its scientific value and raise funds, but they were accustomed to seeing DES as something in the future, not a project in need of their direct and immediate contributions. The starting date of the survey was now just a few months away and the collaboration was still far from ready to validate and use the flood of data that the Dark Energy Camera (DECam) would begin to deliver.

Gary Bernstein recalls:

By May 2012, I had been a DES member for a few years but did not feel I had made any substantial contributions, aside from organizing a DES collaboration meeting at UPenn. I recognized, though, that

I was one of surprisingly few DES members with extensive experience in handling astronomical charge-coupled device (CCD) data. Most of DES members were either hardware and software experts from the high-energy physics world who had little experience with astronomical data, or cosmological theorists and data analysts who never worked directly with telescopes or sky images. I led the construction of the first mosaic CCD camera ever installed on the Blanco (the 'BTC' camera, installed in 1996–1997) and spent most of my research time working with CCD images, so it seemed that I could make a valuable contribution, filling the gap between the instrument-builders and the scientists. I decided to jump into full-time DES work by helping with the SV program, and I've been working full time on DES ever since.

Co-chairing the SV task force with Klaus Honscheid worked out extremely well. I warned him that I knew a lot about CCD imaging in general and could speak the language of the instrument teams, but I knew almost nothing about the specifics of DECam. Fortunately, Klaus knew basically everything about the software running the camera and survey, and the camera construction groups and Cerro Tololo Inter-American Observatory (CTIO) staff were fully engaged to cover any hardware issues during SV. I also had no experience working with a large group like DES, but Klaus's background in large high-energy physics projects made for successful coordination of the large number of people contributing to SV planning and execution.

Given that expected first light was only a few months away yes, the collaboration cut this a bit close and a more extended planning period is recommended for future surveys there was some urgency to set up a strong team and to develop a science verification plan. A request from the SV co-chairs to join the effort was met with great enthusiasm and by the beginning of the summer the SV team was established. The members were G. Bernstein and K. Honscheid (co-chairs), H. Lin, M. March, D. Gruen, P. Melchior, N. Sevilla, M. Soares-Santos, S. Kent, A. Roodman, K. Reil, C. Davis, E. Rykoff, R. Armstrong, A. Plazas, J. Annis, M. Jarvis, V. Vikram, M. Sako, J. Marriner, A. Elliot, E. Neilsen, E. Suchyta, K. Patton, P. Martini, J. Hao, D. Tucker, B. Flaugher, T. Diehl, J. Frieman, D. DePoy, D. Gerdes, A. Sypniewski, L. Buckley-Geer, R. Ogando, R. Nichol, L. Old, N. Mccrann, B. Nord, K. Hoffman, P. Rooney, J. Helsby, A. Fausti, S. Serrano, R. Casas,

W. Wester, A. Walker, D. James, C. Smith, T. Abbott, A. Kunder, R. Gruendl, D. Petravick and K. Paech. N. Regnault (Paris) also contributed.

5.2 The SV plan

The first order of business for the newly established SV team was deciding exactly what we needed to verify. First, we assembled a list of the tests we would require for DECam and the observing procedures to pass before allowing the survey to begin. We also collected a list of goals that were desirable to complete, but their achievement would not stand in the way of starting the survey. We then compiled a list of all the exposures and procedures that were necessary to test these requirements and goals, and collected them into a schedule that would fit into the available twenty-three half-nights of SV time, assuming decent weather. We also inventoried the software that would be needed to process the DECam exposures to determine whether the requirements and goals were satisfied, most of which was yet to be written. The DESDM pipelines were not yet at a sufficient state of maturity to rely on using them for SV analyses.

The final DES SV plan is available in DES document archive (DES-doc 6985) and can be summarized by the following key elements:

1. Perform two visits to each of the supernova (SN) search fields, to put into practice the SN template/detection processes.[1] Additional exposures to extend the depth and gain u-band information on those SN fields that overlapped existing deep spectroscopic surveys were included to enable photometric redshift (see Chapter 15) training experiments.
2. Perform a 'mini-survey' of about 100 square degrees, to the full ten-exposure wide survey depth in all five *grizY* filters, exercising online and offline procedures as planned for DES main survey. The original mini-survey region was chosen to be an equatorial field that covered parts of the Sloan Digital Sky Survey (SDSS) Stripe 82 and a Canada-France-Hawaii Telescope Legacy Survey (CFHTLS) wide field.

---

[1]We look for supernovae by comparing new images with older ones. Each region of the supernova fields is imaged to form a template and then, at a later date, the same area is re-imaged and compared to the template (see Chapter 9).

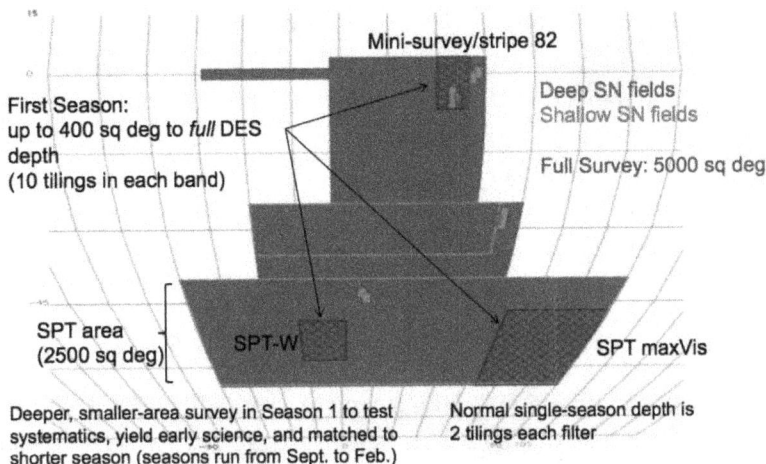

Fig. 5.1 Science verification observing plan overlaid on the Dark Energy Survey (DES) footprint.

As described in Figure 5.1, our SV test survey ended up in the southeast corner of our footprint.

3. Observe two areas of rich galaxy clusters, already covered by deep Hubble Space Telescope imaging, to test star/galaxy separation and photometry in crowded fields.

4. Take standard-star exposures in *ugrizY* filters at beginning, middle and end of each photometric night, to explore the photometric stability of the system.

5. Execute numerous specialized exposure sequences to facilitate astrometric and photometric calibration, check out DES optical quality, confirm signal and noise levels and test system behaviour to fault conditions.

6. Characterize the filter bandpasses using DES calibration (DECal) system during cloudy nights.

The SV Plan was (intentionally) overambitious: we recognized that achieving all of the goals would require far more software development than would be possible in time for SV, and more analysis than we would have person power for during the brief SV period. Indeed, many of the goals were not achieved during SV, even though we eventually extended the

SV period to three months rather than three weeks. Some of the issues raised in the SV plan took several years to fully understand, such as knowing how stable the DECam calibrations are over time. Nonetheless, it was useful to list all of the characterizations of the camera and survey that we should do eventually. And the list of SV requirements would prove critical in identifying problems with the camera and telescope that needed to be resolved before the survey could begin.

5.3 The 'eyeball squad'

At the end of commissioning, DECam images were still plagued with occasional artefacts that warranted further investigations. An 'eyeball squad' was initiated by Kathy Romer and led by Marissa March and Daniel Gruen. Over the first five weeks of the SV period they visually inspected over 3000 DECam frames for defects and anomalies, uncovering several important effects and providing a catalogue of images containing ghosts and scattered light. A few examples are shown in Figure 5.2, and terms such as edge bleed, zigzag lines and barcode stripes were added to the collaboration's vocabulary.

5.4 Early SV phase

DES SV observing nominally started on November 1, 2012, although the lines between SV and commissioning are blurred. The hope was that if it went well, we would start the main DES survey in late 2012.

The weather was excellent throughout this nominal SV period, with only a few hours of cloud over the full twenty-four nights, and a wealth of SV exposures were obtained. Instrument uptime was also very high. Unfortunately, the telescope pointing, a measure of how accurately the telescope can slew to a new position on the sky, was still poor, with typical errors greater than 100 arcseconds. More seriously, stellar images were noticed to often be highly elongated or even bimodal, a feature dubbed guider jumps (see later), degrading the point spread function (PSF)[2] size and ellipticity

---

[2] The PSF is a measure of the blurring of the image of a point source.

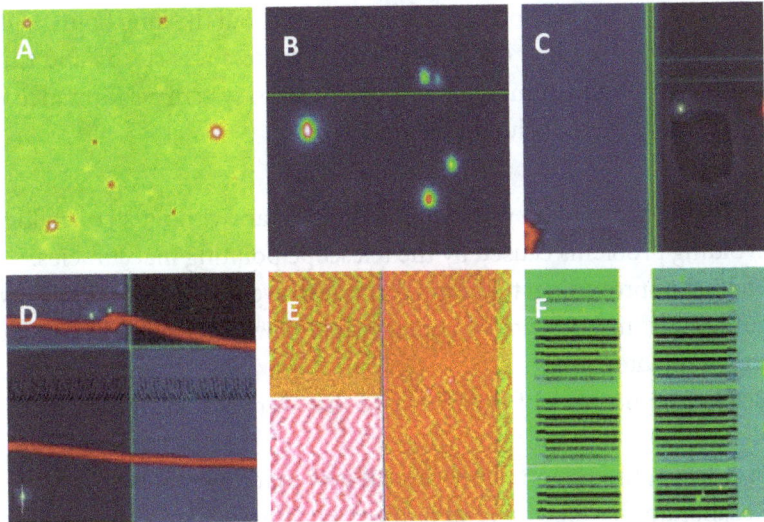

Fig. 5.2 Image artefacts found by the eyeball squad. A: normal point spread function (PSF); B: elongated PSF (tracking); C: tape bumps (charge-coupled device [CCD] mounting); D: herringbone; E: zigzag lines; F: bar codes (electronics noise and synchronization issues). The eyeball volunteers who provided this valuable service were Leon Baruah, Emma Beynon, Heather Campbell, Diego Capozzi, Mark Carter, Lucy Clerkin, Chris D'Andrea, James Etherington, Daniel Gruen, John Katsaros, Rebecca Kennedy, Lyndsay Old, Andreas Papadopoplous, Andrs Plazas, Malagn Rhys, Poulton, Tom Rigby, Kathy Romer, Philip Rooney, Harry Wilcox and Rob Williams.

well beyond DES requirements on many exposures. At the same time, some early exposures showed excellent image quality with seeing of 0.7 to 0.8 arcseconds full-width half-maximum (FWHM) across the full field of view. (For an explanation of FWHM see Chapter 3, footnote 5.)

Addressing these issues was a major focus at the start of SV, with observers on the mountain running nightly tests and the rest of the team trying to make sense of the data during the day. It soon became clear that the inaccurate telescope pointing rendered the baseline design of the DECam guider ineffective. We needed Santiago Serrano, the developer of the guider software, to come to CTIO, but he was in Europe. On very short notice, he dropped everything he was doing in Barcelona and, instead of going home from his office, flew to Chile. Thirty-six hours later, with missed connections and little sleep, he walked into the Blanco control room and announced: 'I am here let's fix it'. This positive attitude was shared by the entire SV team. Nobody ever complained about the long hours and the seemingly never-ending list of problems. Progress was steady and during the early SV phase, a number of improvements to system performance were implemented. These included the following:

- Changing the guider system to find guide stars without star catalogues, avoiding problems caused by the telescope pointing inaccuracies.
- Further improvement to the guider code to avoid selecting cosmic rays as guide stars. Interference of guider readouts with science-array readouts was eliminated.
- Blacken some parts of the telescope that were scattering light onto the CCD array.
- Improve pointing performance with new pointing maps.
- Recalibrate the focus and alignment system and use it to refine both passive and active hexapod position control. The hexapod is a device consisting of six adjustable legs, which enables adjustments to the focus and to the alignment between the optical corrector on DECam and the primary mirror on the Blanco Telescope (see Figure 2.1).
- Various telescope control system bug fixes.
- Development of new tools for studying image quality.
- Improved procedures throughout because of operational experience.

At this stage, image quality was improved and many of the artefacts found by the eyeball squad in early SV data were no longer present. Nevertheless, a few times per night exposures showed double star images, and it required some detective work by Michael Warner and Kevin Reil to get to the bottom of these guider jumps. Using telemetry data recorded by the new Blanco Telescope control system they found that each guider jump was preceded by a spike in the force measured at one of the primary mirror support structures, indicating a faulty actuator. Once that unit was replaced, the guider jumps were gone.

After the guider changes and the repair of the primary mirror support system, it was apparent that oscillations in the telescope's hour angle servo were still common, leading to exposures with elongated PSF (see Figure 5.2). The situation was especially bad north of zenith. Unfortunately, most of the SV targets – the SN fields and the mini-survey – were north of zenith and were impossible to observe with reliably good image quality.

Near the close of this early SV period, we targeted three known rich clusters and one blank field south of zenith in a successful attempt to obtain full-depth imaging in *grizY* filters with sub-arcsecond image quality. A bright moon during the observing period meant that these fields did not achieve expected DES depth in the *gri* bands, but these exposures still provided a valuable test bed for the DESDM processing codes, and were used for initial science investigations.

One major event of the initial SV period was the sudden failure of one of the sixty-two science CCDs on the morning of November 7, 2012. It was known from laboratory testing at Fermilab that the DECam CCDs could be damaged by exposure to bright light, so attention focused immediately on the fact that the camera was operating further into twilight that morning than on any previous night. But the light levels were still orders of magnitude lower than those noted to cause damage to CCDs in the lab and no other cause could be identified, so the failure of N30 remains a mystery. As a precaution, CTIO edict now prevents observation of twilight with DECam. A photosensor on the camera that automatically triggers a shutdown in bright light was adjusted to lower thresholds. It was quickly decided that we would not attempt to replace CCD N30 with one of the spares: the risk of damaging the array was deemed not worth the potential recovery of N30 (plus the half of CCD S7 that is unusable due to an unstable amplifier). In fact, another CCD (S30) failed mysteriously in November 2013, leaving 59.5 functional science CCDs, and we still did not attempt a repair. Even more mysteriously, S30 came back to life in December 2016!

5.5 The SV phase is extended

Poor tracking was known to be degrading some indeterminate fraction of the images' quality. The CTIO differential image motion monitor (DIMM) was also broken, which meant we could not compare DECam- delivered image quality to the best possible angular resolution given the atmospheric conditions at the site (astronomers call this 'seeing'). Given these unre-solved questions about telescope performance and image quality, DES and CTIO decided not to begin the survey in December but instead to extend the SV period over the full season. Between December 2, 2012, and February 23, 2013, an additional twenty-three full nights and thirty-three firsthalf-nights were allocated to DES.

Observers at the telescope write a 'night summary' at the end of each observing session, and from this record we find that during February 2013, out of 69 hours of scheduled observing time, 53 hours were spent actually observing the sky; 9.25 hours were spent observing with the specific purpose of testing the camera, the telescope, or improving a procedure; 6 hours were lost due to telescope of infrastructure failure; half an hour was lost due to camera systems failure and only fifteen minutes were lost due to bad weather.

Efforts to improve observing efficiency and image quality continued through the extended SV period. Diagnostic imaging and engineering tests were done on designated engineering nights but also during DES extended SV time whenever this would accelerate progress. A series of improvements to the telescope system were made over this time, leading to dramatic improvements in tracking performance:

- The guider software was upgraded to reduce timing jitter and delays in guider updates to the telescope control system.
- Faulty primary mirror support elements were replaced; these were causing occasional tilts of the primary mirror that recorded as jumps in the image positions on the detector.
- The horseshoe damper motor was recommissioned.
- Cable wraps were modified to reduce friction and sudden load changes.
- Tuning of servos.

In addition, active control of the hexapod position – first in focus, then in all five degrees of freedom – was successfully implemented using the focus and alignment 'donut' data (see Chapter 4).

By the close of the extended SV period, the telescope tracking performance had improved to the point where we were getting the image quality we would need for weak lensing and the other DES science projects. Because most of this period was subject to tracking oscillations in the north, we abandoned the equatorial mini-survey and settled on an observing program dominated by:

- Observations of the SN fields whenever seeing was poor or when a field had not been visited in five days. The SN fields were observed between seven and fourteen times during extended SV, an average of once per week per field.
- Main-survey-style observing in two fields, 'SPT-E' and 'SPT-W', located in the South Pole Telescope (SPT) field well south of zenith. The SPT-E data would be the primary resource for code tests and early science on DES main survey style data.
- Observations of the Cosmos Evolution Survey (COSMOS) field in *ugrizY*, in collaboration with Arjun Dey's community program.

The cause of poor telescope tracking in the north was finally traced to an improperly adjusted oil scraper on the telescope hour angle hydraulic bearing, which was not fixed until late January 2013.

5.6 SV results and accomplishments

The SV period was an extremely intense and productive period for DES. Long daily teleconferences between CTIO and DES home institutions and many SKYPE sessions during night-time observing were the norm during November and December as we discovered and diagnosed problems, and planned the agenda for each day and night. By the end of extended SV, the daily check-ins had become brief and routine. SV was highly successful, above all, in transitioning DES from a construction project to an observing project (it would take a little longer to transition into a science project). A long list of technical achievements during SV includes the following:

- Establishing the very high quality of the DECam detectors, electronics and optics: careful measures of the signal and noise levels of the camera showed these to be slightly better than the nominal predictions. The fact that there were some exposures with excellent image quality showed that the optics and filters were properly fabricated and installed.
- Identifying the glitches and defects of the camera: through the work of the eyeball squad, we were able to catalogue even the rare types of flaws in DECam images. Most were remedied through adjustments to the electronics, changes in procedure or judicious application of black

paint. In retrospect, some issues with the detectors could have been identified during laboratory testing, namely the low-light-level nonlinearities observed on some of the CCDs and problems with two of the focus and alignment devices. But because DES does not observe at low light levels, these issues do not affect DES science, and the requirements that guided the lab testing did not emphasize performance in this regime.

- Demonstrating and refining the DECam and DES observing schemes: innumerable fixes were made to the software that controls the components of the camera and telescope, to the obstac code (see Chapter 8) that chooses DES targets, to the software that reports the quality of the images and to the system that keeps the camera in alignment. We succeeded in altering our observing protocols to be unaffected by the unexpectedly inaccurate pointing of the telescope. By the end of SV, we had a system that we knew would take the images we wanted for the survey, with little wasted downtime and taking full advantage of the sky conditions.

- Bringing DES SN search up to speed. We chose the ten fields where we wanted to search for SN and succeeded in taking the observations we wanted of each, running the image subtraction pipelines to find the things that changed in these fields and determining which of them were high-redshift SN. DES SN working group went through many late nights and pizza deliveries to accomplish this.

- Producing a miniature version of DES wide survey. In the last weeks of extended SV, we successfully surveyed about 150 square degrees of the sky at the south-eastern edge of DES footprint, with the same number of exposures in each filter as we expect for the completed survey. In the closing month of SV, the telescope tracking issues had been largely resolved and the quality of this data was close to what we wanted for the full survey. This SPT-E survey, plus the images of selected galaxy cluster fields, would fuel the development of the DESDM pipeline and science analyses for the following two years.

The largest piece of unfinished business at the close of SV was the optimization of the image quality. This is always one of the hardest things to do for a ground-based optical observatory, because the seeing is fundamentally limited by air turbulence in the atmosphere and even within the dome. Like other kinds of weather, the seeing is variable and unpredictable, so even in the best of conditions it is difficult to tell whether an improvement in seeing is due to a change in the instrument, or just a random fluctuation in the weather. It can take months or longer to discern cause and effect. SV efforts at improving image quality were further stymied by the blurring caused by the telescope's tracking problems, by the absence of a functional instrument (DIMM) to monitor the atmospheric turbulence, and by our learning curve on the telescope alignment system. Efforts to improve the Blanco image quality and optics control continued well past SV, including major observatory changes such as installing a new chiller system to cool the dome air during the day.

Aside from the 1.5 non-functional science CCDs, the only shortfall of the DECam/Blanco system from the design specifications that had long-term impact on DES data was that the telescope motion was slower than expected, due to the added weight of DECam. The pre-survey plan was to have typical gaps of twenty seconds between exposures; during SV, this was more like thirty seconds, later shaved down towards twenty-five seconds. While five seconds of extra time doesn't seem like much, this adds about 5% to the time of a typical DES exposure, costing about twenty nights over the course of the entire survey. On the other hand, we learned that we could gain some time by observing closer to twilight than originally planned. At the end of SV, we knew that the productivity of DECam would be within a few percent of the design plan, a rare and impressive achievement.

5.7 The impact of SV data

DECam has proven to be a refreshingly boring instrument, nearly free of quirks that would complicate or compromise the analysis of its images. An early demonstration of the power of the DECam/Blanco combination was the discovery by community observers S. Sheppard and C. Trujillo, during the first week of SV, of 2012 VP113, the solar system object with the most distant known perihelion.

By the end of SV, DECam had collected enough data to prove that DES could discover high-redshift SN efficiently, which was important for obtaining time for spectroscopic follow-up observations on other telescopes (see Chapter 14).

The SV images taken in the southern parts of DES footprint served as the development platform for the DESDM data processing pipelines, leading eventually to the 'SVA1' release of processed images and catalogues to the collaboration in September 2013. Further exacting processing was necessary to conduct weak gravitational lensing measurements from the data. The first science paper using DES observations was an analysis of the gravitational lensing fields and galaxy distributions around four massive galaxy clusters, led by Peter Melchior and others (2015). Figure 5.3 shows the distribution of dark matter inferred around the cluster RXC J2248.7-4431, as well as the DECam SV image of this field.

The SV data, particularly the approximately 150 square degrees of survey-quality imaging in the SPT-E field, fueled the demonstration of an impressive array of cosmological survey methods using DES data. Most of these relied upon careful measurement of the galaxy positions, shapes and photometric redshifts across the SPT-E survey, requiring the development of many scientific analysis codes layered atop the DESDM output catalogues. Cosmological constraints based on these analyses were mostly published over the summer of 2015, with the results of SV cosmic shear measurements becoming the first science 'Key Paper' from DES. The

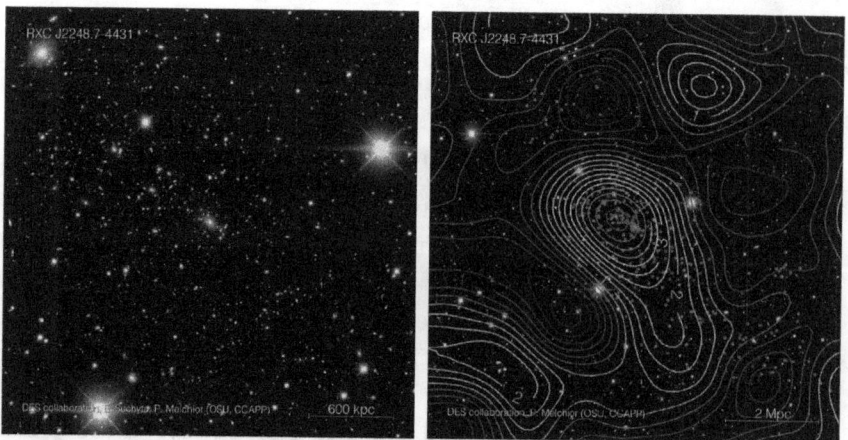

Fig. 5.3 Dark Energy Survey (DES) science verification (SV) image of galaxy cluster RXC J2248.7-4431. Left: the inner 10 × 10 arcminutes. Right: the mass distribution and red-sequence galaxies within 30 arcminutes.

two-year interval between the end of SV observations and the publication of cosmological constraints is actually shorter than that which had been achieved by previous weak lensing surveys. In the long run, DES SV was successful not only in showing that the camera, telescope and operations were ready to begin the survey, but also in showing where we needed to take more time to improve performance and in collecting publication-quality cosmological data that gave the collaboration a head start in developing the science analysis pipelines for the survey data to come.

Reference

Melchior, P., Suchyta, E., Huff, E., Hirsch, M., Kacprzak, T., et al. (2015). Mass and galaxy distributions of four massive galaxy clusters from Dark Energy Survey science verification data, *Monthly Notices of the Royal Astronomical Society* **449**, pp. 2219–2238, doi:10.1093/mnras/stv398.

CHAPTER 6

DES as a Big Data Machine Part I: The Dark Energy Survey Data Management System

Robert Gruendl & Donald Petravick

We are in the age of big data, with massive amounts of unstructured data being continuously generated through science alone. Astronomy is not yet dealing with the same level of information processing as particle physics experiments, but the volume and complexity of the data set is ever-increasing. Besides which, the particular nature of data produced by an astronomy survey calls for unique and creative solutions. This chapter and the following describe some of the ways in which the Dark Energy Survey (DES) is evolving to understand and negotiate these challenges.

6.1 What is the Dark Energy Survey Data Management?

The Dark Energy Survey Data Management (DESDM) team comprises astronomers, computer scientists and system engineers, who work together at the intersection of scientific research and the computing systems that enable it.

The DESDM system provides the infrastructure and computational power needed to store the survey data and to process it into high-quality images and catalogues that can be analyzed by scientists. The Dark Energy Camera (DECam) takes about 400 very wide images per night. Since DECam is a mosaic camera, each raw image contains sixty-two charge-coupled device (CCD) images and is a gigabyte in size. This means as many as 25,000 separate images, a total of 2.5 terabytes of data, may be collected in a single night. These images are sent via a microwave link from the Blanco Telescope on Cerro Tololo to the nearby coastal city of La Serena. From here the images are transferred via a number of national research networks and stored at the National Center for Supercomputing Applications (NCSA) at the University of Illinois in Urbana-Champaign (UIUC). From NCSA, data are immediately farmed out to large computing clusters at both NCSA and Fermilab.

DESDM performs two types of nightly processing as the images arrive at NCSA:

Because DES uses a ground-based telescope, images may be degraded by clouds and other effects. Uniform nightly visual inspection of tens of thousands of images for these defects is impractical. One set of nightly processing tasks uses software to assess whether the images taken meet the quality standards needed to produce a scientifically useful data set. Images that do not meet the data quality requirements of the survey are passed back to the observatory during the daytime, for re-observation. It is important the assessment happens promptly, before the corresponding part of the sky rotates over the horizon. In addition to image quality assessment, this initial data is made available to the DES collaboration, where it has been used to find trans-Neptunian objects and dwarf galaxies that orbit our own Milky Way galaxy.

DES also observes special Supernova Fields, and DESDM supplies additional nightly processing for these observations. Images from the night are compared to a template image derived from previous observations. The comparison detects new, bright objects in and around galaxies. Observations from various nights are compared and a light curve is constructed. The light curves of type 1a supernovae have a distinct shape and for these objects a distance to the galaxy hosting the supernova can be determined. Detection of a supernova may trigger a follow-up spectroscopic observation on another telescope, so it is important that supernova processing occurs promptly and reliably.

In addition to the nightly processing, DESDM re-processes all data and produces a variety of data products in the form of annual releases. Images that pass all the initial tests are cleaned (reduced) to remove artifacts produced by the camera and to subtract signatures from the Earth's atmosphere. Information about the stars and galaxies are catalogued and images from the same part of the sky are combined in order to increase image depth. In this process, data from all years of the Dark Energy Survey are used to provide the best estimate of an object's brightness (photometry) and position on the sky (astrometry). Due to the variable conditions under which the images were taken, this stacking or 'co-addition' process is not trivial.

DESDM then releases these clean, stacked and science-ready images to the rest of the DES collaboration so that scientists can make discoveries. The annual releases are a key data product enabling the majority of the studies characterizing Dark Energy. In its first four years, DESDM has produced over two petabytes of data. Scientists may want to see one or many images. They can download raw or cleaned data corresponding to the CCD sensors in a single exposure of DECAM, or co-added images built from combined data.

Because of the volume of data, most of the science work involves querying catalogues of detected objects. What does this mean? Software, primarily the AstroMatic software, scans the images, which DES has taken. When the software detects a star or galaxy, the software calculates a number of parameters describing the object. Example parameters are the probability the object is a star or a galaxy, measurements of how bright the object is and parameters that characterize the shape of the galaxy. Altogether, about 200 parameters are collected for each object detected. A scientist formulates and executes a Standard Query Language (SQL) query that specifies the objects of interest, and so obtains the data needed for a study. When studying this catalogue data, the scientist may want to confirm that the objects selected by a query look as expected in the images. So DESDM provides a cutout server, which makes small object cutouts corresponding to the objects that are being studied.

Public release: raw data from the survey is made public one year after it is taken and reduced data is released publicly at intervals throughout the survey.

DECam takes images for DES for about half of the year, during the South American summer. This is when conditions are best for a cosmology survey. In the South American winter the Milky Way, that is, the stars and dust within our own galaxy, is mostly overhead. During that time, the DESDM team updates and make improvements to the DESDM system. This involves improving scientific algorithms (how scientific values are measured), fixing bugs in the codes, upgrading the computing and storage infrastructure, running numerous tests and validating image and science quality. Images taken with DECam by astronomers during the non-DES observing period are processed through the DECam Community Pipeline (which was produced by DESDM) at the National Optical Astronomy Observatory (NOAO), and then made available for analysis.

6.2 Early DESDM

In any large survey such as DES, there is continuous learning about the optimal data products, the optimal production cadence and the optimal infrastructure for receiving, storing, producing and serving data products. This is a complex software system, which has to be ready to receive a fire hose of data once the survey starts to observe. The processing system needs to adapt to changing requirements, and to the production schedule itself.

Work on understanding DESDM began at the University of Illinois in 2004, almost ten years before first light. We expect that data service to the collaboration will not end until 2022. A scientific project of this length has to cope with technological change and also a change in the people constructing and running it. This is in stark contrast to a lone principal investigator obtaining a small amount of telescope time and producing a result for a single paper.

The DESDM system was originally envisioned as two monolithic pipelines: one that would reduce the data acquired on an individual night and another that would combine image products from the nightly pipeline to create stacked images. An independent pipeline (with initial steps similar to the nightly pipeline) was also developed to search for super-novae using difference-imaging techniques. Initial funding was provided to develop and test these DES pipelines and the community-use pipeline that would be deployed at NOAO. Provision of a community pipeline was a pre-condition for DES being funded by the National Science Foundation (NSF) and Department of Energy (DOE) in the first place. It could be used by any astronomer who applied for time on the Blanco during the six months of the year when the telescope was not taking images for DES.

A number of architectural decisions were made that have stood the test of time: DESDM would be built on the AstroMatic Tool kit as the major component for cataloguing, astrometry and image co-addition. Catalogues of detected objects would be directly loaded into an Oracle relational database, and scientists would query that database to obtain the subset of all detections needed for their work.

The Oracle database would also keep catalogues from tests runs and prior data releases. This basic capability enables scientists to measure how pipelines are improving by providing the ability to easily compare the quality data products as the processing improves, but also supports the sustained science exploitation of any one release. Deep understanding of survey data is often a multi-year effort by a doctoral student. Other aspects of the architecture needed to evolve with time and technology, experience and effort and the focus needed to build a system that would function on a daily basis for about ten years.

How do you build and test a processing system without the actual camera mounted on an actual telescope? There are two ways: use software simulators to produce images similar to those DECam would produce, and reduce real data from other cameras, even if the camera is not as capable as DECAM. In the end, DESDM did both.

In the early years, tests on the DESDM system were made using simulated DECAM data. These simulations were produced at Fermilab, using computational resources on the Open Science Grid. This data contained simulated imperfections, designed to mimic the point spread function (PSF) and electronic noise, but it was actually very clean data compared to that from real observations. For instance, it is difficult to model shifting weather patterns, but in reality the seeing, the brightness and the noise change from one exposure to the next. This lack of real data was potentially a problem for the community pipeline in particular, because it is supposed to handle any data an astronomer chooses to gather. The DES data are being taken from 5000 square degrees of sky outside the plane of the Milky Way galaxy, because this area is most useful for dark energy measurements, but the community pipeline needed to be able to cope with data from the telescope pointing anywhere on the sky. On the other hand, since we know how the simulation was made, we could compare the objects, which were detected by our software to the objects the simulation placed in the images, and keep score on how well the software worked.

The technical infrastructure supporting DES processing and serving was conceived such that the data reduction and initial measurements would be carried out using distributed databases (a network of supercomputers) instantiated at different sites in the United States and Europe. This process was coupled to a database that tracked the image products and the individual measurements of astronomical objects within those images. The database was to be replicated at secondary and tertiary sites within the Dark Energy Survey collaboration. In addition, a Science Portal to provide analysis computing was to be developed by the LIneA group in DES-Brazil.

However, there was an orchestration problem with sending the data to the network, using the machines and getting the data back quickly. The file systems on these supercomputers are typically made to handle problems involving large amounts of data: the computer reads the data and then many hours later it will write out a result. Our computation jobs are not really that large (compared with data from a particle physics experiment, for example), but the pipelines are full of codes that run in short bursts of twenty or thirty seconds, such that the input for each successive code depends on the output of the previous code. If you have thousands of these running incrementally on the data-sets at once, you run the risk of knocking out the file system, that is, running the file system so heavily that the inputs and outputs are all colliding, trying to access the same resource. There is too much data going in and out of the processors too rapidly. Furthermore, each one of these twenty or thirty second jobs is working on a single CCD, but there are points in the pipeline where you need to know information about all sixty-two CCDs at once (one exposure) and so you have to orchestrate all the jobs as an ensemble. Supercomputers have interconnects between the cores so there are ways for them to share the resources to do this, but the effort needed to synchronize and maintain database instances was more than the project could sustain. A single failure on a CCD or exposure could halt the entire pipeline.

In October 2011, we initiated the DC6B data challenge, to process ten nights of simulated observations (forming a mini-survey) and in the end it took roughly one month to process this data. It was clear that the existing pipelines would have to be altered in order to meet the processing requirements for the survey.

6.3 Intermediate development of DESDM

When we realized the pipelines were not going to be able to run day and night in the way they had been laid out so far, it was essential to come up with a solution as quickly as possible because DECam was coming on line within a year. A large number of different groups were relying on there being a working Data Management System in place for the commissioning period, including the people upgrading the telescope at CTIO, the people at Fermilab building the camera, the funding agencies, and the scientists joining the collaboration.

6.3.1 Splitting up the pipelines

In 2011, there were essentially two monolithic pipelines: one was the nightly production pipeline and one was the co-add pipeline. The nightly pipeline was built to take all data from the night and process it together, but if something failed in the middle of the night, perhaps it was cloudy and one of the exposures was bad, this would halt the execution of the pipeline for everything. Our intermediate solution was to split up the pipelines into fundamental pieces that do specific jobs. That way you can decide whether or not each job is executing correctly, and if you need to re-run it, you can re-run that piece and not the entire monolith again. The original system also assumed that it would always be able to choose the best calibration ahead of time for how to run that night's data, and we discovered that was really not the case.

We need to calibrate two fundamental quantities: one is the amount of light measured from an object (photometry) and the second is the position of an object on the sky (astrometry). The pixels in the CCDs are not all uniformly sensitive to light and not exactly the same size, so you need to make corrections to the measurements for these systematic distortions due to the instrument. As the data is coming in you don't know what the best calibration is going to be, because it's only after all the data is in that you can find any problems that occurred. Prior to some point your calibration worked and after that point you need a different calibration, and we can solve for that after the fact by running the data once through the nightly pipeline and then running it again later on. We call this process First Cut and Final Cut. First Cut takes the last estimate of corrections to be made to the data and applies them again. Final Cut is informed by what happened when First Cut ran. It has prior knowledge about the observations and the data and this is put into what the pipeline is told to do. In this way the pipeline is continually being upgraded. Furthermore, First Cut and Final Cut are built to work on a single exposure, which is the fundamental unit of the survey, so if one exposure fails and you have 500 of them processing, the other 499 keep going.

This method gives us the best result and is actually very close to what used to happen ten years ago, when an astronomer sat at a desk working on data. If that astronomer had been observing at Kitt Peak or CTIO, they would have brought their data home and put it through a set of routines to reduce it. Then they would have looked at the results and thought, 'Hmm, I think I need to change this'. They would change it and run it again, then they would look at the new results and say, 'Yes, that's better, but now I can see this other problem'. So they would run it again. It is making an informed decision about what you're doing.

At first light, in 2012, nightly computing ran locally on the NCSA Private Sector Program Cluster, as did the first initial Annual Releases. As the body of DES data grew along with the sophistication of the workflow system and codes, the DESDM workflow system expanded to include the capability of running on the Open Science Grid at Fermilab, the Blue Waters Supercomputer at NCSA, the Illinois Campus Cluster and the DOEs National Energy Research Scientific Computing Center (NERSC).

6.3.2 A stronger storage solution

In the short term, this solution took the form of standing up computer nodes using refurbished computers (new processors and memory) with the iForge/Private Sector Partners (PSP) Cluster at NCSA. This cluster had a much more robust file system on it than the grid computers. In addition, DESDM worked with the Storage Enabling Technologies group at NCSA to set up a 'storage condominium' that could serve as a central file store. This group was already preparing to provide a large storage facility for many different projects at UIUC and when DESDM began looking for a stronger storage solution, it gave them the final incentive to start providing a large service to a lot of groups.

DESDM now uses a large ORACLE Real Application Cluster instance at NCSA, which is a conceptually simpler system than the grid model. The storage condominium is a shared resource serving multiple petabytes of data and DES occupies roughly a tenth of it. There are administrators who make sure that it stays up and the data is not lost. We then have a second spinning disc store as a backup for that, and the data is also saved on tape at Fermilab. It's a purposely redundant system and it ensures that if there's a catastrophe at one site, the data will survive at another site. One other factor that enabled this decision was the rise of competent high-performance networks supporting research, for example, ESNET and Internet II in the United States, and GEANT in Europe. These networks support access to a centralized system in the United States, almost as if the system was local.

The concept of data service for the DES collaboration changed as well. To speed up implementation of the new system, the requirement that the database be replicated at secondary sites and accessed through a system of web-based portals was dropped. In such a system, it's important to have a high degree of understanding of what the scientist wants. Get it wrong, and all you have done is to make it harder for the scientist to obtain the data needed for a study. However, the Science Portal continued to be successfully developed by LIneA in Brazil (see Chapter 7). The centralized storage and database services could have resulted in a data hoard giving the DES collaboration (and eventually the general public) limited ability to work with the DES processing results. Rather than hold the data hostage, the principle of democratic access through lightweight tools developed first at NCSA was espoused. It became clear that the scientists within the DES collaboration could be expected to use query languages and diverse tools of data intensive science to do their work. Essentially, any collaboration member has equal access to any file or database item. The bonus that comes with this is simple: if you provide access and simple tools, scientists are very likely to join your effort to improve those tools simply because they want the improvements. Simple web-based (or more technically RestFul) techniques provide access to the files, and simple query extensions make for a happier collaboration.

6.3.3 Realistic test data for the community pipeline

We had to provide the community pipeline – it was one of the criteria for the survey to exist – so we began to source real images. A prime source of real data was data from the Mosaic II imager, the camera that DECam would replace on the Blanco Telescope. DECam is a wide-field camera, with a field of view many times larger than the apparent size of the moon, whereas Mosaic II had only eight CCDs and sixteen amplifiers. Nevertheless, working with Mosaic II data provided experience of working with some similar effects we would see with DECam. There is a huge archive of images taken by astronomers using Mosaic II. In that archive, you can find observations taken within the galactic plane, observations of the Magellanic clouds, images with nebulae spread throughout them or images crowded with millions of stars. We wanted to see what would happen if you took these observations and tried to make the pipeline run. Then we could go to NOAO and discuss what worked and what didn't, and what needed to be done.

When the community pipeline was developed enough, it was handed over to NOAO to run and maintain. NOAO facilitate and support open access to an ensemble of US-operated observatories. We remain responsible for an underlying software pool and when we make improvements to that pool we make sure they understand what those changes are and they can decide whether or not they want to use them. They have taken some of our solutions, but sometimes they decide it's not necessary for what they're doing, because in the end they're not trying to produce a large, well-calibrated survey. They're trying to process groups of a few nights of data that have been taken by an investigator from a small institution, and they want to produce a product that that investigator can use as a starting point for their scientific work.

6.3.4 Taming hostile codes

At DESDM, we run a number of codes that are written by other people. This includes, for instance, the weak lensing codes. When someone writes a code, they have a specific idea in mind about how they're going to make it execute, but we have to embed that code inside a larger system. There are pre-conditions to how the code is going to find the data it will run on, and what the outputs are going to look like. So we have to rewrite parts of these hostile codes to fit them into our system.

By the time of DECam first light in September 2012, all the above measures had been put into practice. However, by the end of the period of Science Verification observations, it had become apparent that further changes would be necessary. The initial plan for the DESDM pipeline was very ambitious: it would not only measure the brightness and position of all the objects on the sky, and clean and stack the data, it would also produce weak lensing results and photometric redshift results, and this was not yet the case.

6.4 DESDM at present and looking to the future

The early processing framework (software middleware that managed the actual processing jobs, assembled inputs and curated outputs) was written with the early pipelines in mind. With the changes in the pipelines to handle the data in more flexible units, it was clear that this middleware needed to be updated. We took advantage at this point to change two principles that underlay how the system worked. One was a module called file-ingest that 'discovered' the type of each pipeline output. This was replaced by a much simpler notion, according to which the user must declare the 'type' of each output. This relieved the middleware of a problem and allowed the user to make changes and simply declare them to be so (rather than write an algorithm to identify each new data type). Along with this we imposed a rule that all filenames be unique, which simplified finding data products and also naturally allowed a means to express processing provenance within the database using the Open Provenance Model and to further tie the processing provenance to quality assessment.

In addition, we began to modernize the actual pipeline codes. Primarily this has been achieved by switching to Python (with wrapped C when necessary). This choice was driven by the fact that Python has been emerging as the language of choice for many younger astronomers. The payoff has been that a majority of the DES collaboration can contribute ideas, insights and sometimes even code, to help improve the pipelines.

We have now reached a point where the nightly processing runs with minimal intervention. The data comes in and within about twenty minutes it starts going through a machine to produce a result; then about three hours later that result is done. The supernovae data takes a little longer to process, but in general, the whole night's data is processed within six hours. Unless there's a problem, the most significant thing the telescope operators have to do now is keep watch on the weather. For example, if cloud cover becomes very heavy, they may have to prevent an exposure from being put through the pipeline. They also need to look at the results to confirm the processing has been done and has finished successfully, but overseeing the pipeline is fairly straightforward now.

As the survey proceeds, we have gradually been adding to what DESDM does. As of May 2017, we are probably 80% of the way to what was originally promised and we have found alternative methods that will better serve the collaboration in the long run. For instance, in the end the Weak Lensing group wrote their own codes for the shear pipeline (see Chapter 11), but DESDM is now running those codes, as we said we were going to do. While the Weak Lensing people were developing their codes, they needed to work very closely with the data and scrutinize what it was doing. This intimacy with data is in the collaboration's best interests because the results they are starting to get are phenomenal. In the last year, those codes have reached the point of maturity where they think they can run them as a production job, that is, they no longer need someone to constantly monitor the results.

The collaboration produces a 'gold catalogue' as a way of addressing shortcomings in the processing. It contains all the most reliable data about objects in the survey. DESDM produce an initial catalogue and then the producers of the gold catalogue essentially look at the systematics in what we have produced and remove regions with too many errors. Additional information is then added to this catalogue, such as the photometric redshifts. Responsibility for the gold catalogue has been rotating through a different person for each data release.

We also hope to run the photometric redshift code sometime soon, at which point we can say we've run everything we said we were going to run. However, the collaboration has been discovering as we've been going along that often the solution to a problem is not definitive. We continually change and update our system and we are always trying to add something new. The knowledge we have gained is proving invaluable as we come to the final stages of the Dark Energy Survey and begin to consider the generation of surveys that will succeed it.

CHAPTER 7

DES as a Big Data Machine Part II: Source Extractor and the Dark Energy Survey Science Portal

Emmanuel Bertin, Luiz da Costa & Angelo Fausti

This chapter introduces the Dark Energy Survey (DES) Science Portal, an innovative collection of software systems developed by the DES-Brazil Consortium. The Science Portal provides scientists with straightforward, efficient methods to evaluate, explore and analyze the vast amounts of data collected by Dark Energy Camera (DECam). In Section 7.2, Emmanuel Bertin describes his Source Extractor (SExtractor) package, which is used to build a catalogue of objects from an astronomical image. For obvious reasons, this is of fundamental importance to astronomy.

7.1 Analyzing large data sets: The Dark Energy Survey Science Portal solution *(Luiz da Costa and Angelo Fausti)*

Astronomical surveys generating large amounts of data and metadata, such as the Dark Energy Survey (DES), are setting new challenges to enable comprehensive and timely analysis in a collaborative environment involving up to hundreds of astronomers. Furthermore, the computational resources required to handle the volume of data involved go beyond what can be done with a personal computer and call for special procedures and hardware to streamline the process.

To address these needs, members of the Laboratório Inter-institucional de e-Astronomia (LIneA), the technical arm of the DES-Brazil Consortium, have developed software systems to evaluate the quality of the raw data produced by Dark Energy Camera (DECam), tools for data exploration allowing the visual inspection of images and catalogues processed by DES Data Management (DESDM) at National Center for Supercomputing Applications (NCSA) and an innovative, web-based framework, designed to meet the data analysis and processing needs of the DES science collaboration.

7.1.1 Monitoring the quality of the raw data with Quick Reduce

The Quick Reduce (QR) development started in 2011 and a robust version of the system was ready in September 2012 for the science verification phase of DECam. QR was designed to monitor the quality of the DECam exposures in real time. From the QR interface, the observer can monitor the DECam exposures assembled by the telescope control system, Survey Image System Process Integration (SISPI), and select a set of charge-coupled device (CCDs) to be processed using a simplified version of the Data Management algorithms. The QR pipeline includes overscan, bias and flat field corrections and produces a source catalogue using the Source Extractor (SExtractor) algorithm (Bertin and Arnouts, 1996). Several quality checks are performed on a sample of high signal-to-noise ratio and good-quality-flagged objects. QR presents a complete log showing the raw and reduced images, the variation of the point spread function (PSF), which describes the blurring of the image of a point-like object by the telescope and the atmosphere, along the focal plane, the PSF distortion, number counts and sky brightness tests for each exposure. The median, mean and RMS (Root Mean Square – see Chapter 8, footnote 1) values for each CCD are stored in a database and can be visualized through interactive plots as new exposures are observed. Alarms can be configured with predefined limits for each test and sent to the SISPI alarm system. QR results are transferred from Cerro Tololo Inter-American Observatory (CTIO) to Fermilab on a daily basis, allowing members of the DES collaboration to follow the quality of the data over different periods of time, as shown in Figure 7.1.

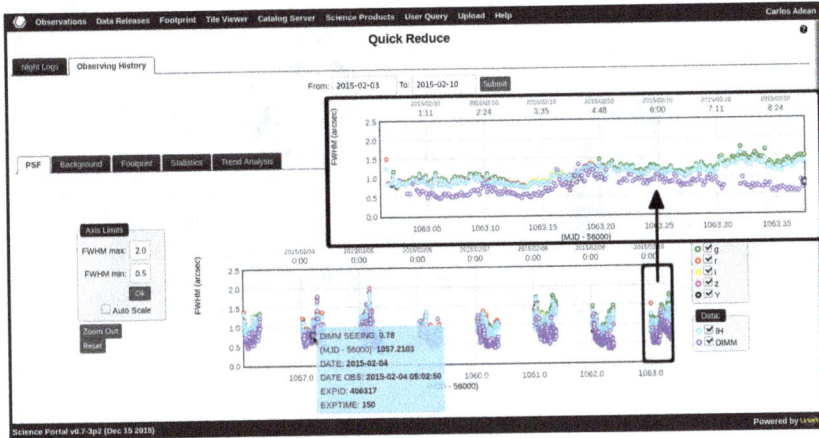

Fig. 7.1 Quick Reduce (QR) results exported to the DES Science Portal at Fermilab. The plot shows the variation of the full-width half-maximum (FWHM) (arcsecond) in a selected period of time. The points in the inset correspond to measurements on each exposure. QR measurements (green) are compared with Image Health (cyan) and the differential image motion monitor (DIMM) seeing (violet) for reference.

7.1.2 Exploration of DES processed data

With the first data products released by DESDM in 2013 (Science Verification data) and in 2014 (the first DES annual data release), there was a clear need for data exploration tools. These tools should enable the members of the collaboration to visualize in an efficient manner the large number of images and catalogues produced in each release, search for defects in the data and evaluate the impact of changes in data processing algorithms in the different data processing campaigns run by DESDM. In 2014, the so-called DES Data Server was made available by LIneA at Fermilab and the system was used by the collaboration for evaluating the first DES data releases. At that point, the Tile Viewer tool was developed to navigate through the co-added images observed in different filters and RGB colour composites, and to overlay the detected sources and masks for inspection (see Figure 7.2). In the Tile Viewer, the scientist can add annotations on specific positions of the image and share interesting findings with other members of the collaboration and, if necessary, mark defects on the images, producing at the end an overall summary of the different problems to help

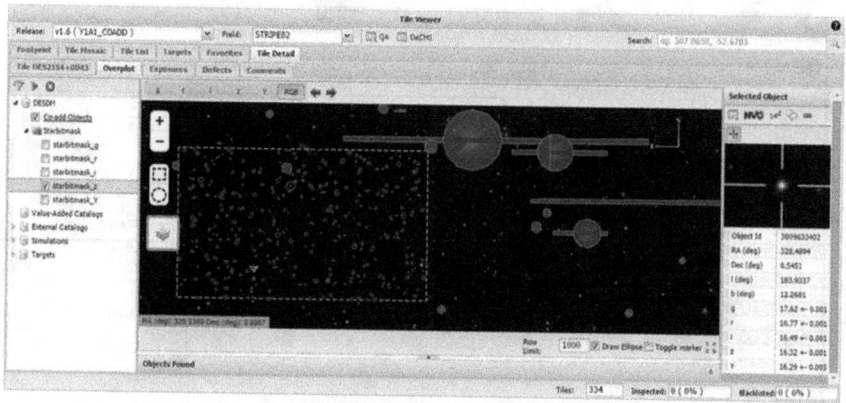

Fig. 7.2 The Tile Viewer showing an RGB colour image for one of the Dark Energy Survey (DES) tiles in the first annual (DES Y1A1) data release. Overlaid on the image we can see detected objects and masks. The selected object appears on the right panel in the 'detail', where the main properties of that particular object are listed.

improve the data reduction algorithms. Suddenly, there was a large interest on the DES Data Server for visual inspection of Strong Lensing and Galaxy Cluster candidates as they were discovered from the first analysis carried out by the DES collaboration. This need drove the development of specialized tools to upload lists of objects to the DES Data Server database and to automatically create postage stamps (small images carved out from larger images) around them for visualization in the Target Viewer tool (see Figure 7.3). In the Target Viewer, the scientist can inspect images of hundreds to thousands of objects, divided into several pages, presented as a mosaic or a list of properties from which the scientist can accept/reject and rank their candidates. From there, it is also possible to inspect the complete list of properties obtained from the co-added objects catalogue and in some cases inspect system properties such as the cluster radius and the cluster members in the case of galaxy clusters, and so on. Key features for this 'searching' process that are implemented in the DES Data Server are: (1) user friendly interfaces for uploading and managing the lists of targets, (2) automatic creation of postage stamps, (3) the ability to save the user feedback on a database, (4) the ability to create reports to do collaborative work and (5) create reports summarizing the results. Finally, this version of the DES Data Server had prototype tools to carry out database queries and access science products. Over the last three years of operation (2015–2017), there have been about 4200 visits to the DES Data Server by over 220 users.

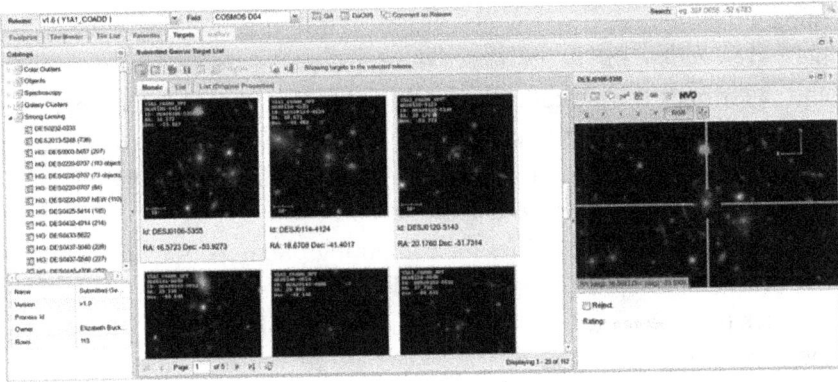

Fig. 7.3 The Target Viewer tool displaying a list of Strong Lensing candidates found in the Dark Energy Survey Year 1 annual (DES Y1A1) data release (mosaic view).

While useful for exploring the science verification and first-year data releases, this solution was not scalable as it required the transfer of images and catalogues from NCSA to Fermilab and relied on a local database. Also, the DES Data Server was developed as a monolithic application and as we added more features, it became very difficult to maintain. To address these limitations, the development of the DES Data Server was frozen in early 2016 to implement a new system, with newer technologies integrating third-party tools for data visualization. More importantly, the new system is integrated with the DES science database and the DES science archive at NCSA, eliminating the need to move data and significantly improving its readiness when new data releases are produced. The so-called Science Server will be one of the data release interfaces at NCSA, providing access to the first public data release of DES data accumulated during the first three years of operation, scheduled to occur in late 2017.

7.1.3 End-to-end processing

The main data products released by DESDM include stacked (co-add) images, mangle masks and object catalogues. For some studies, these data are enough for the science analysis, and the combination of queries on the DES Science Database and the aforementioned data exploration tools are the main mechanisms for data access. However, more sophisticated analyses, such as those needed for dark energy studies, require well-defined statistical samples. For these cases, ancillary products are needed to take into account the variations of observing conditions across the sky, which impact the depth reached by observations. Additional information is also needed to enable the separation of the detected objects as stars or galaxies and to estimate the photometric redshifts (photo-zs). This is critical information for large-scale and clustering studies and is also essential in order to compute the basic properties of objects classified as galaxies, such as absolute magnitude, mass and spectral type.

The Portal hosts pipelines for computing photo-zs, starting from the creation of spectroscopic samples and training sets used to train neural network photo-z codes, to the final estimation of photo-zs for all objects in the photometric catalogue (Gschwend et al., 2018).

Creating consistent astronomical catalogues from large multi-wavelength photometric surveys is undoubtedly a complex task, involving a large number of steps and parameters. It is, therefore, critical to be able to efficiently produce them to, among other things, evaluate the impact that different choices of algorithms and parameters may have on the final results of any given analysis. Also critical is to have a system in place that allows one to keep track of how these catalogues were created, so that they can be reproduced at any time, especially those catalogues that are ultimately to be used in publications and distributed to the general public.

The process for creating catalogues in the portal is divided into stages. First is the creation of masks that remove defects leading to spurious detections, bright objects that may affect the measured properties (shape, flux) of nearby objects and maps showing the spatial variation of the observing conditions and the depth reached by the observations in each passband. This step also includes possible small corrections to the measured magnitudes and a first characterization of the data that may guide the user in the preparation of a catalogue for a given application. Secondly, different algorithms can be used to classify detected objects as stars and galaxies and also to estimate photometric redshifts.

The capability of creating science-ready catalogues quickly and with full control of the input data and the configuration used is an important asset for supporting science analysis using data from large astronomical surveys. It is important to emphasize that the portal incorporates algorithms available in the public domain or developed by the collaboration at large. Its main contribution is that it provides an environment where these algorithms and procedures are all integrated and the processes are

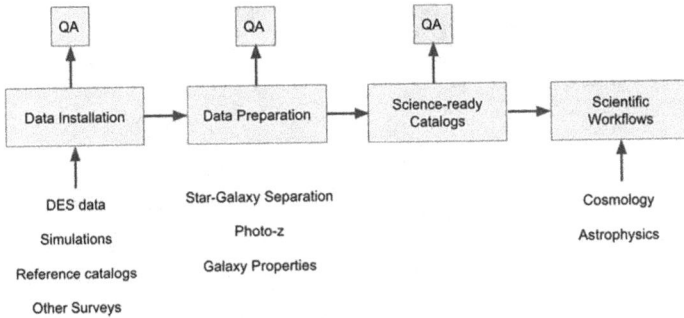

Fig. 7.4 Conceptual view of the end-to-end process in the Dark Energy Survey (DES) Science Portal.

streamlined. To address these requirements in a structured way, a novel system for creating science-ready catalogues and running scientific workflows has been devised and developed by LIneA to automate all steps involved in the science analysis. The conceptual view of this framework is shown in Figure 7.4. This system is currently running at the Laboratório Nacional de Computação Científica (LNCC) in Brazil.

At the time of the writing (spring 2017), there are thirty-six workflows implemented in the portal to execute tasks in the different stages of data installation, data preparation, science-ready catalogues and science analysis. Figure 7.5 shows a dashboard interface used to monitor the execution of these workflows and follow the chain of processes involved in the creation of a given data product, the inputs and configurations used as well as additional information for each process. There are three science analysis workflows operational in the Portal. Other workflows are in the process of being integrated by members of DES-Brazil, covering galaxy archeology, galaxy evolution and methods to estimate cosmological parameters based on cluster counts and angular correlation function measurements. In addition, the Portal supports the creation of special samples and other utilities such as comparing photo-zs, catalogues and galaxy cluster catalogues, making it an extremely useful scientific tool. It is important to note that the output of analyses that generate samples of special objects such as galaxy or stellar clusters can be exported to the DES Science Server running at NCSA for visualization and validation of the results.

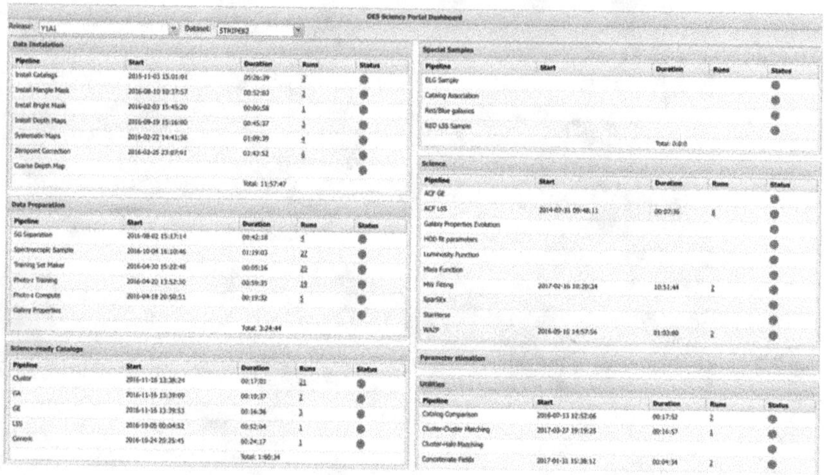

Fig. 7.5 Dashboard interface showing the different stages of the end-to-end processing and all the workflows currently available in the portal.

The development of the systems described earlier is the result of a team effort bringing together professionals from astronomy (e.g. A. Carnero, J. Gschewend, R. Ogando) and computer science (e.g. C. Adean, R. Campisano, J. G. Dias, R. Loureiro, L. Nunes, C. Singulani, G. Verde) to address the data analysis challenges imposed by large astronomical surveys. It is hoped that it serves as a model for future long-term projects to enhance the efficiency of extracting scientific results from massive data volumes in a secure and reproducible way. While the Portal has been designed with DES needs in mind, the framework is generic and can be adapted to other long-term projects involving large teams spread out geographically and massive volumes of data.

7.2 SExtractor and image analysis *(Emmanuel Bertin)*

Like many other large survey pipelines, DESDM uses several low-level data processing packages of the AstrOmatic software suite. These packages were developed over the years by Emmanuel Bertin, while working on different experiments. Emmanuel, who works in France, formally joined the DESDM team in 2007 as an external collaborator, since no French laboratory is part of the consortium. Emmanuel travelled a couple of times to Urbana Champaign in Illinois and to Munich in Germany to visit DES team members, although much of his work was done from France.

7.2.1 SExtractor

The SExtractor (Source Extractor) package is in charge of generating 'raw' catalogues of sources from the processed images. SExtractor[1] was originally developed by Emmanuel in the early nineties, while working on his PhD thesis at the Institute of Astrophysics in Paris (IAP), to extract sources from scans of large photographic plates. Computer memory was extremely limited at the time – the most powerful workstations had about 128 megabytes – but the scans were already about two gigabytes in size. Hence the usual procedure for doing computer measurements on those data was first to deal only with the pixels that make up the brightest part of the objects, and then to split the images into many small tiles for further analyses. The implementation was generally complex and required a lot of bookkeeping. SExtractor replaced that procedure with a single call to an executable that would carry out all of the processing by *streaming* the content of the large image in only two internal passes. Sources were now analyzed and written to the output catalogue on-the-fly.

Around 1994, another PhD student at IAP, Stéphane Arnouts, was also struggling to create a catalogue of sources. However, the images this time were from a survey done with CCD detectors, namely the European Southern Observatory-Sculptor Faint Galaxy Redshift Survey (Arnouts et al., 1997). Stéphane and Emmanuel worked together to modify SExtractor to work with images from digital detectors (Bertin and Arnouts, 1996). The software was then handed out to colleagues working in United Kingdom and United States labs, where it was extensively applied to deep, high-resolution images coming from the recently refurbished Hubble Space Telescope to extract catalogues of faint stars and galaxies (Kneib et al., 1996; Méndez et al., 1996). From this point on the software rapidly spread throughout the community. Over the years, features were added to accommodate the requirements of various studies and imaging surveys, most notably the European Southern Observatory Imaging Survey

---

[1]The name of the package came out as a student joke; SExtractor was initially meant to be the optical counterpart of the 'Extractor' software that had been developed at the Infrared Processing and Analysis Center (IPAC) for cataloguing infrared data. The name is considered controversial in some US laboratories where it is generally pronounced 'S-Extractor', highlighting cultural differences between both sides of the Atlantic.

(da Costa et al., 1999), the Canada-France-Hawaii Telescope Legacy Survey (Cuillandre et al., 2012) and obviously, DES.

Like all other AstrOmatic packages, SExtractor is open source. As a consequence, improvements and fixes to the code that are considered reasonably stable in the DES version are propagated to the general public version within a few months, and vice versa. The SPREAD_MODEL star/galaxy classifier and the MAG_DETMODEL optimized colour estimates are examples of new SExtractor features that were originally developed for DES and are now used by the general community.

7.2.2 Image stacking

Among the products delivered by DESDM are stacked images. Stacking (or co-addition) entails averaging overlapping exposures that are warped to a common pixel grid; this process is taken care of by the SWarp package (Bertin et al., 2002). Stacks are a convenient product, on which astronomical sources can easily be detected and measured down to the nominal survey limit. Nevertheless, co-addition has a bad reputation in the weak lensing community; the resampling and trimming of pixel values, which are part of the stacking procedure, can induce subtle but often systematic effects on the shapes of objects and distort the underlying lensing signal. In DES, a much more severe issue is the presence of abrupt changes of the image quality across the stacked image. In imaging science, the image quality can be defined by the PSF, which represents the normalized image of a point source (e.g. a star). The PSF of ground-based instruments varies significantly from exposure to exposure because of atmospheric turbulence. Those variations, combined with the DES tiling strategy and the presence of gaps between the detectors, are responsible for the abrupt changes across the stacked images. Unlike smooth variations, they cannot be mapped easily. The tiling strategy used for surveying the sky is one of the specificities of DES. In imaging surveys made with mosaic cameras, one would generally try to maximize the overlapping area between successive exposures covering a given sky area, with only the minimum amount of shifting required to fill the coverage gap between detectors. Not in DES. Instead, the tiling pattern is shifted by about half the field of view in any of six possible directions at each new pass of the telescope across the same region of the sky. This tiling strategy is a huge benefit to the experiment, because it allows scientists to homogenize the response of the instrument very precisely and meet the tight DES requirements on calibration over a wide range of angular scales. For image stacks, however, it has the disadvantage that the PSF, which represents the normalized image of a point source (and which defines image quality), changes abruptly within regions that are much smaller than the DECam field of view.

The initial approach taken by the DESDM team to address this issue was to homogenize the PSF prior to stacking. To this aim, the PSFEx package (Bertin, 2011), AstrOmatic's PSF modelling software, was modified to compute homogenization filters that would be applied to individual exposures prior to stacking. Those filters were computed in a way similar to that used for image subtraction in transient sky studies (Alard and Lupton, 1998). However, in DES, the target PSF was synthetic and defined as a perfectly round star image. Tony Darnell, who was in the DESDM team at the time, successfully implemented the PSF homogenization and co-addition pipeline (Darnell et al., 2009).

One downside of the filtering process is its impact on the estimation of measurement uncertainties, which becomes much more complex on the homogenized images than on regular stacks. This is a critical issue for DES, as many scientific analyses require accurate measurement uncertainties. Specific pixel uncertainty maps were attached to the PSF-homogenized stacks, which would allow measurement uncertainties to be estimated with good accuracy for point sources. For resolved objects, the estimation would be more approximate. Although PSF homogenization was successfully applied to a large volume of data, including DES image simulations and the Blanco Cluster Survey (Desai et al., 2012), the added complexity for deriving uncertainties, especially limiting magnitudes, and the presence of ringing artefacts around bright stars convinced people in the DESDM team that this was not the right approach. PSF homogenization was therefore eventually turned off in the image stacking process, and measurements that would depend on the PSF were performed on individual exposures instead of the stacks.

7.2.3 Moving to multi-exposure image analysis

The ngmix code, which had been developed by Erin Sheldon and Matt Becker for weak lensing purposes, was designed from the ground up to deal with individual exposures simultaneously and not with stacks. As such it did not have issues with abrupt PSF variations, and would advantageously replace SExtractor's PSF-dependent measurements (galaxy model fitting), while SExtractor was being re-factored to manage multiple exposures.

Transforming the SExtractor code to perform multi-exposure measurements proved extremely difficult for Emmanuel. The original SExtractor was written in the C language, but C is not abstract enough to easily manage the combinatorial explosion that occurs when configuring multiple measurements on multiple exposures, observed in different channels and at different epochs. Moreover, dealing with many images made it more memory intensive. Hence fewer instances of SExtractor could run on a compute node, which means that the code had to be parallelized to take advantage of all the central processing units (CPUs). Eventually, Emmanuel focused on a version with a reduced scope specifically for DES.[2]

7.2.4 Improving deblending and the road ahead

The *deblending* of overlapping sources has been a long-standing issue for source extraction codes. In DES, overlapping sources are predominantly found in the core of galaxy clusters. Even with carefully tuned detection parameters, SExtractor often fails to deblend some galaxies in dense galaxy clusters. The GAIN algorithm developed by Yuanyuan Zhang and others (2015) for DES showed notable improvement in these regions. Emmanuel demonstrated an iterative algorithm based on galaxy model subtraction. However both techniques rely on heuristics that require tuning; more aggressive settings unavoidably lead to complex single objects such as large, patchy galaxies being split. These objects are ubiquitous in a large survey like DES.

---

[2]A more general version of SExtractor is actively being developed in C++ for the general community in the context of the Euclid mission at Integral Science Data Centre (ISDC) in Geneva.

Clearly, a more 'intelligent' deblender is required, trained with examples and capable of taking decisions directly from complex image patterns. A category of machine learning algorithm has emerged in recent years that is precisely capable of that: 'Deep' convolutional neural networks (Lecun et al., 2015). ConvNets have revolutionized the field of image processing and have also started being used in astronomical image analysis for addressing complex problems (Dieleman et al., 2015; Lanusse et al., 2018), including deblending (Paillassa and Bertin, 2019). Current training algorithms for deep ConvNets require huge data sets with typically hundreds of thousands of images, but this is obviously not an issue for a big data experiment such as DES. ConvNets are, therefore, likely to play an important role in the final years of DES data processing.

References

Alard, C. and Lupton, R. H. (1998). A method for optimal image subtraction, *The Astrophysical Journal* **503**, pp. 325–331, doi:10.1086/305984.

Arnouts, S., Lapparent, de V., Mathez, G., Mazure, A., Mellier, Y. et al. (1997). The ESO-Sculptor faint galaxy redshift survey: The photometric sample, *Astronomy and Astrophysics Supplement Series* **124**, pp. 1–21, doi:10.1051/aas:1997187.

Bertin, E. (2011). Automated morphometry with SExtractor and PSFEx, in I. N. Evans, A. Accomazzi, D. J. Mink and A. H. Rots. eds., *Astronomical Data Analysis Software and Systems XX, San Francisco:* Astronomical Society of the Pacific Conference Series, Vol. 442, p. 435.

Bertin, E. and Arnouts, S. (1996). SExtractor: Software for source extraction. *Astronomy and Astrophysics Supplement Series* **117**, pp. 393–404, doi:10.1051/aas:1996164.

Bertin, E., Mellier, Y., Radovich, M., Missonnier, G., Didelon, P., et al. (2002). The TERAPIX pipeline, in D. A. Bohlender, D. Durand and T. H. Handley. eds., *San Francisco: Astronomical Data Analysis Software and Systems XI, Astronomical Society of the Pacific Conference Series*, Vol. 281, p. 228.

Cuillandre, J. -C. J., Withington, K., Hudelot, P., Goranova, Y., McCracken, H., et al. (2012). Introduction to the CFHT Legacy Survey final release (CFHTLS T0007), in A. B. Peck, R. L. Seaman, F. Comeron. eds.,*Observatory Operations: Strategies, Processes, and Systems IV, Proceedings of the SPIE*, Vol. 8448, p. 84480M, doi:10.1117/12.925584.

da Costa, L., Arnouts, S., Benoist, C., and Deul, E. (1999). ESO Imaging Survey: Past activities and future prospects. *The Messenger* **98**, pp. 36–45.

Darnell, T., Bertin, E., Gower, M., Ngeow, C., Desai, S., et al. (2009). The Dark Energy Survey Data Management System: The coaddition pipeline and PSF homogenization, in D. A. Bohlender, D. Durand and P. Dowler. eds., San Francisco: *Astronomical Data Analysis Software and Systems XVIII, Astronomical Society of the Pacific Conference Series,* Vol. 411, p. 18.

Desai, S., Armstrong, R., Mohr, J. J., Semler, D. R., Liu, J., et al. (2012). The Blanco Cosmology Survey: Data acquisition, processing, calibration, quality diagnostics, and data release, *The Astrophysical Journal* **757**, 83, doi:10.1088/0004-637X/757/1/83.

Dieleman, S., Willett, K. W., and Dambre, J. (2015). Rotation-invariant convolutional neural networks for galaxy morphology prediction, *Monthly Notices of the Royal Astronomical Society* **450**, pp. 1441–1459, doi:10.1093/mnras/stv632.

Gschwend, J., Carnero Rosell, A., Ogando, R., Fausti Neto, A., Maia, M., et al. (2018). DES Science Portal: I - Computing photometric redshifts, *Astronomy and Computing* **25**, pp. 58–80, doi:10.1016/j.ascom.2018.08.008.

Kneib, J. -P., Ellis, R. S., Smail, I., Couch, W. J., and Sharples, R. M. (1996). Hubble Space Telescope observations of the lensing cluster Abell 2218, *The Astrophysical Journal* **471**, p. 643, doi:10.1086/177995.

Lanusse, F., Ma, Q., Li, N., Collett, T. E., Li, C. -L., et al. (2018). CMU DeepLens: Deep learning for automatic image-based galaxy-galaxy strong lens finding, *Monthly Notices of the Royal Astronomical Society* **473**, pp. 3895–3906, doi:10.1093/mnras/stx1665.

Lecun, Y., Bengio, Y., and Hinton, G. (2015). Deep learning, *Nature* **521**, pp. 436–444. doi:10.1038/nature14539.

Méndez, R. A., Minniti, D., de Marchi, G., Baker, A., and Couch, W. J. (1996). Star counts in the Hubble Deep Field: Constraining galactic structure models. *Monthly Notices of the Royal Astronomical Society* **283**, pp. 666–672, doi:10.1093/mnras/283.2.666.

Paillassa, M. and Bertin, E. (2019). Deblending in crowded star fields using convolutional neural networks, in M. Molinaro, K. Shortridge, and Fabio Pasian. eds., *Astronomical Data Analysis Software and Systems XXVI, San Francisco:* Astronomical Society of the Pacific Conference Series, Vol. 521, p. 382.

Zhang, Y., McKay, T. A., Bertin, E., Jeltema, T., Miller, C. J., et al. (2015). Crowded Cluster Cores: An algorithm for deblending in Dark Energy Survey images, *Publications of the Astronomical Society of the Pacific* **127,** p. 1183, doi:10.1086/684053.

CHAPTER 8

The Dark Energy Survey Strategy and Calibration

James Annis, Eric Neilsen & Douglas Tucker

In this chapter, we discuss the nuts and bolts of the survey: the plan we follow to carry out the Dark Energy Survey (DES) objectives as efficiently and cost-effectively as possible while still collecting data of the highest quality attainable by the Dark Energy Camera (DECam). Integral to this over-all strategy is photometric calibration, which in essence is the process of converting DES data to a standard (externally calibrated) system. Without accurate, unbiased calibration, our science analysis is of no value, so we are continuously refining and improving our plan.

8.1 Constituent surveys within the Dark Energy Survey

The Dark Energy Survey (DES) strategy is designed to support four separate methods of measuring dark energy equation of state parameters. One of the four methods, measurement of distances using type 1a supernova, requires imaging exposures on a small number of fields in many filters following a regular cadence in time. The other three methods all require images over large areas of the sky in multiple filters and can be applied to a common data set. Each of these methods imposes requirements on the data set; some of these requirements are shared by multiple methods, others are unique to specific projects.

8.2 Wide survey

8.2.1 Depth, exposures and image quality

DES uses the Dark Energy Camera (DECam) instrument on the 4 m Blanco Telescope at Cerro Tololo Inter-American Observatory (CTIO) in Chile to collect its science data. The camera has sixty-two science charge-coupled devices (CCDs) arranged in a roughly hexagonal area with a field of view of 3.1 square degrees on the sky. A collection of filters allows images to be taken using light in specific ranges of wavelength (bandpasses). The camera can be read-out in about twenty-seven seconds, and it also takes time to slew the telescope from one pointing to another. Read-out and slew can happen simultaneously, so the time between the end of one exposure and the next is the longer of the read-out time and the slew time. An efficient survey strategy minimizes the number of slews that take longer than twenty-seven seconds.

Different exposures are combined to make 'stacked' or 'coadded' images, which combine data from multiple exposures of the same area of the sky. The limiting magnitude, that is, the brightness of the faintest objects that can be used for science in a field, increases as a function of the total exposure time of all exposures included in the coadded image.

The time to complete the survey is determined by the area covered by the survey, the mean exposure time it takes to complete each exposure in each filter (including read-out and slew time) and the mean number of exposures of the area in each filter.

The science projects driving the wide survey are subject to statistical uncertainty arising from the finite numbers of galaxies measured at different redshifts, and inhomogeneities in the large-scale structure of the universe (cosmic variance). To achieve the dark energy science goals, the wide survey must cover an area of at least 5000 square degrees.

Because DES does not include a spectroscopic component, the different science projects need to approximate redshifts for galaxy clusters using photometry in multiple filters (photo-z). The requirements for effective photo-z measurements drive the selection of four filters (g, r, i and z, corresponding to green, red and near-infrared wavelengths) and impose limits on the minimum needed exposure times in each filter. Additional photometry at longer wavelengths, further into the infrared, can substantially improve photo-z measurements. The DES collaboration therefore entered into an agreement with the VISTA Hemisphere Survey (VHS), an infrared survey being performed with the 4 m VISTA telescope at Paranal Observatory, under which DES provides Y-band imaging data in exchange for imaging in J, H and K bands at a suitable depth from the VHS survey. With the addition of the Y-band imaging collected under this agreement, the total filter set used by DES becomes g, r, i, z and Y.

Measurement of the brightness of objects detected in each exposure requires mapping between flux measured by the camera to the brightness of objects in the sky (the photometric calibration). Because the fraction of light from each object absorbed by the atmosphere varies with time, (depending on the composition of the atmosphere at the time and in the direction of observation) this calibration must be made for each exposure. The uncertainty in the calibration introduces a systematic error in the magnitudes of objects detected in the image. If there are many exposures of the same area of the sky taken on different nights, the errors introduced by imprecision in the photometric calibration can be averaged over the different nights for a smaller total contribution to the uncertainty in the magnitudes of the objects measured. Increasing the number of exposures improves photometric precision relative to the same total exposure time spread over fewer exposures.

Similarly, uncertainty in the modelling of the point spread function (PSF) of each image is an important factor in limiting the precision of weak-lensing measurements of dark energy parameters. The width of the PSF corresponds to the degree of blurring in the image. Turbulence in Earth's atmosphere is a major factor in the PSF and since it varies with time it must be measured separately for each exposure. If there are many exposures of the same object, uncertainties resulting from imprecision in the modelling of the PSF can be averaged over the different exposures: as with photometric calibration, increasing the number of exposures improves the precision of weak-lensing measurements relative to the same total exposure time spread over fewer exposures.

Taking large numbers of exposures on the same area of sky incurs a cost, however: each exposure takes about twenty-seven seconds to read out. So, for example, where a single exposure of 180 seconds takes a total of

$180 + 27 = 207$ seconds to complete, dividing this same exposure time into three different exposures takes $3 \times (60 + 27) = 261$ seconds.

To balance these considerations, DES has adapted a wide-survey strategy in which an area of 5000 square degrees is covered ten times with ninety second exposures in g, r, i and z filters, and six tilings of forty-five seconds each plus two tilings of ninety seconds exposures in the Y filter.

8.2.2 Hexes and tilings

The outline of the area imaged by DECam in a given exposure is roughly hexagonal, with flat sides to the north and south, and vertices to the east and west. Hexagons of uniform size and orientation can be arranged to completely fill ('tile') a flat two-dimensional space. Near the celestial equator, the sky can be approximated as a flat space in which celestial coordinates (right ascension (RA) and declination) approximate rectangular coordinates. In this area, the sky can be efficiently 'tiled' by images of the footprint, and it is possible to generate a collection that image all parts of the sky without either overlapping or leaving spaces between exposures. In areas far from the equator, this approximation breaks down. The wide survey divides the celestial sphere into lunes (areas bounded by lines of constant RA) 30° wide, and tiles each lune using an inverse sinusoidal projection. The result is an arrangement of telescope pointings that are approximately well tiled in a hex-like tiling arrangement except at boundaries between lunes, at which there are discontinuities.

Once an initial set of such pointings has been generated, additional sets can be generated from it by applying offsets to each. To cover the survey area with ten exposures, nine different offsets can be applied to an initial set. The size of these offsets is comparable to the radius of the field of view, so that an individual object on the sky will appear at a variety of points on the focal plane. Figure 8.1 shows the initial tiling, and the next two tilings in red and blue.

8.2.3 Footprint

To accumulate a sufficient sample of galaxies and overcome uncertainty resulting from inhomogeneity due to large-scale structure, the wide-survey footprint covers slightly more than 5000 square degrees on the sky. The

Tiling 1   Tilings 1 and 2   Tilings 1, 2, and 3

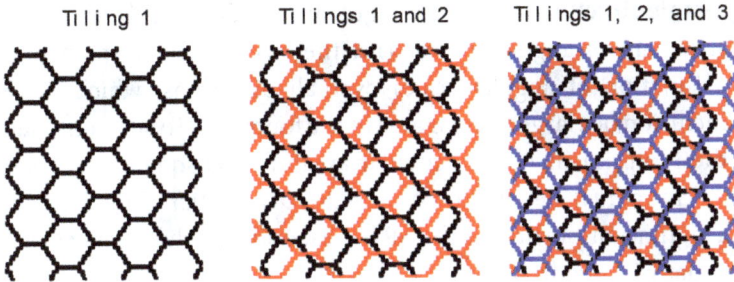

Fig. 8.1 Hex layouts showing different numbers of tilings.

dark energy science projects depend on photometry of galaxies and so are vulnerable to obscuration by dust and stars in the Milky Way. Therefore, the footprint chosen avoids the plane of the Milky Way, excluding parts of the sky where extinction by dust exceeds 0.15 magnitude in i band or where the stellar density exceeds the density at the south galactic pole by a factor of seven or more.

In addition to these general restrictions, each dark energy science project introduces additional constraints:

Clusters Measurement of dark energy parameters using galaxy cluster scale halo abundance as a function of redshift requires mass estimates for detected clusters. These will be obtained from the South Pole Telescope Sunyaev-Zel'dovich (SPT-SZE) Survey, so the DES footprint must overlap the SZE survey footprint as much as possible. This leads to the inclusion of an area bounded in declination, δ, by the limits of the SPT-SZE survey, $-65° < \delta < -40°$; and by galactic stellar density and extinction in RA.

Photometric redshifts Because DES does not include a spectroscopic component, redshifts must be estimated using photometry of galaxy clusters. Calibration of the relationship between photometry and redshift requires a large area of overlap between the DES footprint and area covered by spectroscopic surveys such as Baryon Oscillation Spectroscopic Survey (BOSS) and Extended Baryon Oscillation Spectroscopic Survey (eBOSS). This requirement drives the inclusion of two equatorial quadrangles: one with $317° < \alpha < 360°$, $-2° < \delta < 2°$, and other with $0° < \alpha < 45°$, $-7° < \delta < 5°$.

Large-scale structure Measurement of large-scale structure requires measurement of angular power (i.e. the size of density fluctuations as a function of angular scale) over all directions, which drives the inclusion of a large circular area in the footprint. To accommodate this, the footprint includes a circular area of 3000 square degrees (about $31°$ in radius) centered on $\alpha = 38.8°$, $\delta = -39.5°$. This area overlaps a significant portion of the SPT area, and (nearly) connects it to the equatorial quadrangles.

Weak lensing Measurement of dark energy parameters using weak lensing depends critically on having a large area of images with very small PSFs. To maximize the ability of the survey to take advantage of weather with minimal atmospheric turbulence (which leads to wide PSFs), which occurs most often in December, January and February, the large-scale structure 'circle' is centred to optimize visibility during these months (in so far as it is possible without excessive contamination by dust and stars from the Milky Way), with some additional area to the east between $\delta = -40°$ (the northern boundary of the SPT area) and $\delta = -18°$. The eastern edge of this region is set where few distant galaxies can be well measured because too many stars from the Milky Way fill the images.

With the addition of a small amount of area to improve the connection between the large-scale-structure circle and the equatorial quadrilaterals, the full area of the footprint reaches 5122 square degrees. Figure 8.2 shows the perimeter of the wide-survey footprint.

8.2.4 Quality cuts

All of the scientific programmes rely on being able to perform measurements across the area covered by the footprint. When attempted images are of too poor a quality to enable these measurements (e.g. because of clouds or poor seeing), the images need to be retaken.

Rejection and repetition of exposures relies primarily on two metrics: the faintness of objects whose brightness can be measured to the necessary precision; and the sharpness of the images, as indicated by the width of images of stars in the images. All programmes require precision photometry (measurement of the brightness of astronomical sources), and the

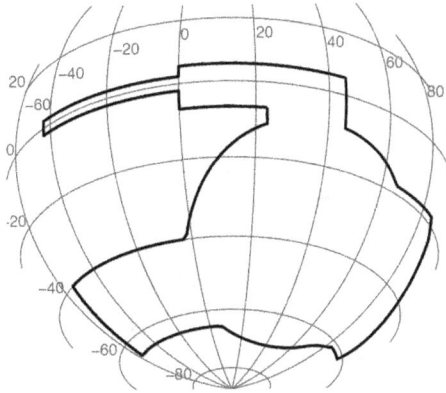

Fig. 8.2 The Dark Energy Survey (DES) wide-survey footprint, with right ascension (RA) and declination both marked in units of degrees.

weak-lensing program depends critically on measurement of the ellipticity of galaxies. The exact criteria used to reject and repeat exposures must balance image quality and survey progress: if the image quality requirements are too demanding, the survey will not make adequate progress. Loose requirements, however, will compromise the quality of the science results.

8.3 Time-domain survey

The DES time-domain survey collects sequences of exposures at ten different telescope pointings, for a total area of a roughly thirty square degrees. It observes eight of these pointings using 'short' sequences of exposures taking a total of twenty minutes of observing time, including exposures in the g, r, i and z filters. The remaining two pointings are observed in longer sequences in each filter separately, taking between fifteen minutes and over an hour of observing time, depending on the filter. DES observers attempt to complete every sequence according to a cadence in which successive completions are separated by no more than one week.

With this regular monitoring of the large area of sky, DES can detect when supernovae begin to explode, and follow the progress of each supernova as it brightens and then fades in each of the four wavelength bands. The shape of the resultant light curve allows the estimation of the intrinsic brightness of each supernova, which in turn allows the distance to be estimated.

To use these distances to measure dark energy equation of state parameters, the redshifts of each supernova are required as well. These can be either obtained from spectra of the supernovae themselves (or their host galaxies) or using 'photometric' redshifts of either the supernova or host. Spectroscopic redshifts are more precise and accurate, but require external sources of data, either from follow-up observing with spectrometers on 8 m to 10 m class telescopes or existing galaxy redshift surveys. The locations of such surveys and the desire that they be accessible to suitable telescopes with spectrographs in both hemispheres drives the selection of the specific pointings used.

8.4 Schedule

At the beginning of each observing season, the DES collaboration provides the observatory time allocation committee (TAC) with a request for specific nights on which DES should be scheduled for observing. The TAC then integrates DES observing and other uses of the telescope to generate an observing schedule. The TAC allocates time in units of half-nights: a p.m. portion staring at sunset and ending at local solar midnight, and an a.m. portion starting at midnight and ending at sunrise.

Several considerations inform the construction of the initial schedule request. Among the most important are:

Accessibility of the footprint Different areas of the sky rise and set at different times on different dates. The DES footprint is optimally visible from CTIO in late October, and is reasonably observable for at least part of the night starting in August and ending in February. The full six-year DES survey is therefore divided into six observing seasons, the first of which began in August of 2013 and ran through until February of 2014, the next starting in August of 2014, and so forth. If the DES footprint can only be observed for part of a half-night, then DES will be able to use this time less effectively than a better scheduled half-night, and time which a different project may have been able to better use the telescope will have been lost. Footprint accessibility limits DES to a.m. half-nights before September, and p.m. half-nights after mid-December.

Bright and dark skies Just as the Sun causes the whole sky to appear blue during the day, the Moon causes the whole sky to glow blue at night when it is above the horizon. This light introduces background noise into astronomical images, limiting the magnitude at which photometry of a given precision is possible. Because this light is blue, images in shorter wavelength bands are more sensitive than those in longer wavelength bands: exposures with the g and r filters can only be usefully taken when the Moon is down, while those in i, z and Y can tolerate some scattered light from the Moon, to varying degrees. When the Moon is within a few days of full, however, even z and Y images are severely degraded. The requested schedule must be allocated sufficient time in each phase of the Moon to allow each filter to be completed.

Time-domain cadence The measurement of supernova light curves depends on measurements of the brightness of the supernova at regular intervals; long gaps between measurements severely degrade the quality of the measurements. Therefore, the DES schedule must limit the duration of periods of time over which there are no scheduled DES half-nights.

Weather and seasonal changes CTIO is in the southern hemisphere, so nights are longest near late June and shortest near late December. However, the effect of this change on available observing time is counterbalanced by seasonal weather variations: there are fewer cloudy nights in December and January than in August or September, such that on average the total clear night time at CTIO is roughly constant over the whole of the DES observing season. However, the 'seeing' (which arises from turbulence in the atmosphere) also varies seasonally. The weak-lensing project in particular is critically dependent on exposures taken when the seeing is good, and since the total time in which the seeing is good enough for weak-lensing is greater later in the observing season, the weak-lensing program depends on observing time later in the season.

8.5 Tactics

DES is allocated a limited amount of observing time that covers a range of conditions: the weather (both cloud cover and seeing) varies on short time-scales; different times experience different sky brightnesses due to the presence of the Moon and different areas of the footprint are visible at different times and dates. The various survey exposures require different conditions. For example, wide exposures taken in the g and r filters should not be attempted when the sky is bright due to the Moon, and wide exposures in r, i and z (with which the weak-lensing survey will be measuring ellipticities) should not be attempted in poor seeing. To allow time for these more demanding exposures to be completed, they need to be observed in preference to less demanding exposures when optimum conditions obtain.

Care also needs to be taken that certain exposures are left to be completed during later nights in the schedule. Late in the schedule, only exposures of the far eastern end of the footprint can be taken. These pointings could also be effectively observed earlier in the year, but if they are, nothing that remains to be completed at the end of the year will be observable.

Completion of the survey requires accurate and timely record-keeping, such that exposures are not repeated unnecessarily and exposures, which do need to be repeated, are made while they are still visible.

DECam records DES images at a rate of roughly one every two minutes. It is impractical for human observers to schedule observing on an exposure-by-exposure basis, so DES uses an automated scheduler called obstac, to implement **observing tactics**.

Before observing each night, the observing team updates a database at the observatory with fields marking the success or failure of exposures taken on previous nights, such that obstac can repeat failed exposures or schedule time-domain sequences when required by the observing cadence.

During observing, the DECam data acquisition executes exposures from an observing queue, taking exposures from the head of the queue and executing them as each successive exposure completes. The queue may be edited either interactively by human observers through a web-based user interface, or programmatically through a message-passing Application Program Interface. Edits may occur either prior to observing, or concurrently with it.

During the night, obstac interacts with the data acquisition system through the API to the queue and uses it in much the same way a human observer would: it monitors the queue of upcoming exposures and adds exposures to the queue as necessary.

Before selecting an exposure to add to the queue, obstac extrapolates the time at which it will be executed using the current contents of the queue and estimates the seeing under which it will be observed using data from recently completed images. Then, it attempts to select an exposure following a sequence of steps that depend on current conditions. The steps are as follows:

- If any observable time-domain sequence has not been observed in the past seven days, add it to the queue; otherwise
- If the estimated seeing is better than 1.1″ (arcseconds) and the Moon is below the horizon, select a wide-survey exposure in g, r or i if any are observable, and add it to the queue; otherwise
- If the estimated seeing is better than 1.1″ and the Moon is above the horizon, select a wide-survey exposure likely to be successful given the current seeing and sky brightness and add it to the queue; otherwise
- If any currently observable time-domain sequences have not been observed in the past four days, select the one completed least recently that is likely to be successful and add it to the queue; otherwise
- Select a wide-survey exposure in g (if the Moon is below the horizon) or Y (if the Moon is above the horizon) and add it to the queue

When it must select among wide-survey exposures, obstac balances two competing criteria. To ensure that pointings in the west are completed before they become inaccessible and that pointings in the east remain to be done when they are the only ones accessible, obstac generally prefers pointings furthest west among those available; and to avoid using observing time slewing the telescope, it prefers pointings close to the current pointing of the telescope.

8.6 Photometric calibration

Photometry is the measurement of brightnesses. In astronomy, brightnesses are typically measured in quantities ('magnitudes') that can be traced to units of specific flux (e.g. 'energy per unit second per unit area per unit wavelength' or 'energy per unit second per unit area per unit frequency'). Photometric calibration is the process of ensuring that these brightnesses are measured consistently and accurately. Furthermore, the concept of photometric calibration itself can be classified into two categories: relative and absolute.

The purpose of relative calibration is internal consistency. DES has a requirement that its photometry must be internally consistent to 2% RMS[1] over the full survey area in each of the five filters. In other words, all else being equal, after relative calibration the measured brightness of a star in DES should not depend on its position within the survey area (to a tolerance of 2% RMS). In DES, this level of relative calibration is accomplished by means of the survey strategy of multiple overlapping tilings of the survey footprint in each filter band, as described previously in the Survey Strategy section. Unless a star is a variable star, its brightness should be the same in any overlapping exposure (of the same filter band) that contains it. If the brightness is not the same, a correction factor can be calculated using the ratio of observed brightnesses in the different exposures. (Of course, for accuracy and precision, not just one but many, many stars in each overlapping exposure are used to calculate this correction factor.) We note that the relative calibration requirement was already achieved during the DES Year 1 data processing campaign, and by the time of DES Year 4 data processing campaign, we were achieving relative calibrations of $\approx 0.5\%$ RMS.

The purpose of absolute calibration is to tie observed count rates (e.g. the number of photons detected from a given star each second) to real physical units of specific flux (e.g. ergs/sec/cm^2/Hz). Absolute calibration can itself be classified into two sub-categories: absolute flux calibration and absolute colour calibration, where absolute flux calibration refers to getting the overall scale right, and absolute colour calibration refers to getting the relative scale right between filter bands. (Note that, in astronomical parlance, the term 'colour' corresponds to the ratio of fluxes in two different filter bands.) As a concrete example, consider a star whose measured

---

[1] RMS, or Root Mean Square, is a measure of the scatter in measurements. For a relative calibration of 2% RMS, one expects c. 68% of measured brightnesses to be calibrated to within ± 2% of a mean uniform calibration.

fluxes in g, r, i, z and Y are all exactly 10% too low. This star's absolute flux calibration is off by 10%, but its absolute colour calibration across these filter wavelengths is perfect. In yet another way to look at it, note that the specific fluxes in the five filter bandpasses of DES constitute a very low-resolution spectrum of the object. Getting the absolute colour calibration for a star right means getting the overall shape of this star's low-resolution spectral energy distribution right; getting the absolute flux calibration right means getting right the overall multiplicative scale – a single number – that converts the measured brightness at each wavelength into physically meaningful units (see Figure 8.3). For DES, the absolute colour calibration requirement is 0.5% between g and r, between r and i and between i and z, and is 1% between z and Y. The absolute flux requirement is 0.5% in i-band relative to a set of well-calibrated spectrophotometric standard stars. As of the DES Year 3 and Year 4 processing campaigns, we were achieving absolute flux and colour calibrations at the $\approx 1\%$ level, based upon a single external reference star, the Hubble Space Telescope (HST) spectrophotometric standard star C26202; for the DES Year 5 and Year 6 processing campaigns, we are adding many other external reference stars in order to improve on the Year 3 and Year 4 results.

The above description provides a general conceptual framework for photometric calibration, but how did the DES team implement this conceptual framework as a practical plan? For any ground-based astronomical survey, not only does the instrument itself need to be calibrated (including the detectors, the filters, and the telescope optics), but so does the transmission of the atmosphere above the observatory (including the effects of clouds and the ease at which light of different wavelengths passes through varying amounts of aerosols and water molecules in the atmosphere). The DES photometric calibration plan takes both instrumental and atmospheric effects into account, and can be summarized in these five points:

1. **Instrumental Calibration (Nightly and Periodic):** Measure the average and wavelength-dependent pixel-to-pixel relative photon-detection efficiency of the camera and filters by observing a large, uniformly illuminated screen within the telescope dome ('dome flats' and 'DES Calibration' ['DECal'] system response scans, respectively), as well as other instrumental effects.

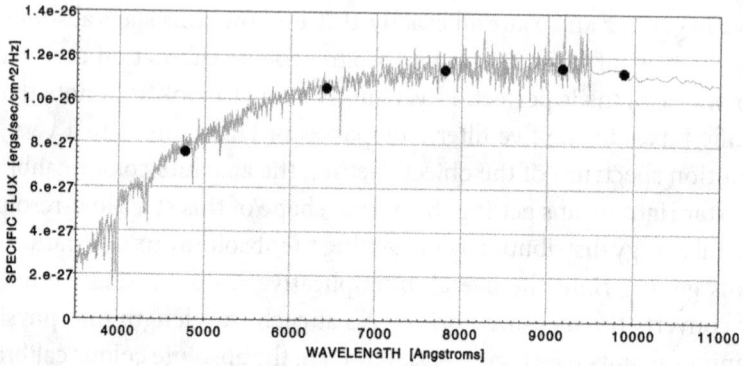

Fig. 8.3 The spectrum of the Hubble Space Telescope (HST) spectrophotometric standard star C26202 in units of specific energy flux per unit frequency (grey line), and the expected values of the mean specific energy flux for this star in the Dark Energy Survey (DES) *g, r, i, z, Y* filters under the assumption of good absolute photometric calibration (filled black circles). Note that the filled black circles, in essence, describe a very low-resolution spectrum for C26202. If the absolute *colour* calibration were off, the filled black circles would not trace the shape of the HST spectrum. If the absolute flux calibration was 10% too low for all five filters, the filled black circles would still trace the shape of the HST spectrum, but they would be offset 10% below the grey line. (We note that this HST spectrum of C26202 was used as the external reference for both the absolute flux and the absolute colour calibration for the DES Year 3 and Year 4 data processing campaigns.)

2. **Photometric Monitoring:** Monitor the sky with a Radiometric All-Sky Infrared CAMera ('RASICAM') to detect clouds, a dual-band GPS monitor ('GPSmon') to monitor the amount of precipitable water vapour overhead and a four-camera narrow-filter-band atmospheric transmission monitor ('aTmCam') to measure the amount of aerosols and precipitable water vapour overhead.[2]

3. **Nightly Calibrations:** Observe fields of previously calibrated stars ('standard stars') with DECam at the beginning and the end of the night in order to measure the overall calibration of the instrument plus atmosphere for each filter band on a night-by-night basis.

4. **Global Relative Calibrations:** Use the extensive overlaps between exposures over multiple tilings and the observations of frequently

---

[2]Precipitable water vapour, which can vary substantially during a night, affects the transmission of light in the *z* and *Y* filters. Aerosols, which vary less dramatically, affect transmission predominantly in the *g* filter.

observed fields (e.g. the supernova fields) to tie together the DES photometry into an internally consistent system across the entire DES footprint.

5. **Global Absolute Calibrations:** Use DECam observations of exquisitely well-calibrated external reference stars[3] in combination with DECal system response scans and measurements of the atmospheric transmission to tie the DES *grizY* photometry to physical units of specific flux ('AB magnitudes', which are a logarithmic measure of specific flux in units of ergs/s/cm^2/Hz).

Finally, we note that photometric calibration is an iterative and evolving process, in which lessons learned in the early part of the survey are applied as the survey progresses. For instance, the photometric calibration strategy in the first couple years of DES depended heavily on RASICAM and the nightly observations of standard star fields, whereas the DECal scans, the GPSmon and the aTmCam were mostly ignored. Starting with DES Year 3, the photometric calibration strategy depended much more heavily on the DECal scans, the GPSmon and the aTmCam, but much less so on RASI-CAM and the nightly observation of standard stars. It is expected that, as more lessons are learned and as more external resources become available (e.g. new data from the *Gaia* satellite), the DES calibration strategy will continue to evolve.

---

[3] For the external reference stars, DES is using HST 'CalSpec' spectrophotometric standards, like the star C26202, combined with a 'Golden Sample' of about 100 pure-hydrogen-atmosphere white dwarfs within the DES footprint.

Dark Energy Science

In this section, we discuss the core dark energy science in more detail, with chapters on Type Ia supernovae, large-scale structure, weak gravitational lensing, galaxy clusters, photometric redshifts, spectroscopy, simulations and theory and combined probes.

CHAPTER 9

Type Ia Supernovae

Chris D'Andrea, Rick Kessler, John Marriner, Bob Nichol &
Masao Sako

The observation of Type Ia supernovae is the most mature technique for investigating dark energy and is one of the four main cosmological probes in the Dark Energy Survey (DES). Unlike the studies of large-scale structure, galaxy clustering and gravitational lensing, the study of supernovae involves measuring the expansion rate of the universe, rather than the formation of structure. In this chapter, we describe what Type Ia supernovae are, why they are useful for cosmology and what new discoveries we are making in DES. In the final section, we reveal the preliminary cosmological parameters we derived from our Year 3 season supernova analysis.

9.1 Introduction to supernovae

For centuries, early astronomers and star gazers looked up to the heavens and saw an unchanging universe, a refuge of stability that contrasted with the temporary nature of life on Earth. This view changed dramatically in the twentieth century when scientists were surprised by the evidence that the universe is not static, but has been expanding since it began 13.8 billion years ago with a 'Big Bang'. Not only is the size of the universe changing but everything within it is in a continuous state of transformation. Stars are born, they burn by the process of nuclear fusion and some end their existence in a violent explosion. These explosions can shine with the brightness of ten billion suns and they are known as supernovae (SN). The observation of supernovae by the Dark Energy Survey (DES) is the subject of this chapter.

9.1.1 Types of SN

There are two generally recognized types of SN explosions, classified as type I or type II. Type II have hydrogen absorption lines in their spectra, whereas type I do not. Depending on the exact shape of their light curves and the nature of their spectra, each of these classifications can be further subdivided. Those designated as Ia (SNIa)[1] are widely agreed to be the result of a thermonuclear explosion of a carbon-oxygen white dwarf (WD) star. This is due to runaway nuclear fusion at extremely high temperatures. In general, nuclear fusion in stars depends on the mass of the star. WDs are the usual final phase of medium-mass stars: about the mass of our Sun or a bit larger. However, if a WD is in a binary system with a second star, it can gradually accrete enough mass (the limit is 1.4 solar masses) to initiate an out-of-control fusion reaction to cause a type Ia SN. Billions of joules of energy are released within seconds. Figure 9.1 shows a typical SNIa light curve from DES data. The light from the explosion rapidly reaches a peak over the course of about two weeks and then dies away over a longer period (the long tail of the graph is not shown). The light from the explosion is mostly trapped within the debris, and thus the SN explosion phase is difficult to observe. Over time, the debris from the explosion expands and becomes more transparent. While the transparency increases with time, the radioactive material powering the light decays away with time. This delicate balance between transparency and radioactive debris results in a peak brightness occurring about three weeks after explosion.

While SNIa occur from 1.4 solar mass WDs, much more massive stars can explode from a different mechanism. Stars with at least ten times the mass of our sun can fuse nuclei as heavy as iron. However, the amount of iron that can be accumulated in the core of these massive stars is also restricted to ~1.4 solar masses (the 'Chandrasekhar limit'). When this limit is reached, gravity squeezes the iron core until it implodes (a process known as core collapse). The core density increases until the neutrons within the atoms cannot be pushed any closer together. Then the in-falling matter rebounds causing a shock wave, and the star violently explodes. This is a type II SN.

---

[1] We will use the acronym 'SN' to denote both supernova (singular) and supernovae (plural). Similarly, SNIa will be used for both the singular and plural forms.

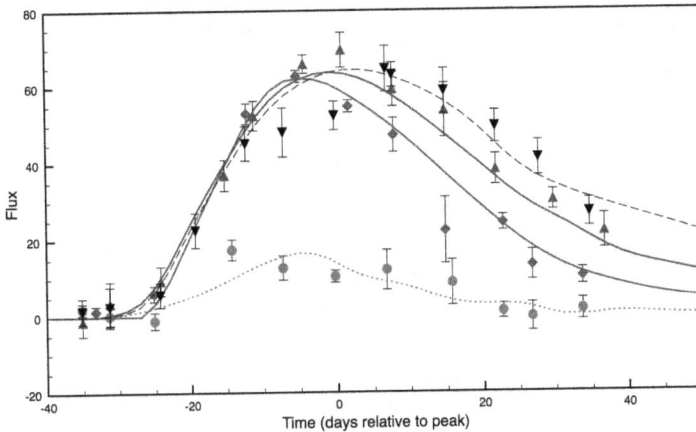

Fig. 9.1 A Dark Energy Survey (DES) light curve showing the measured brightness in the four DES bands: *g* (circles), *r* (diamonds), *i* (triangles) and *z* (inverted triangles). The lines displayed are for fits to an empirical model: *g* (dotted), *r* (dot-dash), *i* (solid) and *z* (dashed). This supernova Ia (SNIa) is at a redshift $z = 0.69$. The units for the flux are not standard: an AB magnitude of 20 corresponds to a flux of 1000 or about 36 micro-janskys.

9.1.2 Superluminous SN

In addition to cosmology, DES is sensitive to other closely related transient phenomena. The collaboration has identified a number of interesting objects known as superluminous supernovae (SLSN). SLSN are rare compared to other types of SN and are thought to result from the explosions of very massive stars, releasing phenomenal amounts of energy. There are different categories of SLSN but the explosion mechanisms are uncertain and different from the relatively well-understood Ia and core collapse SN. SLSN are ten to a hundred times brighter than SNIa and remain bright for up to six months before fading, rather than just a few weeks. This means they can be observed at higher redshifts than SNIa and could provide information on the early universe.

DES also has a program to rapidly identify SLSN and target them for spectroscopic follow-up. The first DES SLSN was reported by Papadopoulos et al. (2015). DES SLSN, denoted DES14X3taz, exhibited an unusual double peak in its light curve (Smith et al., 2016). DES is notable for its sensitivity at high redshifts and has reported a spectroscopically confirmed SLSN at redshift $z = 1.86$ (Pan et al., 2017).

9.2 Cosmology from SN

SNIa have two properties that make them useful for studying the rate of expansion of the universe. The first property is that they are extremely bright. At the peak of the explosion, a SNIa is about as bright as a typical galaxy with 10^{11} stars. Hence, SNIa can be observed in distant galaxies. Second, the range of peak brightness of SNIa is narrow due to the uniform mass of WDs that explode. This uniformity of SNIa allows them to be used as 'standard candles': objects whose intrinsic brightness is known. Comparing this known brightness to the object's apparent brightness (as observed from Earth) allows us to calculate the distance that the light has travelled. Type II SN have also been used as cosmological distance indicators, but so far SNIa have produced overwhelmingly more accurate results.

In addition to measuring the apparent brightnesses of SNIa, one can also measure their redshifts, z. Redshifts are often determined from a spectrum of either the SN or the host galaxy. After the light is emitted, the wavelength grows with the expansion of the universe, causing the light to become redder. The redshift is, therefore, a measurement of the size of the universe or equivalently, the age of the universe when the light was emitted (see Chapter 14 for further discussion of redshifts). By measuring the redshift (time the light was emitted) and the apparent brightness (distance the light has travelled) for SNIa at a variety of redshifts, it is possible to determine the history of the expansion of the universe over the redshift range of the observed SN.

We know experimentally that the effect of dark energy has evolved as the universe has expanded. In early cosmological times, matter was dominant and the expansion of the universe was slowing due to gravitational attraction. In the current era, dark energy is dominant and causes the rate of expansion to increase. The inclusion of a cosmological constant in Einstein's theory results in a very specific expectation for the rate of expansion, and dark energy experiments are often compared to this theory. Other cosmological theories give different predictions, which can also be tested by the data. Since the current understanding is that the effect of dark energy has only become important in past few billion years, there is good reason to believe that the redshift range probed by DES is where most of the action lies. In 1998 to 1999, a plot of apparent brightness against redshift (called a Hubble diagram) for SNIa provided the first direct evidence for cosmic acceleration by two groups: The Supernova Cosmology Project headed by Saul Perlmutter and High-Z SN Search team led by Brian Schmidt.[2]

Despite extensive efforts, SN are not quantitatively understood from first principles. In particular, the details of the explosion mechanism have been the subject of extensive research and debate. As a consequence, SNIa analyses are based on empirical models and often rely on a comparison of the relative brightness of SNIa in nearby (low redshift) galaxies with those in distant (high redshift) galaxies. This type of analysis does not rely on knowing the absolute brightness of the SN, but on the assumption that the brightnesses is the same at low and high redshift.

While the measurement of the expansion of the universe is simple in theory, there are a number of complications in practice. The variation in SNIa brightness is uncomfortably large (more than x2, depending on the SNIa selection criteria). This can be reduced significantly (to about 15%) by carefully measuring the light emitted from the time the SNIa is first detected to the time it fades below the detection limit. It is possible to measure the light from distant SNIa typically over a two-month period even though the time that it is visible varies depending on the distance. The shape of the light curve is also important: broader light curves arise from SNIa that are intrinsically brighter. Accounting for the correlation between light curve shape and brightness can significantly improve the distance measurement. In addition, there are variations in SNIa colour that are correlated with brightness. Colour can be measured with broadband filters, such as the *griz* filters on the Dark Energy Camera (DECam). The fact that the light curve correlations between brightness and shape and colour are invariably used to improve distance estimates is often emphasized by referring to SNIa as 'standardizable' candles.

---

[2] For which they, alongside Adam Riess, were awarded the 2011 Nobel Prize in Physics.

9.3 Strategy for SN observing with DECam

In designing the strategy for DECam SNIa survey, it was necessary to take into account several factors. One factor was that in order to study the effect of dark energy on the expansion history of the universe, we needed to make measurements of SNIa over a wide range of redshifts. Another factor was that the observation of SNIa light curves requires repeated measurements of the same part of the sky. All the other dark energy science objectives rely on the measurement of essentially static properties of galaxies (brightness, clustering and shapes), which could be accomplished with only a few visits to each portion of the sky. Thus, there was compromise between the SN science, which required repeated monitoring (every few days) of a small part of the sky and the other dark energy science, which required measurements over the widest possible area.

While there was a need for a large sample of low redshift (i.e. relatively nearby) SNIa for a cosmological sample, DECam was not well suited for a low redshift survey. This is jointly due to the overhead of the camera readout time and the need to change the telescope pointing frequently. We also considered a survey that would preferentially target medium ($z \sim 0.4$) redshift SNIa, but we felt that such a survey did not capitalize on the characteristics of DECam that enables accurate measurements of colour up to redshifts $z \sim 1$.

SNIa are rare compared to the number of stars in the universe, but a long exposure of a single pointing will typically result in the discovery every night of a new SNIa that is capable of yielding a precise distance measurement. We considered long observations of a single field of about two hours, consisting of a series of ten-minute exposures that sampled light in each of the four *griz* filter passbands. The two-hour total observation time at a single pointing was something of a practical maximum: it is a significant fraction of the approximately eight observing hours per night, and longer exposures would only reduce the statistical noise of the images by a factor proportional to the square root of the total observing time. We also considered shorter observing times per pointing, which would yield more SNIa overall, but at lower redshifts.

As a throwaway thought, we considered a hybrid survey: a survey consisting of two deep fields (long exposures) and several shallow fields (medium length exposures). To our surprise, we found that such a survey strategy was more attractive than we expected. We applied the constraint that the total exposure time of all the shallow fields together should equal the exposure time for a single deep field. So our survey would still have two thirds of the high redshift SNIa, which would be obtained by a survey with three deep fields. The addition of the shallow fields resulted in substantially more SNIa overall, which gave better statistical accuracy in the medium redshift range, and produced a more diverse sample for studying systematic effects. The final SN survey has ten fields – two of which are long exposures with a total area of 27 square degrees.

Early plans for a DES SNIa survey had called for about 10% of the survey time to be devoted to SN observations. However, more time than that would be required to meet the goal of analyzing more SNIa than were obtained by the Supernova Legacy Survey (SNLS). We initially proposed using non-photometric time, that is, nights when there is thin cloud cover that partially absorbs light in a way that is spatially and time dependent, making an accurate photometric calibration difficult.[3] For SN observing, the problem is less severe since a single night of photometric observing suffices to calibrate all the other nights. Although poor 'seeing'[4] reduces the discovery potential for distant SN, the consensus was that giving the SN survey most of the time with poor seeing as well as significant time with good seeing was the best compromise conducive to the overall goal of advancing our knowledge of dark energy with DES.

9.3.1 SN simulations

The different survey strategies were simulated by (Bernstein et al., 2012). These simulations were a key tool in designing the survey strategy, and are also critical for understanding the results. The detection of SN depends on

---

[3] A photometric calibration requires the measurement of a standard star, whose brightness is known, under the same conditions as the objects with unknown brightness. If the observing conditions vary, the calibration is more difficult, if not impossible.

[4] The contribution of atmospheric effects to the size of a point source such as a star.

their brightness, light-curve shape as well as other factors that result in a sample, which is biased relative to the complete population. Simulations help us to understand and factor in these biases. Surprisingly, the result of the simulations was that our figure of merit (a measure of our degree of uncertainty in the process) for sensitivity to dark energy didn't depend strongly on the survey strategy.

9.3.2 SN pipeline

The sequence of operations necessary to discover SN (and other transients) is known as the 'SN pipeline'. The SN pipeline uses standard DES image processing software that calibrates each pixel so that the resulting image is a background-free, calibrated, linear response to astronomical sources. The heart of the pipeline is 'difference imaging', where the most recent image data is subtracted from a DECam image taken a year or so previously. The difference will show the appearance of new sources: the SN (and also moving objects such as asteroids). The objects found through difference imaging are further analyzed to determine whether the detected object has the characteristics of a genuine astronomical source or is (as is often the case) the result of an artefact in image processing. The SN candidates that meet the selection criteria are placed in a database and subjected to further analysis. SN candidates that have appropriate properties (e.g. magnitude, colour and redshift) are selected for spectroscopic follow-up. Although the pipeline was fully functional at the beginning of operations, it has taken three years for the pipeline to mature and for various bugs to be eliminated. The pipeline is described in more detail by Kessler et al. (2015).

9.4 Spectroscopic follow-up of SN

While the need for spectroscopic observations was widely recognized, the initial DES proposal did not address this need. Previously published SNIa surveys had used spectroscopic observations (typically one observation near the peak of the light curve) to determine the type of SN and its redshift. In most cases, with a good spectrum taken near peak brightness, the identification of the SNIa type is unambiguous. A precise measurement of the redshift can also be obtained from the host galaxy, although there can be some uncertainty in cases where the identification of the host is incorrect. It is possible to identify both the type and the redshift from the light curve data alone, but that method is less accurate in general and less robust because it occasionally produces excessively large errors. Similarly, galaxy redshifts can be obtained from measurement of the brightness in the different filter bands (photometric redshifts), but this method, too, is less accurate and less robust than spectroscopic measurements.

To address the need for spectroscopic observations, DES established a task force to coordinate the efforts. The observing time was to be obtained by individual proposals. The SN follow-up observation proposals were tailored towards the interests of the individual principal investigators, the characteristics of the spectrograph and, in the cases where telescope time was awarded on a competitive basis, a science case that was judged to be appealing. Overall, DES has been awarded about 1500 hours of observing time at the Anglo-Australian Telescope (AAT), Gemini, the Gran Telescopio Canarias (GTC), Keck, Magellan Baade, MMT, the Southern African Large Telescope (SALT) and the Very Large Telescope (VLT) array. DES was fortunate to attract the interest of a group from AAT and the resulting collaborative work, OzDES, is further described in Chapter 14. The AAOmega spectrometer at AAT is highly efficient, because it is a multi-object spectrometer, that is, it can measure many objects in the same exposure. A plan was developed to measure a large number of galaxies in DES SN fields; in particular, those which have hosted SN previously detected by DES. In addition, spectra of SNIa are obtained using all the follow-up telescopes by carefully coordinating DES SNIa discoveries with the telescope availability and capability.

Table 9.1 shows the number of SNIa that were discovered during each of the first four years of survey operations. The spectroscopic observations of the host galaxies can occur over several years. The SNIa discovered in the 2012 to 2013 observing season are most complete with respect to host galaxy follow-up, while subsequent years are still acquiring host galaxy data.

The totals shown are intended only as a rough indication of the progress of the SNIa survey and represent SNIa with well-measured light curves, but the final analysis may differ depending on the selection criteria. Looser selection criteria result in more SNIa and higher statistical precision, but stricter selection criteria reduce contamination from non-SNIa light curves. The 'SNIa Discovered' column includes SNIa with spectroscopically measured redshifts and those whose redshifts were determined

Table 9.1 Summary of supernovae Ia (SNIa) discovered by the Dark Energy Survey (DES).

Year	SNIa Discovered	SNIa with Redshift[a]	SNIa with Spectrum
SV 2012–2013	760	581	19
Y1 2013–2014	710	489	63
Y2 2014–2015	611	440	124
Y3 2015–2016	464	190	10

[a]Measured with a SN spectrum or a host galaxy spectrum. Most of the redshifts are from the host galaxy.

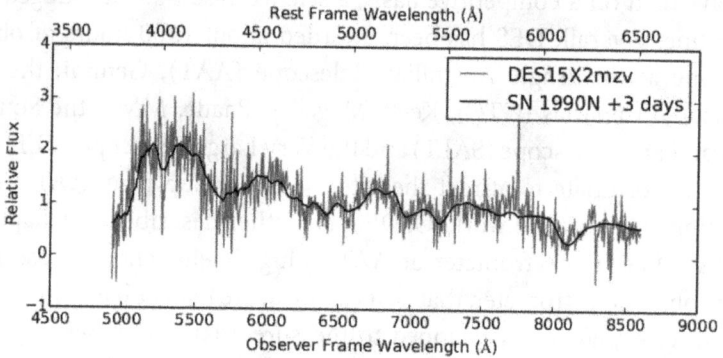

Fig. 9.2 A supernova Ia (SNIa) spectrum obtained by the OzDES collaboration (Yuan et al., 2015).

from the light curve. For the SNIa without a spectroscopically measured redshift, additional selection criteria are applied to reduce background noise, and their number is reduced relative to the sample with spectroscopically measured redshifts.

A SNIa spectrum obtained by the group at AAT (Yuan et al., 2015) is shown in Figure 9.2. The shape of the spectrum is characteristic of SNIa and is created by the wavelength-dependent absorption of light passing through the debris from the SNIa explosion. The curve shows good agreement with a redshifted spectrum of SN1990N, a well-measured low redshift SNIa. Comparison of spectra with a library of templates such as that from SN1990N provides a nearly unambiguous identification of SNIa.

It was clear that there would still not be enough observing time for spectroscopic observations of all the SNIa, especially at high redshift. These require the largest of the follow-up telescopes and are usually too faint for observation by AAT and the other mid-sized telescopes. As a consequence, it was planned that the analysis of the full DES SNIa sample would have to rely heavily on typing using the light curve information only. Successful execution of this plan required DES SNIa analysis to break new ground in producing cosmological parameters with a sample that corrects for mis-identified SN types and inaccurate or missing redshifts.

9.5 Year 3 unblinding

By Year 3, a large enough sample of SNIa had been assembled to plot our own Hubble diagram, and from this to estimate preliminary cosmological parameters. The 334 SN in the sample are made up from 206 SNIa between $0.02 < z < 0.85$ in DES and an external sample of 128 low $z < 0.1$ SNIa.

Producing an accurate measurement of cosmological parameters from SNIa has traditionally been a difficult and tedious task, and DES analysis is no different. While the principle is simple – identify Type Ia SN, plot redshift against magnitude and fit against different cosmological models – the data must be carefully processed to characterize and minimize errors in measurement, calibration and sample selection. In our analysis, we took into account fifty-eight potential sources of systematic error alone!

On December 22, 2017, after tests on 400 DES-like simulations, we were ready to 'unblind' our results (more about unblinding in Chapter 13). More than 30 members of the SN working group around the world switched on their laptops and joined in a remote video conference. As Dillon Brout, who led the unblinding session, reached the moment of truth, he explained:

'I'll press enter and after a few seconds we'll see the result for w and Omega matter taking into account combined statistical and systematic errors and a plot will show in this window. Don't jump to any conclusions! Just give it a few seconds – Are we ready? 3-2-1...'

There was an audible shared nervousness – like the moment in *The Hitchhiker's Guide to the Galaxy* when Deep Thought announces the answer is 42 – as the analysis ran through its processes – and then seconds later, some figures appeared on the screen:

$$w: -1.002 \pm 0.057$$
$$\text{OMEGA MATTER}: 0.314 \pm 0.017$$

A collective 'WOW!' echoed around the conference as we all took in the information. According to our DES SN results, within our uncertainty, we live in a universe where $w = -1$ and dark energy is consistent with a cosmological constant!

We spent the next eleven months writing a set of nine articles detailing with all aspects of the analysis, and including several analysis improvements along the way. A short (few-page) article summarizing the methods and results was recently accepted by the *Astrophysical Journal* (Dark Energy Survey Collaboration 2019). The published results, $w = -0.978 \pm 0.059$, and $\Omega_m = 0.321 \pm 0.018$ (1-sigma errors) are very similar those above announced on the unblinding telecon. These results strongly support the Λ with cold dark matter (ΛCDM) paradigm, and they agree with previous constraints using SNe Ia (e.g. Pantheon, JLA). Our published Hubble diagram and plot of cosmological constraints derived from the combined results of our DES SN and Planck cosmic microwave background (CMB) are shown in Figures 9.3 and 9.4.

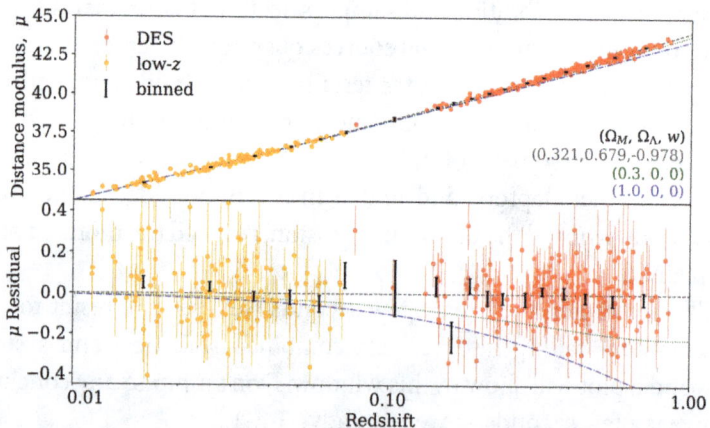

Fig. 9.3 Hubble diagram for the Dark Energy Survey (DES) supernova (SN) 3YR sample. Top: Distance modulus (μ) from a fit (black bars, which are used for cosmology fits) and for each SN (red, orange circles). The dashed gray line shows our best fit model, while the green and blue dotted lines show models with no dark energy and matter densities $\Omega_m = 0.3$ and 1.0 respectively. Bottom: Residuals to the best fit model; 1-sigma error bars show 68% confidence. From Dark Energy Survey Collaboration (2019).

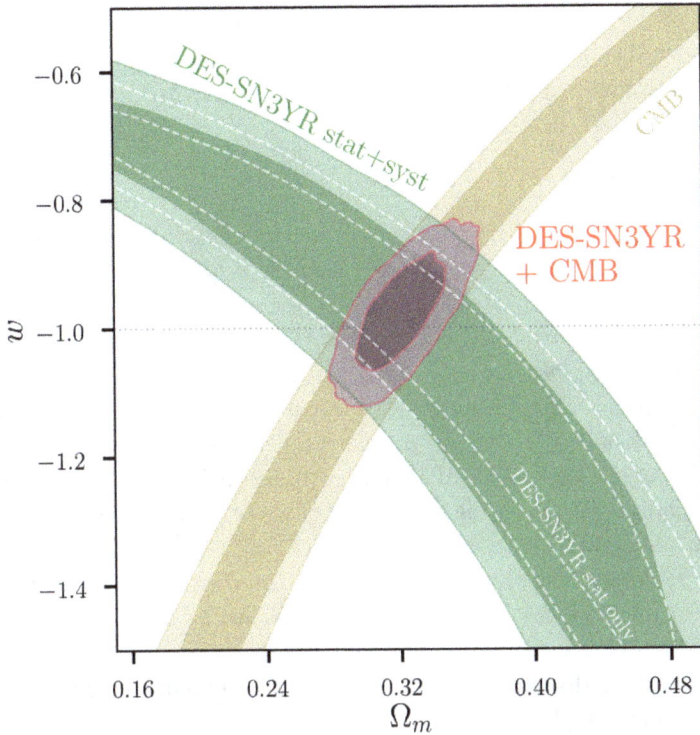

Fig. 9.4 Constraints on $\Omega_m - w$ for the flat wCDM model (68% and 95% confidence intervals). Supernova (SN) contours are shown with only statistical uncertainty (white-dashed) and with total uncertainty (green-shaded). Constraints from cosmic microwave background (CMB) (brown) and the Dark Energy Survey (DES) SN and CMB combined (red) are also shown. From Dark Energy Survey Collaboration (2019).

When our findings were circulated, Nick Suntzeff, who founded the High-z SN Search team, sent a congratulatory email:

'What a great result! This is indeed precision and accurate cosmology. The unblinding has been done almost precisely twenty years since the acceleration of the universe was discovered by the SCP and HZT... While many many want to see w not equal to −1 which would reveal radical new physics, I think that the present result is more exciting. It's a puzzle that someone is going to explain and probably many pieces will fall into place'.

Peter Nugent who was a member of the Supernova Cosmology Project was more succinct, emailing:

'Congrats!! (such a booooring universe we live in)'.

References

Bernstein, J. P. et al. (2012). Supernova simulations and strategies for the Dark Energy Survey, *The Astrophysical Journal* **753**, 6, p. 172.

Dark Energy Survey Collaboration (2019). First cosmology results using Type Ia supernovae from the dark energy survey: Constraints on cosmological parameters, *The Astrophysical Journal* **872**, 2, p. L30, doi:10.3847/2041-8213/ab04fa.

Kessler, R. et al. (2015). The difference imaging pipeline for the transient search in the Dark Energy Survey, *The Astronomical Journal* **150**, 6, p. 172.

Pan, Y.-C. et al. (2017). DES15E2mlf: A spectroscopically confirmed superluminous supernova that exploded 3.5 Gyr after the Big Bang, *Monthly Notices of the Royal Astronomical Society* **470**, 4, p. 4241.

Papadopoulos, A. et al. (2015). DES13S2cmm: The first superluminous supernova from the Dark Energy Survey, *Monthly Notices of the Royal Astronomical Society* **449**, 2, p. 1215.

Yuan, F. et al. (2016). DES14X3taz: A type I superluminous supernova showing a luminous, rapidly cooling initial pre-peak bump, *The Astrophysical Journal* **818**, 1, p. L8.

Yuan, F. et al. (2015). OzDES multifibre spectroscopy for the Dark Energy Survey: First-year operation and results, *Monthly Notices of the Royal Astronomical Society* **452**, 3, p. 3047.

CHAPTER 10

Large-Scale Structure of the Universe

Martin Crocce, Will Percival, Ashley Ross & Eusebio Sánchez

The Dark Energy Survey (DES) is one of the most important of the current large galaxy surveys and it is investigating the nature of dark energy using four independent and complementary methods. Among them is the precise study of the spatial distribution of galaxies, or large-scale structure (LSS) of the universe. This chapter describes how LSS is sensitive to dark energy and other cosmological parameters and how DES is improving our current understanding of LSS, with the goal of solving the problem of the nature of dark energy.

10.1 Introduction

The most shocking consequence of our current cosmological model, the lambda cold dark matter (ΛCDM) theory, is that everything that has ever been directly observed makes up just 5% of the contents of the universe. The remaining 95% is understood to be formed from two different components: dark matter (25%) and dark energy (70%). The existence of the dark components is inferred from the effect they have on ordinary matter, but they have never been observed in the laboratory and their physical nature remains unknown.

Dark energy is one of the most profound mysteries of physics and understanding its nature is a fundamental problem for both cosmology and particle physics. In 2019, all available observations are compatible with dark energy being a cosmological constant (Λ), a term that Einstein added to his general theory of relativity in 1917 and later retracted. This is a conundrum for physics, because the predicted value of the cosmological constant is many orders of magnitude larger than the observed value. However, more exotic possibilities exist, most of them implying new physics beyond the standard model.

One of the best approaches to tackle these questions is through deep and wide galaxy surveys, where large volumes of the universe at different epochs are observed. These surveys allow us to study the universe's overall structure and evolution. It is not an easy task, since we need to sample a volume that is large enough to be meaningful and representative, and to probe different epochs to learn about the evolution. These demanding requirements are met only by very large extragalactic surveys – for example, the Baryon Oscillation Spectroscopic Survey[1] (BOSS) and the VIMOS Public Extragalactic Redshift Survey (VIPERS).[2]

10.2 Large-scale structure of the universe

One of the most significant achievements of modern cosmology is the measurement of the large-scale structure of the universe. Galaxies are not uniformly distributed in space, they are clustered. On large scales, the universe shows a very rich structure with a hierarchical order. Galaxies form groups and clusters which, in turn, are interconnected into superclusters by 'filaments' and 'walls'. These long strings of galaxies surround vast regions of almost empty space, known as voids. The overall picture of the large-scale structure of the universe looks like soap foam. However, this hierarchy has a limit. If we smooth the picture on large scales (around hundred million parsecs or 300 million light years), it starts to look much more homogeneous. This is expected from the cosmological principle, which states that on the largest scales the universe looks the same at every point and in every direction. We know from the cosmic microwave background that in the initial stages of cosmic evolution, when atoms formed, the universe was smooth to around 1 part per 100,000.

---

[1] http://www.sdss3.org/surveys/boss.php
[2] http://vipers.inaf.it/

Understanding how this structure formed and how it has evolved into what is observed today is the basis of many key questions in cosmology. It may seem a hopeless task to establish the origin and development of such a complex pattern. However, we now have a theoretical description of the origin and growth of structure in the universe and, amazingly, it is relatively simple. It requires a mechanism to create the seeds, in the form of small accumulations of matter, and some physical process that induces the growth of these initial seeds into the hierarchical clustering pattern we see today. This process is called gravitational instability, and is easy to understand. The observed pattern appears to have grown from pre-existing, tiny (quantum) fluctuations that were amplified by the continuous action of gravity, acting mainly on dark matter, over 13.8 billion years.

The fluctuations in the matter density arose in the very earliest universe, a fraction of a second after the Big Bang singularity. The most accepted mechanism for these primordial seeds is cosmic inflation, which explains them as the quantum fluctuations of a field, the 'inflation'. This means that, surprisingly, we need both general relativity and quantum mechanics to explain the formation of structure in the universe. Once the primordial fluctuations appeared, regions with a matter density slightly larger than the average grew due to the purely attractive nature of the gravitational force, resulting in the rich structure we observe today.

The structure of the universe is, therefore, very sensitive to the matter–energy content of the universe. The existence of dark matter and dark energy, and some of their physical features, can be inferred from the statistical properties of the clustering of galaxies. This is because the observed structure is the result of a competition between these two cosmic titans. On the one hand, dark matter exerts attractive gravity and tends to make structures larger. On the other hand, dark energy exerts repulsive gravity and consequently tends to prevent the growth of structure.

The observed large-scale structure (LSS) cannot be explained if the only matter that exists in the universe is the ordinary matter we are made of (conventionally known as baryonic matter, since it is formed from protons and neutrons, which are known in particle physics as baryons), because in that case, structures as large as those we observe today could not grow. Some additional gravitating matter is needed. This is the dark matter. In fact, most of the matter in the universe must be dark, to reconcile the observed structure with the gravitational growth of structure. In addition, the action of a repulsive gravity at late times is needed, to explain the observed sizes and quantity of clusters and superclusters. This is the dark energy. As we can see, the measurement of the properties and evolution of the LSS is extremely sensitive to the dark sector of the universe.

10.3 Baryon acoustic oscillations

In addition to dark matter and dark energy, we know that baryonic matter exists, or we would not be here, and it has its own role in the LSS of the universe.

The early universe was much hotter and denser than today. In these conditions, matter was in the form of a plasma. There were no electrically neutral atoms (the temperature was too high), only electrically charged nuclei and electrons that, due to the electric charge, were tightly coupled to the photons. There were continuous collisions between baryons and photons, so that if baryons and electrons combined to form an atom, a collision with an energetic photon would have immediately destroyed it. Therefore this plasma felt two competing effects: attractive gravity due to its mass, and repulsive pressure due to the violent collisions with the photons. The baryons were gravitationally attracted by accumulations of dark matter, but when they became dense enough, the photon pressure prevented their collapse. The combined effects of these two forces produced compressions and rarefactions, that is, acoustic waves that propagated as spherical waves through the plasma. These are the baryon acoustic oscillations (BAO). Dark matter is unaffected by photons so it remained at the centre of these propagating sound waves.

At the same time, the universe was expanding and cooling. When the temperature cooled to below 3000 K, the energy of the photons became low enough that they could no longer break apart atoms. This is known as the era of 'recombination' and it happened when the universe was about 380,000 years old. Atoms are electrically neutral so the lower energy photons did not interact with them anymore and, as a result, could propagate freely. Therefore, the universe became transparent. The expansion of the universe made the density of matter much lower, so the acoustic waves could no longer propagate and the baryons were left sitting in spherical shells around the initial dark matter densities. The characteristic radius (known as the acoustic horizon) of the spherical shells formed when the baryon wave stalled is imprinted on the distribution of baryons in the universe as a density excess. Through the later effects of gravity, dark matter also clumps preferentially on this scale. Since the density of baryons is much smaller than the density of dark matter, the excess of baryon density is small. It is, nevertheless, very important. The atomic physics is very well-known, and the size of the acoustic horizon at the time of recombination can be calculated using measurements of anisotropies in the cosmic microwave background, in particular from the Planck experiment. This provides us with a known distance scale in the universe. This 'standard ruler' allows us to determine the expansion rate of the universe by comparing this reference distance in the cosmic microwave background to observations of the location of the density excesses today (using galaxy clustering). By studying the expansion of the universe, we can study the geometrical properties of space, which depend on the universe contents, including dark matter and dark energy.

The acoustic horizon scale is very large, at around 150 million parsecs (490 million light years). This is good because it allows us to measure this distance even if it is very far away. But it also complicates the measurement, because in order to observe such a large distance, huge, giga-parsec volumes of the universe must be mapped. We expect to see an enhanced number of galaxy pairs separated by 150 million parsecs.

The LSS is a very powerful way to study the universe, because it contains a wealth of cosmological information in a two-fold way. On the one hand, the statistical properties and evolution of LSS are very sensitive to the matter–energy content of the universe. On the other hand, the BAO scale provides a standard ruler and an independent method to check the contents of the universe, based only on the geometrical properties of space.

10.4 Measuring large-scale structure

The initial density perturbations that acted as seeds for the formation of structure in the universe were a quantum process (the quantum fluctuations of the primordial inflation field) and, as such, cannot be predicted from theory. The precise locations of galaxies, clusters, superclusters and voids are the result of a random process and can only be studied in a statistical way. Huge maps of the universe must be constructed, where as many galaxies as possible are located, and the statistical properties of the galaxy distribution are studied on these maps. What is measured is the excess of probability of finding a pair of galaxies at a given distance when compared to the similar probability calculated using a totally uniform distribution. This excess of probability is known as the correlation function. It can be thought of as a measure of the clumpiness of the universe. Different amounts of dark matter and dark energy, or different properties of dark energy, predict different correlation functions and different evolution of the correlation functions with time. If we are able to measure the correlation function for different epochs in the universe, we will be able to infer its matter–energy content.

Since the BAO produce an accumulation of galaxies at the acoustic horizon scale, this translates into a peak in the correlation function at this scale (there's a higher probability of finding a galaxy pair separated by the BAO scale). Again, if we are able to observe this peak at different epochs in the universe, we can map the expansion of the universe and infer its matter–energy content. The critical issue here is our understanding of how exactly the visible galaxies trace the dark-matter distribution. We can only observe the visible light, but we know that galaxies formed in the high density regions of dark-matter halos. Therefore using galaxies to trace the underlying dark matter introduces an inherent bias, by over-weighting the overdense regions and under-weighting the underdense regions. Some uncertainty also appears on small scales, where the most used approximations (that the inhomogeneities are small) do not apply anymore, and non-linear equations must be used. These equations are extremely complex, and can be solved only numerically or by using new approximations, introducing uncertainties into the description of the correlation function.

Different epochs of the universe translate into different values of a galaxy's redshift (more fully explained in Chapters 14 and 15). Therefore, to map the correlation function at different epochs, we need to select galaxy samples at different redshifts and compute the correlation function for each sample. In this way, we map the evolution of LSS and BAO.

10.5 Large-scale structure in DES

The LSS is a 3D spatial distribution, but the Dark Energy Survey (DES) is a photometric survey so its precision in the measurement of the redshift of galaxies is limited (see Chapter 15). To optimize the DES sensitivity to cosmological parameters, we are measuring the angular correlation function of galaxies (the probability of galaxy pairs being separated by a given angle on the sky) in photometric redshift slices, whose size is adapted to the DES precision in the redshift measurements. This method gives us the evolution of angular LSS with time, allowing the exploitation of the full sensitivity of DES to the contents of the universe. LSS is one of the four key probes in DES, together with supernovae (Chapter 9), the number counts of clusters of galaxies (Chapter 12) and weak gravitational lensing (Chapter 11).

Since LSS is one of the core measurements in DES, the working group dedicated to the analysis and study of the LSS of the universe has been active from the very beginning of the project, developing new analysis methods and studying the best ways to maximize the sensitivity to cosmological parameters. The first data set that allowed us to make an initial measurement of LSS was taken between November 2012 and February 2013, during the science verification (SV) period of observations. These data provided our first observed map of galaxy density, shown in Figure 10.1. The colour shows the number of galaxies found in regions of one arcminute in size, mapping the structure of the universe. The underdense regions are in light yellow and the overdense regions are in dark red. The mapped region covers an area of 116 square degrees (0.2% of the total sky) and contains more than two million galaxies. This is only 3% of the total area that DES will cover, but it allowed us to make an initial measurement of the LSS, which was presented in Crocce et al. (2016).

Figure 10.2 shows the correlation function for galaxies at redshifts 0.7 (left) and 0.9 (right). The DES measurements are the dots, while the expected correlation functions for a universe with 70% dark energy, 25% dark matter and 5% normal matter, as expected from the current standard model of cosmology (ΛCDM), is the solid line. As we can see, the DES data is in perfect agreement with the standard model of cosmology. The dashed

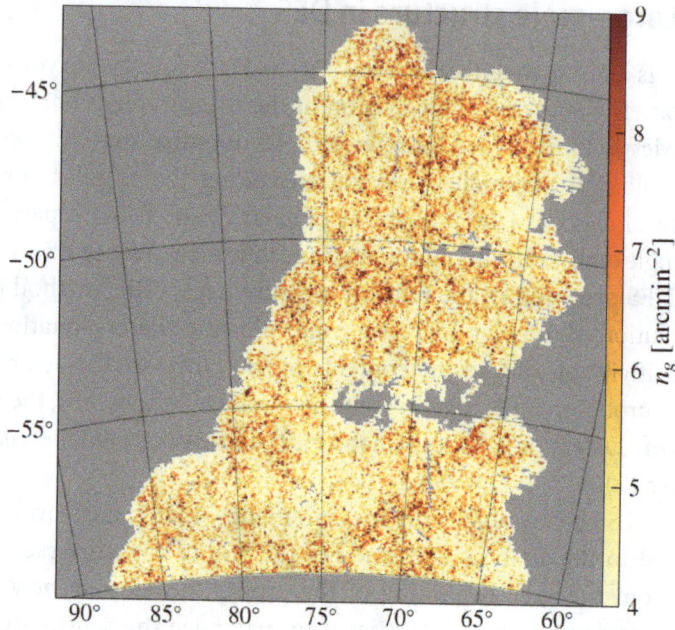

Fig. 10.1 Map of the measured galaxy density using the Dark Energy Survey (DES) Science Verification data set. The horizontal axis corresponds to right ascension and the vertical axis corresponds to declination. The map covers an area of 116 deg^2 in the southern hemisphere and contains more than two million galaxies. The mean density is 5.6 galaxies per square arcmin, with inhomogeneities represented by colours, from underdense regions in light yellow to overdense regions in dark red.

line is an approximate calculation using linear equations, valid only for large scales.

The shape of the correlation functions of Figure 10.2 shows the galaxy clustering of the universe using real data from DES. At large scales they tend to zero, in agreement with the homogeneity of the cosmos. But for smaller scales the correlation functions grow, which tells us that the probability of finding close galaxy pairs is much larger than that expected from a purely random distribution. The exact shape of this function depends on the matter–energy contents of the universe.

The area covered by the DES SV data set was too small to make strong cosmological inferences. By 2017, this changed radically with the analysis of the first year of DES data (DESY1), spanning roughly ten times more area while reaching sufficiently distant galaxies, at redshifts $\gtrsim$ 1.

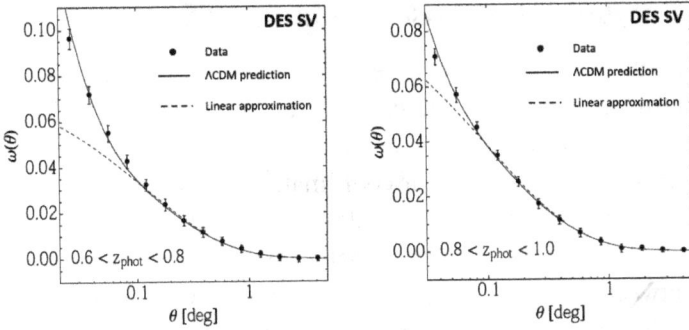

Fig. 10.2 Correlation functions measured in the Dark Energy Survey (DES) Science Verification data set, at two different redshifts, 0.7 (right) and 0.9 (left). The dots are the measured data, while the solid line is the expected correlation function for a universe that contains 70% dark energy, 25% dark matter and 5% ordinary matter, as expected from the current standard model of cosmology ΛCDM. Both are in perfect agreement. The dashed line is an approximate calculation, using linear equations, valid only for large scales.

This enabled high precision measurements of galaxy clustering in five tomographic bins in the redshift range $(0.15 - 0.9)$ using the galaxies with best photometric redshifts, selected with the redMaGiC algorithm. Such measurements provided one of the three different '2pt' functions used to infer state-of-the-art cosmological constrains, as described in Chapter 13.

Several observational and theoretical systematic effects need to be accounted for in order to provide reliable LSS measurements. In particular, one has to disentagle true LSS from fluctuations induced by the changing weather and observing conditions under which measurements are made. Also, in order to try to understand the effects of non-linear growth and galaxy bias, we run large simulations that are specially designed to mimic the DES survey, including all the effects that are present in the real measurements. Using the simulations, one can switch on and off effects such as the bias, the non-linearities and some observational systematics, allowing us to control their effects on the correlation function. There are several simulations, such as Marenostrum at Institut de Ciencies de l'Espai (MICE) or the Blind Cosmology Challenge (BCC) set of simulations, that include different implementations of these effects, especially the non-linearities, that help in the control of the total influence on the correlation functions. This is described in more depth in Chapter 16.

10.6 Baryon acoustic oscillations in DES

The extent of DESY1 data, covering over 1300 deg^2, enabled the first measurement of comoving (angular diameter) distance to redshift of ~ 1 by locating the position of the baryon acoustic feature in the correlation function of galaxies. In angular coordinates, this shows itself as an excess of galaxy pairs at separations of $2° - 3°$ depending on redshift. This analysis required defining an optimal galaxy sample, which is a compromise between (high) number of galaxies and (low) photometric redshift errors. The LSS group also developed a dedicated large set of synthetic galaxy catalogues, the 'HALOGEN' mocks (see Chapter 16), which were key to stress test the analysis, and several theoretical tools. In all, the BAO measurement achieved an error of $\sim 4\%$, was robust to several analysis choices, and matched the accuracy of existing data sets based on spectroscopic data, as shown in Dark Energy Survey Collaboration (2018). Besides being an achievement in itself, it laid the ground for more accurate measurements in the near future with newer and larger DES data sets, and in the longer term with data from LSST or Euclid.

10.7 Looking to the future

As the next seasons of DES data is processed, we will measure the LSS with increasing precision. The upcoming analysis of the first three years of DES data (DESY3), covering three times more area, will allow us to measure BAO with a precision of roughly 2.5% at $z \sim 1$. This has the potential of being the most precise distance measurement to such redshift until DESI arrives. With the improvement from future data, DES will become very sensitive to many important cosmological parameters, including dark matter and dark energy, but also the mass of neutrinos, whose presence alters the properties of the LSS and, consequently, the shape of the correlation functions. This, together with optimal BAO scale measurements and the combination of probes, will allow stringent tests of the geometry of the cosmos and, perhaps, new and exciting discoveries about the properties of the dark side of our universe.

References

Crocce, M., Carretero, J., Bauer, A. H., Ross, A. J., Sevilla-Noarbe, I., et al. (2016). Galaxy clustering, photometric redshifts and diagnosis of systematics in the DES Science Verification data, *Monthly Notices of the Royal Astronomical Society* **483**, pp. 4301–4324, doi:10.1093/mnras/stv2590.

Dark Energy Survey Collaboration (2018). Dark Energy Survey Year 1 Results: Measurement of the Baryon Acoustic Oscillation scale in the distribution of galaxies to redshift 1, *Monthly Notices of the Royal Astronomical Society* **483**, pp. 4866–4883, doi:10.1093/mnras/sty3351.

CHAPTER 11

Weak Gravitational Lensing

Daniel Gruen, Bhuvnesh Jain & Michael Troxel

Weak gravitational lensing is a unique probe of the dark side of the universe. Measurements of the weak lensing-induced distortions of the shapes of distant galaxies provide a direct way to map the distribution of dark matter and to probe dark energy. We discuss how these challenging measurements are made and interpreted. In particular, we describe the work done on weak lensing in the Dark Energy Survey (DES) and the results from the first season of data.

11.1 Introduction

While the Dark Energy Survey (DES) project was being formulated, and before it was funded, it was clear that weak gravitational lensing would be one of the techniques used to study dark energy. What was not clear was that lensing on this scale could work, or that we could convince the funding agencies that this technique was ready. The proposed DES data set would be twenty or thirty times bigger than that of the Canada-France-Hawaii Telescope Legacy Survey (CFHTLS), which was then the largest weak lensing 'mass map' in existence. There was concern about the limitations imposed by systematic uncertainties and whether we would we be able to solve them. Systematic uncertainties, or 'systematics', are caused by something inherent to the system being observed, such as the distortion of the image by the telescope optics and the atmosphere. Our hope was that since the Dark Energy Camera (DECam) would be built from scratch, it would be of the necessary high standard, but the Blanco Telescope was not proven to be capable of the image quality required for weak lensing. At that stage, no weak lensing measurement had ever been made at the accuracy required for DES with any telescope. It all came down to the instruments and how well we could clean the data and model any remaining effects.

11.2 Weak gravitational lensing

Gravitational lensing provides astronomers with a key tool to study the distribution of dark matter in the universe. Crucially, it allows us to turn observed galaxy images into a map of the matter density of the universe. It refers to the process by which the path of light from distant sources in the universe is bent as it travels towards us through the gravitational field of intervening matter. The explanation in Einstein's general theory of relativity is that the presence of matter curves space-time, and the path of a light ray is deflected as a result. We can think of the intervening mass as acting like a giant lens. As with any lens, the shape of the image depends on both the distance from your eye to the lens and on the distance of the lens from the source of light, as illustrated in Figure 11.1. Lensing is, therefore, a way of measuring distances. The shape of the image also depends on the properties of the lens. Therefore, unlike other observational probes that measure distance, such as Type Ia supernovae and baryon acoustic oscillations (BAO), the lensing effect also provides a direct measure of the density of the intervening matter. And it doesn't require any prior knowledge of the composition or dynamical state of the astronomical objects.

When the lens is sufficiently strong, multiple images of the same source are observed. As described in Chapter 19 on 'strong lensing', Einstein rings and multiple images in these rare systems are used to learn about the mass distribution in the inner parts of galaxies and clusters. 'Weak' gravitational lensing is a less dramatic effect: it refers to the magnification and/or 'shearing' of distant galaxy images due to the differential deflection of neighbouring light rays. A shear could, for example, stretch an image along one axis and compress it along the other, so that an initially circular image looks like an ellipse. The distortion is small, typically inducing an ellipticity of order one percent. This is negligible compared to the intrinsic shape of individual galaxies, but it can be measured statistically using the coherence of the lensing shear over millions of galaxies across the sky. Only in the past decades has it become possible to accurately measure

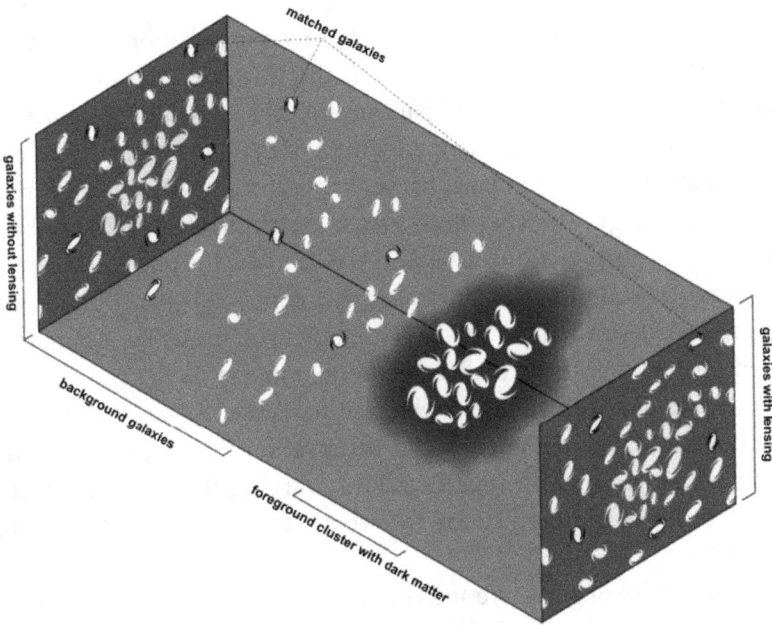

Fig. 11.1 Diagram showing the effects of gravitational lensing in three dimension (3D). Image by Michael Sachs under Creative Commons.

these subtle changes in order to study the distribution of dark matter in the universe.

11.3 Weak lensing and dark energy

In our current standard cosmological model of a cosmological constant Λ with cold dark matter (ΛCDM), the universe began 13.8 billion years ago and almost immediately underwent a vast, exponential expansion (inflation) in a tiny fraction of a second. After that it continued to expand, but at a much less rapid rate. About five billion years ago, the expansion began to accelerate under the influence of dark energy.

We assume[1] that the universe starts off as homogeneous and isotropic (the same in all locations and in all directions) but that there are tiny

---

[1] It's a very general principle that you can't make any inferences without first making a number of assumptions. The important thing is to be aware of what the assumptions are and in what ways they might bias our study.

quantum fluctuations that generate density perturbations, which grow under gravity into all the structure in the universe. The model can be characterized by a number of parameters. These include parameters to describe the expansion rate and curvature of the universe as a whole, and the density of its constituents: baryons, photons, neutrinos, dark matter and dark energy. We also need statistical parameters to describe the nature of the perturbations, and others to describe the physical state of the universe. Typically, a cosmological model contains between five and ten parameters, depending on what assumptions you make. For example, you could assume that the overall shape of the universe is flat (no curvature) and this would remove one of the free parameters. We don't know the exact values of the parameters, but observations have constrained them to lie within a narrow range of values. Varying the parameters in the model changes the predictions for what we can observe, for example, with DES.

In weak lensing, we are interested in how the parameters affect the distribution of matter in the universe. If we can measure the 'clumpiness' of matter at different redshifts, then we can measure how structure in the universe has evolved over time. This relates to dark energy because dark energy is causing the expansion of the universe to accelerate. If space expands very quickly, then matter is pulled apart before it has time to cluster as much. Therefore, if dark energy is more powerful, then the structures are less massive.

As we discuss in the introductory chapter, dark energy is parameterized by a dimensionless number, w, equal to the ratio of its pressure p to its energy density ρ, such that $w = p/\rho$ (the equation of state). If $w = -1$ and the rate of variation of w with time is zero, this is equivalent to Einstein's cosmological constant, Λ: it tells us that dark energy has remained constant over time. However, it may be that the dark energy density has varied over the course of the universe, in which case we might detect a value other than -1. This would imply that we need to revise the current paradigm.

It is astounding that we can presume to describe the entire universe with such a small number of variables, but in reality, things are more complicated because we have to include tens to hundreds of extra 'nuisance' parameters in our analysis. These are numbers that parameterize the systematic uncertainties in (1) our knowledge of astrophysics and (2) the process of making the observations.

11.4 Three regimes of weak lensing

The strongest weak lensing signal occurs when light from distant galaxies is lensed by nearby clusters of galaxies (for more information on galaxy clusters, see Chapter 12). This is called *galaxy–cluster* lensing. Clusters are a particularly sensitive probe because only a slight change in the cosmological parameters makes a big difference to the mass of these extreme objects. If you can measure the mass of the cluster accurately using lensing, you can then compare this to the prediction from your theory. For example, weak lensing studies of large cluster samples play an important role in determining the relation between the observed cluster properties and mass (Chang et al., 2018).

When light from distant galaxies is lensed by individual galaxies closer to Earth, it is called *galaxy–galaxy* lensing. The statistical correlations in the shapes of background source galaxies is weaker than for galaxy–cluster lensing.

The third regime is called *cosmic shear*. This is when the light from background galaxies is distorted by the large-scale structure (see Chapter 10) between the galaxies and Earth, but we don't have detailed knowledge of the foreground galaxies or clusters. Cosmic shear also produces an observable pattern of alignments in background galaxies, but the distortion is much less than that due to clusters or galaxies. The straightforward interpretation of the signal provides us with a measure of the growth of large-scale structure, which is a way to test the effect of gravity on cosmological scales and to infer the properties of dark energy.

11.5 Systematic uncertainties

Precise measurement of galaxy shapes is not an easy task because the observed images have been 'corrupted' in several ways. Even in space-based data, the finite size of the mirror and the complicated telescope optics give rise to a 'point spread function' (PSF). The PSF is a measure of the degree of spreading, or blurring, of a point source. In ground-based data,

Galaxies: Intrinsic galaxy shapes to measured image:

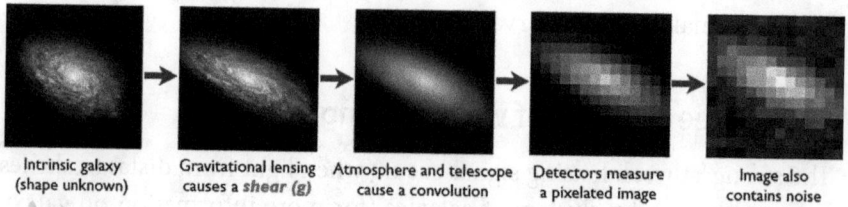

| Intrinsic galaxy (shape unknown) | Gravitational lensing causes a *shear (g)* | Atmosphere and telescope cause a convolution | Detectors measure a pixelated image | Image also contains noise |

Stars: Point sources to star images:

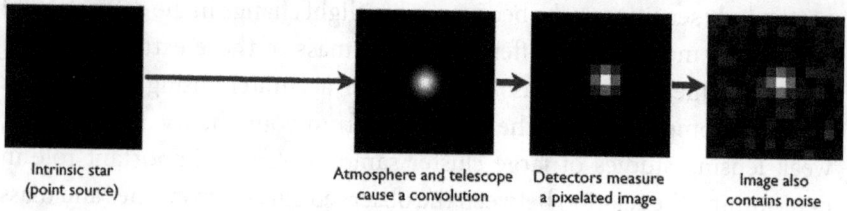

| Intrinsic star (point source) | | Atmosphere and telescope cause a convolution | Detectors measure a pixelated image | Image also contains noise |

Fig. 11.2 The upper panels show how the original galaxy image is sheared, blurred, pixelated and made noisy. The lower panels show the equivalent process for (point-like) stars. Scientists only have access to the right hand images (Bridle et al., 2009).

the situation is made worse by turbulence in the atmosphere (an effect called 'seeing'). Furthermore, the image is sampled in discrete pixels, with a detector that may not be 100% efficient.

Figure 11.2 illustrates the scale of the problem. We need to derive the amount of shear, but we only have access to the fuzzy, pixelated images on the right.

Cosmological inferences of lensing measurements also rely on knowing the redshifts of the source galaxies, which is a measure of how much light from these galaxies has been stretched as it travelled to us through expanding space. The galaxies of interest are too faint to be observed spectroscopically, so weak lensing studies rely on photometric redshifts (photo-z's). See Chapter 15. DES records images using five filters, spanning 400 to 1080 nm, and each filter images the sky in a different wavelength band of light: g, r, i, z and Y (we may only use r,i,z in the wide field for Y3). Measurement of the galaxy colours and other properties using some of these filters provides a coarse estimate of the galaxy's redshift, but uncertainties in the estimated photo-z's can contribute to systematic errors.

In total, the primary sources of systematics are the following:

- Knowledge of the PSF.
- Correction of the PSF and shear calibration: one cannot directly see the weak lensing signal on the sky, since a weakly sheared galaxy appears unchanged due to its much larger intrinsic ellipticity. Therefore, we can never be sure that the recovered lensing signal for any object is cosmological in nature and not dominated by observational distortions.
- Intrinsic alignments. These are due to the tidal gravitational field which can cause the shape of a galaxy to be aligned with another due to direct interactions. It is currently the least well-known systematic uncertainty.
- Photometric redshifts.
- Theoretical model predictions. For example, there may be uncertainties due to nonlinear gravitational clustering and baryonic gas physics.

Each of these errors can be modelled, but small residuals in the corrections of these errors may well be comparable to statistical errors in lensing measurements. We can never reduce the systematics to zero, so our target is to reduce them until they are small compared to the statistical effects. Statistical uncertainties can be lessened by gathering more data, and we have a six-year survey in which to do that.

11.6 From observations to cosmology

To make cosmological measurements from raw galaxy images, we need to follow a sequence of processes called a 'pipeline'. Schematically, the steps are as follows:

1. Object detection: the detection of the faint galaxies that are used in the analysis. This can be performed on the individual exposures (multiple images of the same area of sky are typically obtained) or on a 'stacked' image (the images are added together). In either case, an algorithm to distinguish stars from galaxies is needed.
2. PSF estimation: to deal with the effects of the PSF, a sample of moderately bright stars is identified from the actual data. These are subsequently used to characterize the PSF.

3. PSF correction: having identified the galaxies of interest, the next step is to correct the observed galaxy shapes for the convolution (blurring – see Figure 11.2) by the PSF.

4. Shear calibration: we must estimate, either by fully simulating our survey and measurements and/or by cleverly and systematically modifying the real images, how well our shape measurement algorithms are able to recover a true input shear. We then correct all measurements by this amount. This is the most difficult, yet most important step in the analysis and is one of the most active areas of study and innovation.

5. Measurement of shear correlations: once a catalogue of galaxy shapes is available, we can use it to calculate cosmological statistics. The 'two-point correlation function' gives us a measure of the degree of clustering in the distribution of galaxies. It describes the probability that two galaxies are separated by a particular distance and it can also tell us to what extent they are pointing in the same direction (i.e. how much their shapes are correlated). The two-point correlations of the shear are calculated simply from the galaxy positions, ellipticities and, in some cases, weights that characterize the signal-to-noise of the shape measurement. The 'power spectrum' is the Fourier transform of the correlation function and it gives us an elegant statistical description of the density fluctuations of matter in the universe.

6. The final stage is to use the statistics to find the most likely values of the cosmological parameters. See Chapter 13.

11.7 Weak lensing in DES

In 2003, when DES was still just a possibility, Josh Frieman (University of Chicago) approached a bright young postdoc called Erin Sheldon (then also at the University of Chicago) and asked him whether he would be interested in developing a pipeline of computer code that could deliver cosmological parameters through shear measurement. Erin Sheldon thought it was a great idea. Mike Jarvis and Gary Bernstein at the University of Pennsylvania had developed a code to measure the shapes of galaxies based on a method called 'shapelets' and they, together with Bhuvnesh Jain at U. Penn, also became involved with DES early on. Erin began working with Mike to develop a pipeline using the shapelets code, but they kept hitting a dead end due to noise bias.[2]

In about 2008, Erin Sheldon began working on a new weak lensing code, which he called '*ngmix*'. It worked, in his words, 'pretty well': it was good at modelling the galaxy shapes and it was very fast to run, but it had the same problem with noise that shapelets had. When the DES science verification (SV) period began in September 2012, Erin learned about a technique called 'lensfit', which the CFHTLS group were using. He decided to write his own lensfit and combine it with *ngmix* to deal with the noise bias issue. For a long time, he was working more or less by himself on Long Island (at Brookhaven National Laboratory) and nobody knew what was going to come out of it. You would email Erin and you wouldn't hear back; you would call him and he wouldn't answer the phone. Then you would see him at a collaboration meeting and ask, 'So, are you going to have a code that we want to apply to the data'? And he would answer, 'Not yet, I can't really say. I'm still working on it. I don't know what the bug is but I'm debugging'. Then suddenly he had something that ran. It ran fast and the results were spectacular. He took a risk but that's what he wanted to do and in the end he came through, to the relief of all of us.

At the same time, Joe Zuntz was leading the development of an independent main shear pipeline called '*im3shape*', along with Sarah Bridle and a team including Tomasz Kacprzak, Niall McCrann, Barnaby Rowe, Michael Troxel, Lisa Voigt and others, based in the University of Manchester. Not only was Joe running one of the two pipelines but also helping everybody with their computational issues and at the same time he developed and ran a code for doing parameter estimation (CosmoSIS). For a few years, everybody was relying on Joe.

People were encouraged to develop different pipelines from the beginning because there are dozens of things that can go wrong during the long process of measuring the PSF and measuring the shapes of galaxies, and so on. Real data is incredibly complicated, so it is a huge advantage (almost a necessity) to have two independent codes because they provide an important cross-check on each other. Sarah Bridle and Bhuvnesh Jain

---

[2]Noise bias means that when your signal gets noisy, sometimes the property you estimate from the signal deviates from the truth, even on average.

(University of Pennsylvania) were the first coordinators of the weak lensing group and one of their challenges was to motivate people to work on different codes.

One clear test of different methods is to use simulated data. So we created simple simulated images and applied a known PSF to the data, then tested each method to see how closely it could estimate the lensing signal, that is, the amount of shear applied to the original image. These simulations may lack a systematic effect that is present in real data, but fortunately a number of diagnostic tools can be used to test whether systematics are present in the recovered signal. The tests cannot guarantee whether the recovered signal is free of systematics, but they do often indicate whether systematics are present. This means our ability to correct for shape measurement systematics can be tested using simulated data, which gives lensing an important advantage over other methods.

A series of image analysis challenges using simulations were organized across the weak lensing community. DES members who played a key role in setting these up included Barnaby Rowe, Mike Javis, Gary Bernstein, in addition to Rachel Mandelbaum and Jim Bosch. One consequence of these simulation challenges was the development of an open-source code toolkit called GalSim, which was subsequently used to create large suites of simulated DES-like images to characterize and calibrate weak lensing measurements in real DES data.

Once *im3shape* and *ngmix* became the two main codes, a lot of people worked to test them with simulations and make sure they agreed. This was a huge task in itself. Running a code is relatively easy; checking they work and that you fully understand them is hard. Mike Jarvis coordinated the shape catalogue paper based on data from the SV season of observations, which is when both codes were first put to the test with real data. He and others discovered that the most tricky part in comparing both catalogues on simulations was so-called 'selection effects' – for some galaxies, only one of the codes measures a good shape, and thus you can't directly compare the results. While Mike found some ways of making these selection biases less severe, they are what limited the accuracy of the SV shape catalogues. Other people worked independently on improving the PSF. Making the shape catalogue is the hardest part of the process and takes the most computing hours, and there were a lot of glitches to smooth out. Having good collaborators is essential because they can help you look at a problem differently.

After the Jarvis et al. (2016) paper was published with the SV data, Erin Sheldon began working to improve *ngmix*. What made *ngmix* so good was it was both 'multi-epoch' and 'multi-band', which meant that all the times an object has been imaged, through all the different filters, are processed simultaneously. If you stack the images together, you lose information, about the PSF in particular, so *ngmix* measures the shapes in each exposure, rather than working with the co-adds. Matt Becker, a postdoc (then based in Stanford) realized he could make the code even better if he added 'multi-object' fitting (MOF). The light from galaxies in clusters and stars in galaxies ends up blending together, but he wanted to divide up the light. This idea has transformed the pipeline.

The main reason weak lensing is difficult is because, in contrast to most physics experiments, there is no calibration source. You can never know what the exact shape of a galaxy is before it is sheared, so there is nothing certain you can use to calibrate your measurements. Instead you have to rely on simulations of the data to do your calibration, and if your simulations aren't perfect, then the calibration won't be perfect. Unfortunately, it's extremely hard to do realistic simulations. In 2015, Eric Huff, then a postdoc at the Ohio State University, came up with the idea of 'metacalibration'[3], which meant you could bypass the simulations altogether. The idea is that first you measure the shape of a galaxy in an image, then you introduce a known, fake shear onto the galaxy. Then you measure the shape again and observe how your measurement changes. The assumption is that if the shear is very small, which is for weak lensing, then you can apply this same factor to the real data to find the real shear. Metacalibration uses the images themselves to calibrate the measurement. This was very technically challenging at first but it has gradually become much easier to implement. Eric Huff got it to work with low noise and Erin Sheldon got it to work with real noise. In total, it took about two years to get *ngmix* working with metacalibration and MOF. This new code, *metacal*, was used to produce

---

[3] Eric came up with this independently, but it turned out there was a paper written on a similar idea in 2000.

a catalogue using data from the first full season of observations, Year 1, which covered about ten times the area of sky as the SV season.

To improve the other pipeline, using *im3shape*, we set out to produce much more realistic simulations. Simon Samuroff (University of Manchester) led this work and produced a simulation that looks like our survey in almost all ways imaginable. It may never be possible to simulate reality exactly, so there are limitations to this approach, but it means we have two excellent, independent methods of shape calibration.

11.8 DES Year 1 weak lensing catalogues

Michael Troxel (Ohio State University) and Daniel Gruen (Stanford University) took over as coordinators of the weak lensing group in 2016 and they have continued to push the field forward. Joe Zuntz worked with Michael Troxel in Sarah Bridle's group in Manchester and describes him as, 'the hardest working person I've ever met'. Daniel Gruen pretty much invented 'trough lensing', which involves comparing the gravitational shear around underdense regions of the sky with the shear around overdense regions in the sky.

By 2017, DES was setting the world standard for weak lensing analyses. Using these two fully independently calibrated and very different shape measurement methods, *metacal* and *im3shape*, the Weak Lensing Working Group used the DES Year 1 (DESY1) data to produce two complementary weak lensing catalogues covering around 1300 square degrees of sky and including 34.8 million galaxies. At the time, this was the largest weak lensing map ever created.

DES Data Management produces a 'gold' galaxy catalogue, made up from the galaxies about which we have the most reliable information. This is used to produce a *shape* catalogue and a photo-z catalogue.[4] The statistical properties of the mass distribution are inferred from the lensing "two-point" function described in Section 13.2.3 (Cosmic Shear). Combining the information in both these catalogues we are able to generate, so-called, '3 × 2pt' (combining weak lensing with large scale structure) functions, which are used as a probe of cosmology itself and are described in more detail in Chapter 13.

---

[4]Once you have these catalogue, you can branch off in many directions. For example, Tomasz Kacprzak (ETH Zurich) wrote a paper on 'peak lensing'.

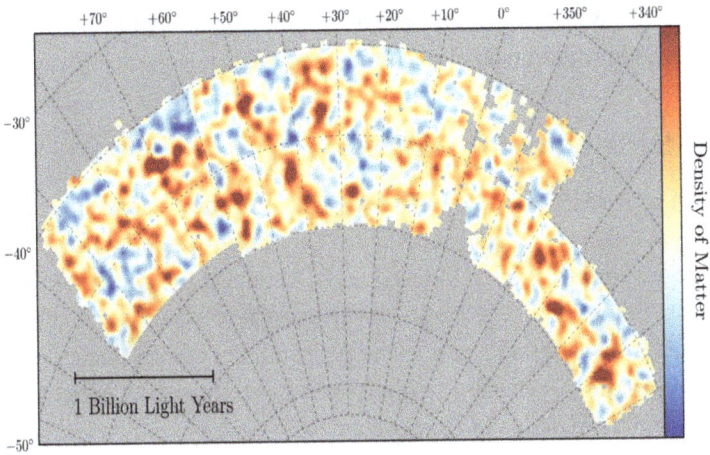

Fig. 11.3 Mass map from the weak lensing analysis of Y1 data on more than twenty-six million background galaxies in Dark Energy Survey (DES). The colours represent the density of dark matter. The blue patches are voids and the dark red patches are clusters. From Chang et al. (2018).

So as not to potentially bias our work towards an 'expected' result, we conduct a 'blind analysis' using the weak lensing data. We did this in two ways: firstly, by taking our carefully calibrated shear catalogues and scaling the calibration by a random factor and secondly, by shifting the cosmology results by an unknown amount in a random direction. We changed both *metacal* and *im3shape* in the same way and cosmological constraints still overlapped, so we knew they were consistent even before we unblinded.

It is a big step when you unblind because it's the culmination of several years of work and you can't go back. However, many times you check everything, it is still an anxious moment.

The unblinded DESY1 3 × 2pt results are revealed and described in more detail in DES collaboration (2018) and in Chapter 13.

11.9 DES Year 1 weak lensing mass map

The distribution of the shear over the sky can be used to obtain the surface density of the mass distribution, that is, to make an 'image' of the dark matter distribution, as shown in Figure 11.3. This allows us to visualize the data. A detailed study of this map can tell us more about both dark matter and dark energy. The DESY1 mass maps are the largest weak lensing mass maps to date constructed from galaxy surveys. The map from the full six-year survey will be based on data from 300 million galaxies.

11.10 Looking to the future

We have proved that weak lensing is a viable probe of dark energy and we have developed powerful tools in DES that can be applied to future surveys (e.g. Large Synoptic Survey Telescope [LSST] and Euclid), and we still have more data from the rest of the survey to analyze. DESY3 covers three times as large an area as DESY1. Using the potential data from this alone, as well as in combination with external data sources, such as the CMB, can further enhance our knowledge of the cosmos.

References

Bridle, S., et al. (2009). Handbook for the GREAT08 Challenge: An image analysis competition for cosmological lensing, *Annals of Applied Statistics* **3**, 1, pp. 6–37.

Chang, C., Pujol, A., Mawdsley, B., Bacon, D., J. Elvin-Poole, J., et al. (2018). Dark Energy Survey Year 1 results: Curved-sky weak lensing mass map, *Monthly Notices of the Royal Astronomical Society* **475**, 3, pp. 3165–3190, doi:10.1093/mnras/stx3363.

Dark Energy Survey Collaboration (2018). Dark Energy Survey Year 1 results: Cosmological constraints from galaxy clustering and weak lensing, *Physical Review D* **98**, 4, pp. 043526.

Jarvis, M., Sheldon, E., Zuntz, J., Kacprzak, T., Bridle, S. L., et al. (2016). The DES science verification weak lensing shear catalogues, *Monthly Notices of the Royal Astronomical Society* **460**, pp. 2245–2281, doi:10.1093/mnras/stw990.

CHAPTER 12

Galaxy Clusters

Tesla Jeltema, Tim McKay, Chris Miller,
Joe Mohr & Kathy Romer

In the previous chapter, we explored the importance to the Dark Energy Survey (DES) of weak gravitational lensing. In this chapter, we see how combining this knowledge with measurements from galaxy clusters allows us to learn about the formation of structure and the amount of matter in the universe, which helps us to measure the effects of dark energy on the expansion history of the universe.

12.1 Galaxy clusters

Galaxy clusters are the largest gravitationally bound and virialized struc tures in the universe. They can contain thousands of galaxies bound together within a giant, invisible halo of dark matter. The Milky Way, itself hosting 100 billion stars, is part of a relatively small cluster of galaxies called the Local Group, which also contains the spiral galaxies Andromeda and Triangulum (M33) and at least fifty smaller galaxies. Our nearest large galaxy cluster, the Virgo Cluster, is situated about fifty-five million light years away, in the direction of the constellation of Virgo. Even larger structures – clusters of clusters – can be identified, called superclusters, but most superclusters are not bound by gravity.[1]

Only 1% or 2% of the mass of a galaxy cluster is in the form of the visible galaxies. A further 10% to 15% is hot X-ray emitting gas and all the

---

[1]Both Virgo and the Local Group are members of a vast supercluster recently dubbed Laniakea, which contains approximately 100,000 galaxies stretched out across 520 million light years. The name Laniakea means 'immeasurable heaven' in Hawaiian.

rest (up to 90%) is thought to be of some form of 'dark matter'. The idea of dark matter was first posited by Fritz Zwicky in 1933. He estimated the mass of the Coma galaxy cluster using the velocity of the galaxies near its edge, and found a value that was about ten times greater than an estimation of the cluster's mass based on its brightness and number of galaxies. Vera Rubin and Kent Ford found further strong evidence for dark matter in the 1970s, by measuring the orbital speeds of stars in spiral galaxies. They learned there wasn't enough visible matter in the galaxies to prevent them from flying apart; so either the galaxies must contain a large amount of 'dark' mass or Einstein's theory of gravity needs some modification. As of 2017, dark matter has still not been detected directly. It is possible that a Dark Energy Survey (DES) group might discover weakly interacting massive particles (WIMP) dark matter in ultra-faint dwarf galaxies orbiting the Milky Way (see Chapter 20) via the detection of gamma rays produced by the annihilation of dark matter particles and anti-particles.

12.2 Cosmology from galaxy clusters

Our central aim in DES is to understand dark energy. Galaxy clusters are a useful way to do this because the abundance of clusters is sensitive to cosmological models. The cosmic microwave background (CMB) radiation, which has been travelling towards us since about 380,000 years after the Big Bang, is uniform to about one part in 100,000. But if we look at the distribution of mass in the (observable) universe today, we can see that it is no longer so uniform. Over time, as the universe expanded, slightly denser regions with stronger gravity attracted more mass and structure began to form. Matter gradually condensed into a web-like structure with long filaments of galaxies strung between clumps, or halos, of dark matter, surrounding large voids containing very few galaxies.

The evolution of large-scale structure is highly sensitive to the exact nature and relative amounts of dark matter, dark energy and atomic matter in the universe. In particular, the abundance of galaxy clusters is very sensitive to how clumpy the universe is. If the universe is clumpier, then it is more likely that small clumps will merge together to form galaxy clusters. Dark energy stretches spacetime apart, acting in opposition to structure formation. Galaxy clusters formed only recently in cosmological terms: they have 'only just' managed to gravitationally collapse. But once structure is gravitationally bound, it decouples from the Hubble flow and dark energy is no longer strong enough to pull it apart. With the help of simulations (see Chapter 16 for more details), we can see how the universe would develop given a particular cosmological model in which the constituents and other parameters can be varied. Using powerful supercomputers, the observed structure can be simulated almost exactly.

Dark matter halos sit at the nodes of the cosmic web, where the walls and filaments of visible galaxies intersect. It is at these nodes, where the density is highest, that galaxy clusters form. The number of halos per unit volume per unit mass, or the 'halo mass function,' changes over time. In the early universe, there were no high-mass halos; as the universe became clumpier, more and more developed, and the structure we see today evolved. Therefore, in order to study dark energy, we have to count and weigh the dark matter halos at different redshifts, as far back in time as possible. This allows us to infer how the clumpiness of the universe is evolving. We compare our observations to the model predictions from cosmology simulations to put constraints on what the universe is made of. Figure 12.1 shows how different values for the dark energy equation of state, w, affect the abundance of clusters across a redshift range.

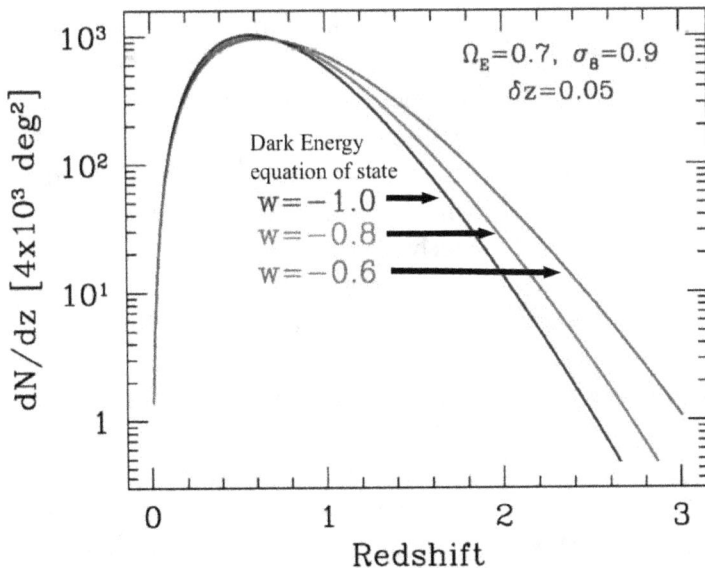

Fig. 12.1 Figure showing how the abundance of clusters is sensitive to cosmology models.

12.3 Galaxy clusters in DES

When it comes to counting dark matter halos, the main problem is that dark matter is invisible: it neither absorbs nor emits electromagnetic radiation of any wavelength. However, at the centre of dark matter halos we can find clusters of visible galaxies. These act as tracers for the unseen dark matter.

We find clusters by using properties of galaxies that are colour-similar. Galaxies at a similar distance will be redshifted by the same amount and should look approximately similar in colour. Also, since they are in the same neighbourhood, they should look fairly chemically similar as well.

A statistical measure of the number of galaxies in a cluster is referred to as 'richness'. Rich galaxy clusters contain hundreds of thousands of galaxies, whereas a cluster with only a few galaxies is a poor cluster. Rich clusters are at the centres of the biggest nodes in the cosmic web. Most of the galaxies in rich clusters are elliptical old galaxies in which star formation ceased billions of years ago. Since light from elliptical galaxies looks red, this set of old galaxies in a cluster is called the 'red sequence'. Young spiral galaxies look bluer, and are collectively known as the 'blue cloud' of a cluster.

We can sort out all the galaxies in a cluster on a colour-magnitude diagram, where we plot the 'colour' of each object against its brightness (apparent magnitude) through a particular filter. Because colour is partly a subjective judgement, astronomers define the colour of an object as the difference in its brightness when viewed through two different colour filters - the DES filters being g (green), r (red), i, z and Y (infrared). When a cluster's member galaxies are plotted on such a diagram, the red-sequence galaxies form a fairly tight line, and can be discerned from the blue cloud galaxies, which are more widely scattered (Figure 12.2).

DECam, the Dark Energy Camera, has a very wide field of view, of three square degrees, which makes it very useful for finding clusters. Figure 12.3 is an image taken by DECam of the Fornax cluster of galaxies.

The clusters look like tiny specks of light buried in the images, but they can be efficiently picked out by their colour. We use an algorithm called The 'red-sequence Matched-filter Probabilistic Percolation', or redMaPPer (Rykoff et al., 2014), to search through the data for tight clumps of red galaxies. The redMaPPer cluster finder was developed jointly by Eli Rykoff

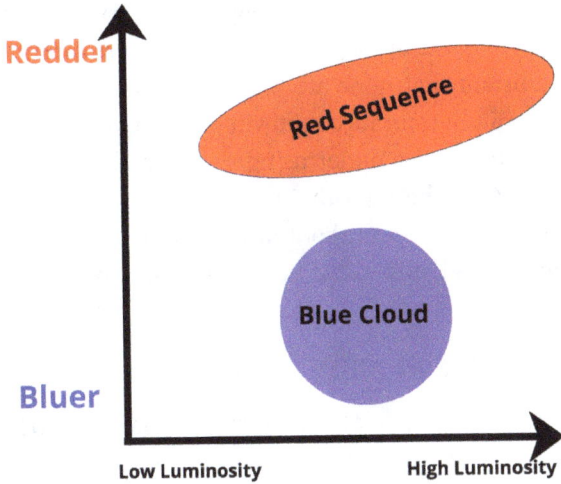

Fig. 12.2 A cartoon colour–magnitude diagram that shows when we plot the colour of each galaxy, a red-sequence of cluster members can be distinguished from non-member blue cloud galaxies.

Fig. 12.3 Full Dark Energy Camera image of the Fornax cluster of galaxies, which lies about sixty million light years from Earth. The center of the cluster is the clump of galaxies in the upper portion of the image. The prominent galaxy in the lower right of the image is the barred spiral galaxy NGC 1365. Photo credit: Dark Energy Survey Collaboration.

and Eduardo Rozo specifically for large photometric surveys. Up to a certain redshift, limited by the size of the telescope, redMaPPer can find all of the clusters. It is important that none are missed, and that there are no false detections, because an inaccurate count will give us inaccurate cosmology. Therefore we test our cluster finder using simulations, to make sure it is finding everything. After the algorithm has identified a cluster, we measure the redshifts of the galaxies to find out how far away they are. In the DES Year 1 data set we have identified 6453 clusters with richness > 20.

12.4 Measuring cluster mass

We wish to know the number of galaxy clusters as a function of mass. The best technique to estimate the mass is through weak gravitational lensing (see Chapter 11). The problem, however, is that it is very difficult to estimate the mass of a large number of clusters using weak lensing – we are finding hundreds of thousands of clusters using redMaPPer! Fortunately, since rich clusters are bigger than poor clusters, we can use the richness (roughly equivalent to the number of galaxies) of a cluster as a proxy for its total mass. This gives us a good approximation, because even though the galaxies themselves are just a small fraction of the total mass, their number in the cluster closely correlates to the total mass. We then use lensing to calibrate the richness-mass relationship. This is why galaxy clusters and weak lensing are complementary techniques to investigate dark energy in DES. Dark matter acts like a giant lens and as a result the images of the background sources are slightly stretched. The more mass there is in the lens, the more the images are distorted. In weak lensing, the distortion is tiny compared to the intrinsic shape of individual galaxies, but if we look at millions of lensed galaxies, we find that their shapes are correlated. From Earth, distant galaxies look like small elliptical blobs pointing in random directions, but if a galaxy cluster is in front of them, their ellipticities align such that they form an arc. We look for these arc patterns around the galaxy clusters as shown for example in Figure 12.4. We understand the bending of light very well from general relativity, so by measuring the amount that the galaxies are aligned, we can estimate the size of the cluster's dark matter gravitational content causing that tangential alignment. From this, we infer the mass of the galaxy cluster. If the cluster lens has more mass, the galaxy

Fig. 12.4 Example of an arc patten of galaxies that have been caused by lensing due to a foreground cluster. In this case the cluster RXC J2248.7-4431. Image by Eric Suchyta.

ellipticities will be more tangentially aligned. Cluster analysis very much depends on work done in the weak lensing working group.

12.4.1 X-ray emission

Other DES groups, in collaboration with external researchers, are working on two alternative techniques to study galaxy clusters: X-ray emission from hot electrons and the Sunyaev-Zel'dovich effect. These techniques provide important cross checks on the work done using redMaPPer and weak lensing.

Galaxy clusters contain extremely hot (at a temperature of millions of kelvins) ionized hydrogen and helium gas, enriched with metals such as iron. A hot gas emits X-rays via the bremsstrahlung process: high energy electrons are deflected by atomic nuclei and rapidly decelerate, producing X-ray photons. Astronomers can find galaxy clusters using an X-ray telescope to search for concentrations of X-rays. (The dark matter halo extends much farther than the central core where X-rays are detectable.)

The XMM Cluster Survey (XCS) is a project to find galaxy clusters using archival data taken from the European Space Agency's X-ray space observatory, XMM-Newton. Kathy Romer is a DES member who is also principle investigator in XCS and she leads a group at Sussex University who work with this data set. Tesla Jeltema (University of California, Santa Cruz) works with data from NASA's Chandra X-ray Observatory. XCS and Chandra data in the DES footprint provide a check on the completeness and sensitivity of redMaPPer allowing us to validate the simulations.

Studies of X-ray emission from clusters are also essential for checking the weak lensing measurements. Lensed galaxies form a ring around the galaxy cluster (again as shown in Figure 12.4) and in order to measure the mass of the cluster, it is important to know the location of the centre of mass. Galaxy clusters usually have a galaxy sitting in the centre – often the biggest one, but not always. We do the best job we can to find the centres using only the DES data and redMaPPer, but the X-ray data allows us to better understand them. If the centre from redMaPPer doesn't agree with the centre from X-rays, then at least one of those two analyses must be wrong, and perhaps both are wrong. It is very important to know how the centres of galaxy clusters behave so as to properly interpret the weak lensing measurements since mis-centering tends to systematically underestimate cluster richnesses.

As well as the issue of mis-centering, other systematic errors that we need to take account of include the weak lensing shear measurement, modelling, cluster triaxiality, line-of-sight and membership of cluster.

We can also analyse the X-rays in order to estimate the temperature of the cluster and use this as another estimate for cluster mass.

12.4.2 The Sunyaev-Zel'dovich effect

Clusters can also be identified by the Sunyaev-Zel'dovich effect, which occurs when photons from the CMB are scattered to higher energies by electrons in the hot gas within a galaxy cluster (or any other reservoir of hot plasma). As a result, the apparent brightness of the CMB is distorted in the direction of a cluster. The change of brightness is independent of the redshift of the cluster, so the effect is a valuable probe of the structure of the universe on the largest scales.

Alex Saro (University of Munich) is finding clusters with the Sunyaev Zel'dovich effect using data from the South Pole Telescope (SPT), a ten metre telescope located at the Amundsen-Scott South Pole Station, Antarctica. The SPT makes observations in the microwave region of the electromagnetic spectrum, and its survey footprint covers the same area of the southern sky as DES. Providing optical follow-up to SPT observations of clusters was one of the central arguments for the formation of DES in the first place (see Chapter 1).

12.5 Clusters beyond cosmology

Finding galaxy clusters and the masses of galaxy clusters has many other uses aside from investigating dark energy. Clusters can also be used to answer questions such as, 'How do galaxies live in clusters?' 'How do they evolve in time?' and 'How do the galaxies in large and small clusters differ?' By combining Sunyaev-Zel'dovich data with X-ray data, we can learn a lot about how the mass in galaxy clusters is related to the gas in the clusters, and about what sort of processes influence the gas. Observations of galaxy clusters across the electromagnetic spectrum (gamma rays, radio waves and X-rays) can also be used to test models of dark matter.

Reference

Rykoff, E. S., Rozo, E., Busha, M. T., Cunha, C. E., Finoguenov, A., et al. (2014). redMaPPer. I. Algorithm and SDSS DR8 Catalog, *The Astrophysical Journal* **785**, p. 104, doi:10.1088/0004-637X/785/2/104.

CHAPTER 13

Theory and Combined Probes

*Scott Dodelson, Tim Eifler, Wayne Hu, Dragan Huterer,
Elisabeth Krause, Eduardo Rozo & Jochen Weller*

Multi-probe analysis is the most promising way to constrain cosmology. In Chapter 11, we described how light from a distant source is 'deflected' by the 'lens' of intervening mass on its journey towards us. As a result of this, a distant 'source' appears distorted by the mass of the foreground underlying dark matter structures – a process known as 'weak lensing'. In Chapter 10 we described how the clustering pattern of galaxies can constrain cosmological parameters. Here we use the weak lensing and galaxy position catalogues, which were produced from the first full year of the Dark Energy Survey (DES) observations, to describe and illustrate the DES 'combined-probe' analysis theory and method. We will describe the three two-point ('3 × 2pt') functions, the importance of understanding the correlation between their parameters and how we fit cosmological models to the data – as well as why we do this blind. In so doing, this chapter charts the journey from 'data' to 'knowledge' and understanding the dark side of the universe.

13.1 From data to knowledge

The Dark Energy Survey Year 1 (DESY1) data set (Drlica-Wagner et al., 2018) was published in August 2017. This covers the period that the Dark Energy Camera (DECam) was taking images between August 31, 2013, and February 9, 2014.[1] The analysis was based on a 1321-square degree region, which is about ten times more area than the science

---

[1] Yes, it really takes three years to analyze and process all the raw images!

verification (SV) region, albeit shallower, and provides the most up-to-date snapshot of the large-scale structure (LSS) of the universe. Using this data set allowed DES create two galaxy-related catalogues.

The first catalogue contains positional information about 650,000 lens galaxies, selected using the 'redMaGiC' algorithm (Rozo et al., 2016), each with a robust photometric redshift (photo-z) estimate.

The second catalogue contains information about the positions and shapes of twenty-six million source galaxies from the METACALIBRATION shear catalogue (Huff and Mandelbaum, 2017).

Generating this amount of data has been a truly astronomical effort. If you have read through the preceding chapters, you will have gained a sense of just how huge an achievement it has been and how the many challenges were overcome by some of the brightest and most original thinkers in astrophysics, cosmology, engineering, computing, optics, mathematics and diplomacy!

The ultimate goal of DES, however, is to increase our knowledge and understanding of how the universe looks as it does today and, specifically, the nature of dark energy. This doesn't reveal itself through *simply* cataloguing the distribution and shape of galaxies. That would be like a detective collecting evidence at the scene of a break-in and neatly filing all these evidences away without further work. Analysis is needed to learn what the evidence means, how the clues combine together, what uncertainties lie in the evidence and how we can ensure that we are being impartial and not led by unrelated previous events. If we can do all this, then we can come up with the best theory as to the cause of the crime and test how well our evidence fits that theory.

We begin by cross-correlating the information contained within the weak lensing (WL) galaxy catalogues. Chapter -1 and Section 11.3 introduce the cosmological model and how observables such as lensing test its parameters.

13.2 Introducing the 3 × 2pt correlation function

By 'auto-correlating' and 'cross-correlating' the data in the **two** catalogues, we generate **three** 'two-point' (2pt) functions, namely:

- A 'galaxy–galaxy' auto-correlation, which compares the **position** of each galaxy with the **position** of every other one (both from the positional catalogue).

- A 'galaxy–shear' cross-correlation, which compares the **position** of each galaxy (from the positional catalogue) with the **shape** of every other one (from the shape catalogue).
- A 'shear–shear' auto-correlation, which compares the **shape** of each galaxy with the **shape** of every other one (both from the shape catalogue).

We can look at each of these two-point correlation functions to get an understanding of what they physically mean.

13.2.1 Galaxy clustering

Galaxies tend to form where there is dark matter and so we are more likely to find galaxies (and indeed dark matter itself) 'clumping' together to form structure, rather than spread around at random. Using statistics, we can quantify the likelihood of finding neighbouring galaxies close together rather than separated by large distances. We measure this statistical correlation between galaxy separation with a function $w(\Theta)$ that gives us an idea of how 'clumpy' the universe is (Θ is the angular separation between galaxies). For example, a value of $w(\Theta) = 0$ for all values of Θ would represent a completely uniformly distributed universe – you'd have no idea about the likelihood that a galaxy has a neighbour at any separation. The higher the value of $w(\Theta)$, however, the more clumpy the universe will be. Therefore, by measuring the positions of galaxies and correlating them with the positions of all other galaxies, we can derive a value for $w(\Theta)$ at different angular separations Θ.

13.2.2 Galaxy–galaxy lensing

The second two-point function is the 'galaxy–shear' correlation, which compares the **position** of each galaxy with the **shape** of every other one.

Again, we start with the idea that galaxies tend to clump together, resulting in regions where there is a higher concentration of mass, which creates the foreground lensing effect. This cosmic lens distorts the shapes of background source galaxies – they appear tangentially sheared (i.e. along a line connecting lens and source galaxy). The amount of shear is quantified by a function (γ_t) that is sensitive to the mass associated with the foreground lens galaxies. We will find larger distortions in the shapes of galaxies behind regions containing large amounts of matter.

13.2.3 Cosmic shear

The third two-point function is the 'shear-shear' correlation, which compares the **shape** of each galaxy with the **shape** of every other one. We intuitively expect that we are more likely to observe a distorted galaxy closer to another distorted galaxy, since their light path has been distorted by the same lens. In other words, the **shapes** of neighbouring galaxies are more correlated than pairs of galaxies at greater separation. Although this is slightly more complex to measure, the functions $\xi_+ (\Theta)$ and $\xi_- (\Theta)$ capture the shape information[2] relevant to each galaxy pair, as a function of angular separation.

13.2.4 Three is the magic number – 3 × 2pt

We can combine these three two-point functions together – referred to as '3 × 2pt' (illustrated in Figure 13.1) – to maximize the use of all the

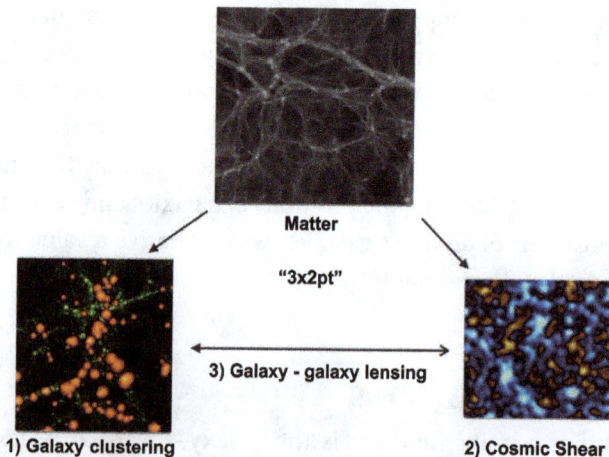

Matter

"3x2pt"

3) Galaxy - galaxy lensing

1) Galaxy clustering

2) Cosmic Shear

Fig. 13.1 Combination of these three probes maximizes use of large-scale structure information and jointly and robustly constrains astrophysical and systematic parameters in the analysis. Image credit: Michael Troxel.

---

[2]The sum and the difference of the product of the tangential and cross component of the shear, measured with respect to the line connecting each galaxy pair.

information. This allows DES the ability to make much more accurate predictions of the underlying cosmology than by using a single two-point function, and increases our understanding of systematic parameters.

13.3 Models, parameters and nuisance

We now wish to take the individual and the combined 3 × 2pt correlation functions and assess the probability that the data was taken from a proposed underlying cosmology, and to derive values of the parameters associated with that theory.[3] We can ask, for example, whether our data is a good representation of a Λ with cold dark matter (ΛCDM) universe. If so, what are the parameters associated with that ΛCDM universe that have enabled us to observe this data?

The DESY1 data was fit with two models. In the first model, ΛCDM, there are six parameters we can estimate. These are the matter, baryon and massive neutrino energy densities (Ω_m, Ω_b and Ω_ν), the Hubble parameter H_0 and two parameters that describe the amplitude and spectral index of the primordial scalar density perturbations, A_s and n_s (considered as the 'seeds' of all structure in the universe).

DES also wanted to consider the possibility that dark energy is not a cosmological constant, that is, that $w \neq -1$. Therefore, a second model, wCDM, was tested in which a seventh parameter w was allowed to vary.

Along with these six cosmological parameters, in the case of ΛCDM, or seven in the case of wCDM, DES also needed to take into account twenty 'nuisance parameters'. These account for uncertainties in our measurements of, for example, photometric redshifts, shear calibration, bias between galaxies and mass and the contribution of intrinsic alignment to the shear spectra. Some of these parameters, those that characterize shear calibration and photometric redshifts (e.g. mass and Δz), have informative priors that we can take into account; but others, such as galaxy bias, do not.

13.4 Covariance

The likelihood that our data is indicative of a cosmological model is dependent not just on the observables that we have measured from our

---

[3] This is the Bayesian approach to statistics that is all the rage in cosmology today.

probes (taking into account the nuisance parameters), but it is influenced by two additional factors: first, our measurements have statistical uncertainties and second, all measurements in the 3 × 2pt data vector are highly correlated.

In order to understand the first aspect, statistical uncertainties, better, imagine a coin toss experiment where you as the experimenter are testing the hypothesis that the chances of heads or tails are truly fifty-fifty. If you flip the coin just twice, you do have a measurement and you can make a statement on your hypothesis, but with very poor statistical accuracy. In other words, the error bars on this two-times coin flip measurement would be very large. If you flip the coin a trillion times, your statistical accuracy is very high and consequently your error bars are very small. The same is true for cosmological measurements. If you observe lots of galaxies (aka a large volume of the universe), your statistical error bars are very small, your measurement accuracy is high. Determining the statistical errors for the DES 3 × 2pt data vector is somewhat more involved than just counting the number of galaxies. However, as a tendency, it is true that the statistical uncertainty will shrink further the more galaxies DES observes in the five years of its total operation.

The second aspect mentioned earlier, the fact that all our measured data points are correlated, is also straightforward to understand. Cosmic shear, galaxy–galaxy lensing, and galaxy clustering all trace the same underlying matter density field, so they are inherently interdependent. As a consequence, the data points cannot be treated as independent measurements where each one contributes independent cosmological information, instead we have to understand and account for these 'data correlations'.

It is important to note that some cosmological probes, such as supernovae 1a or cosmic microwave background, can be treated as independent sources of cosmological information. In this case, the difficult computation of the data correlations can be avoided and the individual measurements can be easily combined.

In case of the DES 3 × 2pt analysis, however, both aspects, the statistical uncertainty and the interdependence of information, must be accounted for. Corresponding information is captured in the so-called 'multiprobe covariance matrix'. This matrix is a critical ingredient of the DES likelihood analysis since it determines the error bars of our measurement and ultimately it determines the confidence with which we reject or affirm a cosmological model.

13.4.1 Enter the matrix

A 'covariance' is a measure of how the uncertainties of two observable properties (or parameters) behave as a pair. A positive covariance indicates a positive relationship, and vice versa. For example, the daily temperature across a month will have a positive covariance with the value of ice cream sales each day, but a negative covariance with the sales of woolly scarves. A covariance equal to, or near to, zero indicates no correlation. The temperature during the day has no correlation with the fractional number of blue cars on the road that day.[4]

Determining the covariance matrices of just one probe with a few data points is fairly simple – a probe with a data vector of only two points has a 2×2 covariance matrix, a three-point data vector probe has a 3×3 matrix, and so on. Each cell in the matrix describes the covariance between two data points. It gets more complex when we try to construct the covariance matrices for multi-probes – the greater the size of the data vector the larger the size of the martix. How, for example, does the uncertainty on the fourth observable from probe A covary with the seventh from probe B?

The covariance matrices that we computed for analyzing the DES data have a dimension equal to the number of data points and the DESY1 data vector contains 457 points. This means that the covariance matrix contains 208,849 elements (457×457) all of which had to be computed accurately. It took the superhuman focus of Elisabeth Krause, a postdoc at Stanford, to derive all equations with pen and paper and turn them into science analysis code. Together with Tim Eifler (NASA-Jet Propulsion Laboratory [JPL]), she designed the corresponding covariance software modules and implemented them in the CosmoLike analysis code. Tim subsequently parallelized these complex computations and deployed them on a high-performance computing system at NASA-JPL.

---

[4]Well, it might, but that is more than likely to be a coincidence.

The result can be seen in Figure 13.2, which shows the full DES 3 × 2pt covariance associated with the galaxy–galaxy, galaxy–galaxy lensing and the cosmic shear probes.

Although determining this multi-probe matrix is a complex task, it is important to do so since, ultimately, it leads (via its inverse, the 'precision matrix') to determining the errors on the cosmological parameters that we are trying to quantify. The community is currently exploring several approaches in estimating the covariance matrix, all of which have different advantages and disadvantages:

- From numerical simulations. The idea here is to create lots of simulations of our survey and study the variations between the different realizations. Remember, that in the coin toss experiment more coin tosses are better than fewer? The same idea applies here, meaning we would have to create lots of (and here lots means more than one million) simulated DES surveys. This is hugely expensive in terms of compute time and not feasible

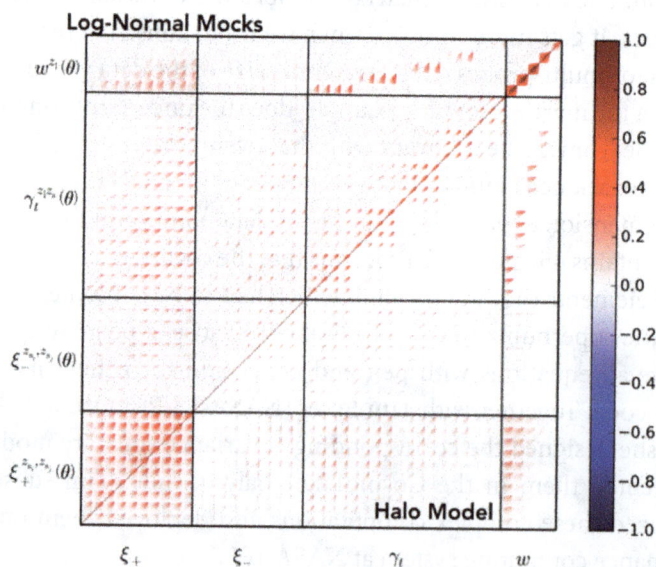

Fig. 13.2 The covariance between the parameters associated with the galaxy clustering, galaxy–galaxy lensing and the cosmic-shear two-point (2pt) multi-probes. The darker the cell, the more correlated the associated parameters. The values of the cells in the upper-left and lower-right halves of the matrix have been derived according to two different methods, as described in Krause et al. (2017).

with the current computing resources available to the community. This statement already holds when used in single-probe analysis and it is not appropriate for the multi-probe analysis that we aspire to.

- From survey data itself. The idea is very similar to the simulations concept, but instead of simulating DES many, many times, we are chopping the existing survey data into smaller pieces and assume that each piece resembles the full DES survey. For this method to work, one needs lots of small DES pieces in order to have good statistics (again see the coin toss analogy), but one also wants each DES piece to be large enough that it still resembles the full survey. Unfortunately, one cannot have both, large pieces and lots of them given that our survey area is limited and consequently this method was discarded due to large uncertainties.

- From analytical modelling. This was the approach we used to derive our matrices. It requires meticulous work to derive all equations, to code them up and to test and validate them. The covariance validation team around Oliver Friedrich, a graduate student in Munich, deserves the credit for testing the final covariance matrices and for making sure that the DES parameter constraints were not impacted by inaccuracies in the covariance calculations.

13.4.2 Two heads are better than one

We now have the input data from our 3×2pt correlation functions, with associated covariance matrices. We have defined our ΛCDM and wCDM models (with associated nuisance parameters) to which we want to estimate how well that data fits. Finally, we wish to derive the cosmological parameters from that fit. What we need is a super 'cosmological parameter estimation' (CPE) pipeline that will do all those things. Better still – let's get two!

Having two analysis pipelines is a huge advantage as any anomalies in results can be quickly cross-checked, to see whether there is an error in one of the software scripts or in the data. Accordingly, DES has built two independent CPE pipelines: CosmoSIS (Zuntz et al., 2014), a code that embodies a modular architecture, built by a joint Manchester-Fermilab-Chicago collaboration headed by Joe Zuntz, and CosmoLike (Cosmological Likelihood Analyses for Photometric Galaxy Surveys), built by Elisabeth Krause at Stanford and Tim Eifler at JPL (Krause and Eifler, 2017).

Over a two-year period, these CPEs were put through their paces by testing them on simulated data from mock DES catalogues (MacCrann et al., 2018). (See Chapter 16 for more details on simulations.)

All the steps in the analysis of the simulations, from measuring the relevant two-point functions to extracting cosmological parameters, were performed as the six ΛCDM (seven wCDM) cosmological and the twenty nuisance parameters were varied. Since the 'true' cosmology dialed into the simulations was known, it was a huge relief when we demonstrated that the two analysis pipelines did indeed recover the correct cosmological parameters, and agreed extremely well. This hugely important comparison ensures the accuracy of DES analysis.

Now to use the real DESY1 WL data, first, close your eyes.

13.5 Blinding

A crime has been committed, the police have made an arrest, the suspect taken for trial and a jury chosen. However, to avoid bias, the jury members are kept unaware of any previous convictions – they are genuinely blind to previous knowledge as it is important that they base their verdict purely on the current evidence. However, despite our best efforts to remain objective, even the most honest researchers are susceptible to biases that lead them to 'prefer' results that agree with what has been found previously. 'I won't include that data because it not what others researchers have found', is a possible thought.[5]

In order to minimize unconscious bias of this nature, before revealing the real results, DES introduced a blinding process. For the WL project, this meant:

- Rescaling the galaxy shapes by a random amount that was not known to any of the scientists doing the analysis.
- Never plotting theory expectations and data on the same plots so that we could not 'see how we were doing' along the way.

---

[5]That said, if we got a result that disagreed with ΛCDM, we might 'want' that to be true because it would be very exciting, but we would check our result a million times.

- Offsetting the extracted parameters by another unknown amount whenever the analysis codes were run, so we could see how small our error bars were but not what our final answers were.

At last, once you are convinced that the pipelines are accurate, that the models, nuisance parameters and covariance matrices are robust and that you have been impartial throughout, you can take off your blindfold.

13.6 DESY1 3 × 2pt results

On July 7, 2017, the results were unblinded, and twenty-seven days later released to the public (Dark Energy Survey Collaboration, 2018).

The analysis combined galaxy clustering and weak gravitational lensing data from Year 1 (1321 square degrees), utlizing three two-point functions (hence referred to as '3 times 2pt'): (1) the cosmic shear correlation function of 26 million source galaxies in four redshift bins, (2) the galaxy angular auto-correlation function of 650,000 luminous red galaxies in five redshift bins and 3) the galaxy–shear cross-correlation of luminous red galaxy positions and source galaxy shears. The headline results strongly support the ΛCDM model of the universe. The DESY1 data combined with Planck, baryonic acoustic oscillation measurements from Sloan Digital Sky Survey (SDSS), 6dF, Baryon Oscillation Spectroscopic Survey (BOSS) and Type Ia supernovae from the Joint Lightcurve Analysis (JLA) data set give $w = -1.00^{+0.05}_{-0.04}$ (68% CL), a spectacular agreement with a cosmological constant. Assuming ΛCDM, matter density is $\Omega_m = 0.298 \pm 0.007$ and the amplitude of mass fluctuations $S_8 = \sigma_8(\Omega_m/0.3)^{0.5} = 0.802 \pm 0.012$. The DES results present a great achievement of its LSS and WL probes, with precision similar to Planck. This allows important comparison of the very early and late universe. The future analyses of the first three years and then the entire six years of data will provide better constraints on the time-varying equation of state of dark energy, modified gravity and neutrino mass.

The headlines from the results are as follows:

- We find consistency within the ΛCDM model at current statistical precision between DESY1 LSS probes and the cosmic microwave background

(CMB). Figure 13.3 shows the 1 and 2- σ regions for Ω_m and S_8,[6] from individual and combined 2pt probes.

- We find no evidence for a wCDM model either from DES alone or when combined with other probes, that is, CMB and baryon acoustic oscillations (BAO).
- Combining DES with Planck shifts the preferred Hubble constant by $> 1\sigma$ towards local H_0 measurements.
- DES relaxes the previous upper limit on the neutrino mass density by 20% when combined with external probes.
- Combining DES LSS constraints with other low-z probes: BAO plus JLA of Type Ia supernova, as well as the high-z CMB measurements from Planck, gives us the tightest constraint ever placed on the ΛCDM parameters most closely related to structure in the Universe, as shown in Figure 13.4.

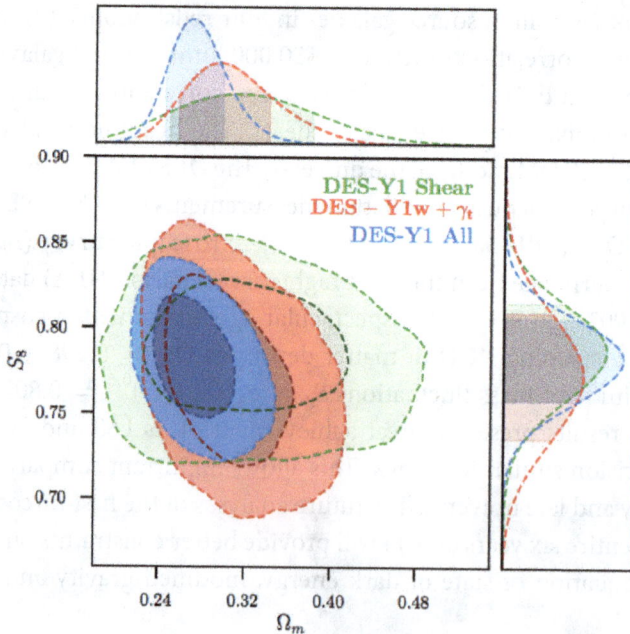

Fig. 13.3 Constraints on the three cosmological parameters S_8 and Ω_m in Λ with cold dark matter (ΛCDM), from the Dark Energy Survey Year 1 (DES Y1).

[6]S_8 is a combination of σ_8 and Ω_m. σ_8 is (roughly) the amplitude of the scalar perturbations.

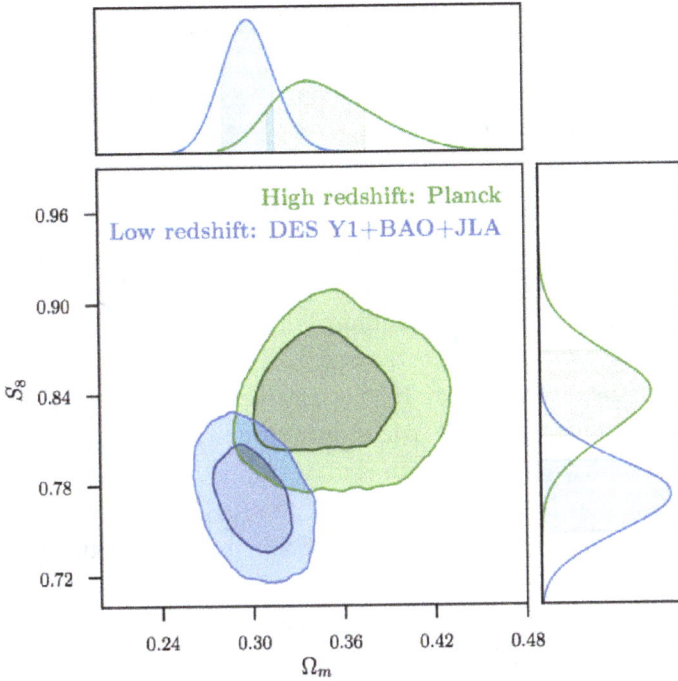

Fig. 13.4 Constraints on the three cosmological parameters S_8 and Ω_m from multiple low-z (the Dark Energy Survey Year 1 [DESY1] + baryon acoustic oscillation [BAO] + Joint Lightcurve Analysis [JLA]) and high-z probe Planck (without lensing).

13.7 Beyond 3 × 2pt

In this chapter, we have shown how, by using the DESY1 galaxy lensing catalogues, we have been able to generate the multi-probe 3 × 2pt functions from which we have estimated cosmological parameters. But why stop at three?

Information from the CMB radiation can also be used to measure lensing. Comparing CMB studies with our DES results is an incredible test of ΛCDM at two extremely different stages of the universe, six billion years apart. Furthermore, CMB lensing systematics are completely different from those associated with galaxy lensing.

So the next stage is to combine the DES 3 × 2pt probes with data from CMB lensing maps. This will generate two further probes - DES galaxy density correlation with CMB lensing maps and DES galaxy shear correlation with CMB lensing maps – and thus a 5 × 2pt function (with an even more complex covariance matrix).

Following earlier detections of cross-correlations between galaxy number counts and CMB lensing using The Wilkinson Microwave Anisotropy Probe (WMAP) and National Radio Astronomy Observatory Very Large Array Sky Survey (NVSS) (Smith et al., 2007; Ho et al., 2008) and later on with higher significance using various galaxy surveys and Planck (Ade et al., 2014), Pablo Fosalba and Tommaso Giannantonio proposed this additional probe to extend the nominal DES 3 × 2pt multi-probe analysis. The first analysis, using DES SV data combined with data from the South Pole Telescope and Planck, led to the first tomographic detection of CMB lensing in several photometric redshift bins and up to $z = 1.2$ (Giannantonio et al., 2016). Donnacha Kirk (University College London [UCL]) also led work to cross-correlate the WL signal from DES galaxies with the WL of CMB maps from the South Pole Telescope team (Kirk et al., 2016). This cross-correlation is now a standard probe included in the so-called 5 × 2pt analysis of DES (i.e. for Y1 data and beyond [Omori et al., 2018]).

13.8 Coming soon

The next round of cosmological analyses of DES data will include data from the first three years of the survey (DESY3), which cover more than three times as much area to greater depth than Y1, and will incorporate constraints from clusters, supernovae, and cross-correlation with CMB lensing.

In the coming years, both COSMOSIS and CosmoLike will be programmed to work on different cosmological models. These include universes with a non-zero curvature (the ΛCDM model that we used in the DESY1 results assumed a 'flat' universe), as well as models that include evolving dark energy, or those enabling us to test 'modified gravity'.

Both the new data and the new analysis techniques will shed more light on dark energy and cosmic acceleration.

Ultimately, and most excitingly, what will we find from the full six years of DES data? Is the case closed on a ΛCDM universe? Or will new observations, different models and results from the other cosmological probes surprise us yet?

References

Dark Energy Survey Collaboration (2018). Dark Energy Survey Year 1 results: Cosmological constraints from galaxy clustering and weak lensing, Physical Review D **98**, 4, p. 043526, doi:10.1103/PhysRevD.98.043526.

Drlica-Wagner, A., Sevilla-Noarbe, I., Rykoff, E. S., Gruendl, R. A., Yanny, B., et al. (2018). Dark Energy Survey Year 1 results: Photometric data set for cosmology, ApJ Sup., **235**, 33.

Giannantonio, T., Fosalba, P., Cawthon, R., Omori, Y., Crocce, M., et al. (2016). CMB lensing tomography with the DES science verification galaxies, *Monthly Notices of the Royal Astronomical Society* **456**, pp. 3213–3244, doi:10.1093/mnras/stv2678.

Ho, S., Hirata, C., Padmanabhan, N., Seljak, U., and Bahcall, N. (2008). Correlation of cmb with large-scale structure. i. integrated Sachs-Wolfe tomography and cosmological implications, *Physical Review D* **78**, p. 043519, doi:10.1103/PhysRevD.78.043519.

Huff, E. and Mandelbaum, R. (2017). Metacalibration: Direct self-calibration of biases in shear measurement, *arXiv*:1702.0260.

Kirk, D., Omori, Y., Benoit-Lévy, A., Cawthon, R., Chang, C., et al. (2016). Cross-correlation of gravitational lensing from DES science verification data with SPT and Planck lensing, *Monthly Notices of the Royal Astronomical Society* **459**, 1, pp. 21–34, doi:10.1093/mnras/stw570.

Krause, E. and Eifler, T. (2017). cosmolike - cosmological likelihood analyses for photometric galaxy surveys, *Monthly Notices of the Royal Astronomical Society* **470**, pp. 2100–2112, doi:10.1093/mnras/stx1261.

Krause, E., Eifler, T. F., Zuntz, J., Friedrich, O., Troxel, M. A., et al. (2017). Dark Energy Survey Year 1 results: Multi-probe methodology and simulated likelihood analyses, *arXiv*:1706.09359.

MacCrann, N., DeRose, J., Wechsler, R. H., Blazek, J., Gaztanaga E., et al. (2018). DES Y1 results: Validating cosmological parameter estimation using simulated Dark Energy Surveys, Monthly Notices of the Royal Astronomical Society **480**, 4, pp. 4614–4635, doi:10.1093/mnras/sty1899.

Omori, Y., Baxter, E. J., Chang, C., Kirk, D., Alarcon, A., et al. (DES and SPT Collaborations) (2019). Dark Energy Survey Year 1 Results: Cross-correlation between Dark Energy Survey Y1 galaxy weak lensing and South

Pole Telescope+Planck CMB weak lensing, *Physical Review D* **100**, p. 043517, doi:10.1103/PhysRevD.100.043517.

Planck Collaboration, Ade, P. A. R., Aghanim, N., Armitage-Caplan, C., Arnaud, Ashdown M., et al. (2014). Planck 2013 results. XVII. Gravitational lensing by large-scale structure, *Astronomy and Astrophysics* **571**, A17, doi:10.1051/0004-6361/201321543.

Rozo, E., Rykoff, E. S., Abate, A., Bonnett, C., Crocce, M., et al. (2016). red-MaGiC: Selecting luminous red galaxies from the DES science verification data, *Monthly Notices of the Royal Astronomical Society* **461**, pp. 1431–1450, doi:10.1093/mnras/stw1281.

Smith, K. M., Zahn, O., and Doré, O. (2007). Detection of gravitational lensing in the cosmic microwave background, *Physical Review D* **76**, p. 043510, doi:10.1103/PhysRevD.76.043510.

Zuntz, J., Paterno, M., Jennings, E., Rudd, D., Manzotti, A., et al. (2014). CosmoSIS: Cosmological parameter estimation, Astrophysics Source Code Library.

CHAPTER 14

Spectroscopic Redshifts

Chris D'Andrea, Tamara Davis, Janie Hoormann, Alex Kim &
Chris Lidman

When we examine an image of the night sky on our computers, it is easy to measure the positions of objects that we see on the screen. However, without making further assumptions, we cannot tell if one object is further away than another. Measuring distances in astronomy is hard. Even the concept of distance is somewhat 'flexible' when the universe is constantly changing size. Redshifts are far easier to measure and they are closely related to distances in a way that is dependent on the cosmology. In this chapter, we give a qualitative explanation of redshift and how we measure it, and discuss spectroscopic redshifts in the Dark Energy Survey (DES) – in particular the work of OzDES at the Anglo-Australian Telescope.

14.1 Spectroscopic redshift

The wavelength of light from an approaching astronomical object is compressed, or 'blueshifted'. Conversely, light from an object that is receding from us is stretched, or 'redshifted'. This 'Doppler shift' resulting from the relative motion of objects is analogous to the pitch change of a siren as an ambulance passes by.

As Edwin Hubble first discovered, almost all galaxies are receding – the light from them being redshifted. This is due almost entirely to the expansion of universe, which is far more important than the small local motions that occur because of the gravitational pull of other galaxies in their neighbourhood. In an expanding universe, a galaxy with a higher redshift, z, is therefore further away and from this we can work out how much the universe has grown since that light was emitted. This information is essential to our understanding of how the universe and its contents have evolved.

The most precise way of measuring the redshift of a galaxy is to obtain a spectrum of it and compare the wavelengths of features in the spectrum with the wavelengths these features have in the laboratory (the 'rest-frame' wavelength). Galaxy spectra have numerous sharp features from the gas and stars within them, so obtaining a precise redshift is relatively easy, if one can get enough signal. Redshifts that come from these galaxy spectra are called *spectroscopic redshifts*.

Redshifts can also be derived by measuring the brightness of light from galaxies using a set of well characterized colour filters. These *photometric redshifts* (or photo-zs) are derived from DECam – hence the Dark Energy Survey (DES) is described as a photometric survey. Photometric redshifts are less accurate than spectroscopic redshifts, but they are much quicker and easier to measure and therefore much less expensive to obtain. Photometric redshift in relation to DES is described in Chapter 15.

14.2 Spectroscopic redshifts in DES

The cosmological constraints that come from weak lensing, galaxy clusters and large scale structure – three of the four DES cosmological probes – can be realized using photometric redshifts, if they are sufficiently accurate and unbiased. One way of evaluating the amount of bias and assessing the accuracy – in effect, calibrating the photometric system – is to obtain spectroscopic redshifts of galaxies that also have photometric redshifts. Having a certain amount of uncertainty is acceptable if we understand that uncertainty. The cosmological constraints that come from the fourth probe, Type Ia supernovae, can only be realized with spectroscopic redshifts – it is not possible to determine accurately enough how fast a supernova is moving away from us using only photometric redshifts. Spectroscopic redshifts are therefore necessary, either directly or indirectly, for all four DES probes. They are also very important for the working groups studying non-dark energy science, such as quasars, strong lensing and galaxy evolution. Therefore the question early on became, how can we obtain as many spectroscopic redshifts as possible for as many of our interesting targets as possible? OzDES was created in response.

14.3 OzDES

In 2007, the Supernova Legacy Survey (SNLS) was coming to the end of its five-year study. SNLS used the MegaCam imager on the Canada-France-Hawaii Telescope to discover and spectroscopically confirm over 500 Type Ia supernovae in four one-square-degree fields spread evenly across the sky. At the same time, the Sloan Digital Sky Survey (SDSS) was finishing a similarly ambitious program with the SDSS telescope in New Mexico. The combined SNLS-SDSS sample, known as the Joint Lightcurve Analysis (JLA) sample, provides some of the tightest constraints yet on the dark energy equation of state parameter.

Both surveys discovered more supernovae than they could confirm with the spectroscopic facilities that were available at that time, and were therefore very interested in using wide-field, fibre-fed spectrographs that could take spectra of hundreds of galaxies simultaneously. The basic idea was to obtain the redshifts of the galaxies that have hosted Type Ia supernovae, after the supernovae have faded from view. It was becoming apparent that it was going to be even more difficult to perform real time confirmation of supernovae in future surveys such as DES and the Large Synoptic Survey Telescope (LSST), which would produce an order of magnitude more supernovae than SNLS and SDSS combined.

In 2010, Karl Galzebrook at Swinburne University organized a conference[1] titled 'Synergistic Surveys of the Southern Sky'. Attending that conference were Bob Nichol and Masao Sako, who were at that time deeply involved in preparing the supernova programme for DES. Also attending was Chris Lidman, who gave a talk on plans to obtain redshifts of SNLS supernova host galaxies with the 2dF fibre positioner and AAOmega spectrograph on the Anglo-Australian Telescope (AAT) (see Figure 14.1). The seed that led to OzDES was planted at this conference.

The 2dF fibre positioner enables astronomers to obtain spectra for almost 400 sources simultaneously from the same exposure. At the top of the telescope, the 2dF robot places 400 fibres with a field of view that is two degrees across (four times the diameter of the full moon). After travelling billions of years through mostly empty space, light from distant

---

[1] http://astronomy.swin.edu.au/S4/home.html

Fig. 14.1 An image of the Anglo-Australian Telescope on the far left, with the centre of the Milky Way setting in the west. Photo credit: James Gilbert.

objects enters these fibres and then passes down 40 m of fibre cable to the AAOmega spectrograph. It takes light one-ten-millionth of a second to travel this last 40 m.

The SNLS host galaxies were observed with the AAT in 2011. They showed that if one were prepared to observe each object for long enough, one could obtain redshifts of very faint galaxies, even with a 4-m class telescope at an average observing site. And while measuring redshifts of the supernovae, one could also obtain redshifts of other sources at the same time. The results were presented at the DES meeting at the University of Pennsylvania in the autumn of 2011.

During the course of 2012, the OzDES collaboration started to take shape. A pilot run with the AAT targeting the DES deep fields took place during December 2012, when DECam was going through its science verification phase. By happy coincidence, the patrol field of the 2dF fibre positioner is a close match to the field of view of DECam, as shown in Figure 14.2. This makes 2dF and AAOmega the ideal instrument to follow targets in the ten DES deep fields.

The pilot run was successful. In addition to obtaining redshifts of faint galaxies, it also demonstrated that one could observe and spectroscopically identify transients – astronomical phenomena such as supernovae that are visible for a period of time before disappearing from view. The first DES supernova was confirmed with the AAT on December 13th, 2012. Named DES12C2a, it was a Type Ia supernova at redshift z = 0.3035. Its spectrum is shown in Figure 14.3.

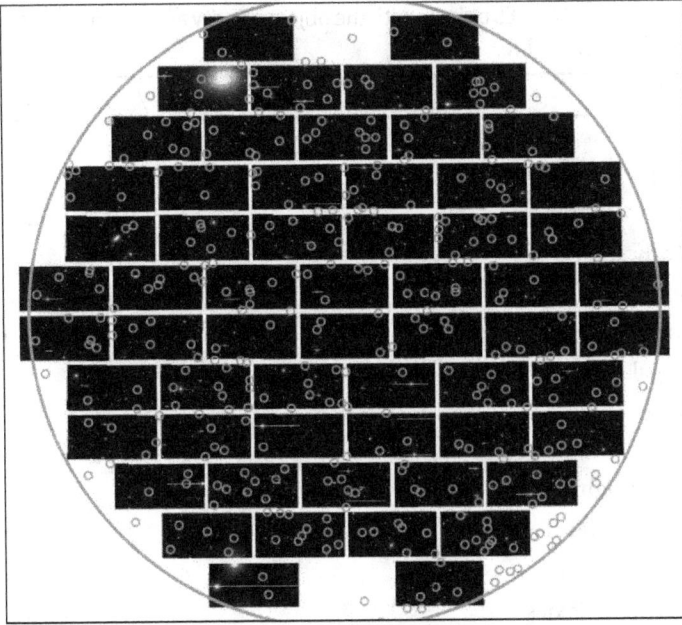

Fig. 14.2 The patrol field of the 2dF fibre positioner (the large circle) together with the position of 400 2dF fibres (the small circles) overlaid on an image taken with Dark Energy Camera (DECam). From Yuan et al., 2015, MNRAS 452(3), 3047.

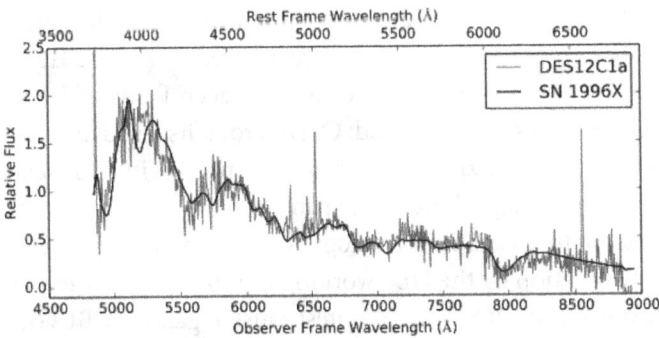

Fig. 14.3 The first spectroscopically confirmed supernova from the Dark Energy Survey (DES). The broad absorption line at 8000 Å from silicon and the W feature at 7000 Å from sulphur unambiguously identify the supernova as a Type Ia. The sharp spikes in the spectra are emission lines from the host galaxy and lead to a precise redshift of $z = 0.3035$.

Table 14.1 List of OzDES objects with the object priority and the maximum number that are allocated a fibre.

Target Type	Priority	Quota
Transients	8	Unlimited
Active galactic nuclei (AGN)	7	100
Supernova (SN) hosts	6	200
White dwarfs	6	3
Strong lens	6	3
Cluster galaxies I	6	10
Radio galaxies I	6	25
Sky fibres	5	25
F stars	5	10
SpARCS brightest cluster galaxies (BCGs)	4	20
Cluster galaxies II	4	40
Radio galaxies II	4	25
RedMaGiC galaxies	4	100
Luminous red galaxies (LRGs)	2	50
Photo-z sources	2	Unlimited
Bright galaxies	1	Unlimited

In 2013, OzDES was awarded one hundred observing nights on the AAT over what was initially a five-year period (2013–2017) to obtain redshifts of thousands of objects in the ten deep fields of DES. This was later extended to a six-year period. OzDES took its last data in early 2019.

As with DES, the OzDES observing season starts in August and ends in December. On average, there is one run a month and each run lasts about four nights. Before each run, catalogues of potential targets are uploaded to a central location by the DES working groups. Some targets come from groups external to DES (e.g. brightest cluster galaxies, BCGs, from the Spitzer Adaptation of the Red-Sequence Cluster Survey (SpARCS) survey). OzDES observes a wide variety of objects. The full list, which evolves over time, is shown in Table 14.1. Also listed is the object priority and the maximum number that can be allocated a fibre in a single configuration. Highest priority goes to transients, followed by active galactic nuclei (AGN), supernova host galaxies and so on.

AGN (also known as quasars) make up about a quarter of all the objects observed by OzDES. We repeatedly observe the same AGN to get time-lapse spectra. This lets us watch how they evolve and use their changing behaviour in a technique known as 'reverberation mapping' (described in Chapter 18).

As a general rule, most targets are uploaded a couple of weeks before each run. Transients are the one exception to this rule. Since a transient can occur at any time, the OzDES observing tools have to be flexible enough to allow transients to be uploaded in real time. It is not uncommon to receive the coordinates of the transients a couple of hours[2] before they are observed.

During each run, a pair of astronomers plan the observations, run the instrument and process the data. A number of other astronomers, the 'redshifters', measure redshifts. They in turn are supervised by the chief 'redshift whip'.

Redshifts are estimated with the Manual and Automatic Redshifting Software (Marz)[3] and added into the OzDES database within forty-eight hours of the observations.

Transients are classified as the run progresses. It usually takes a few days for all the transients to be classified. Usually most transients are Type Ia supernovae. We have confirmed other kinds of supernovae as well, some of which are quite rare, for example, super-luminous supernovae. These supernovae are even brighter than Type Ia supernova, and DES can detect them out to redshift four when the universe was just over a billion years old. A spectrum of a super-luminous supernova taken by OzDES at the AAT is shown in Figure 14.4.

Although, most of the spectroscopy comes from OzDES, DES has also been awarded observing time at the Gemini South Telescope, the Gran Telescopio Canarias (GTC), the Keck Telescope, the Magellan Baade Telescope, Multiple Mirror Telescope (MMT), the Southern African Large

---

[2] It takes about forty minutes for the 2dF robot to configure a field, so an hour is close to the minimum amount of time between receiving the coordinates of the transients and observing them.

[3] http://samreay.github.io/Marz

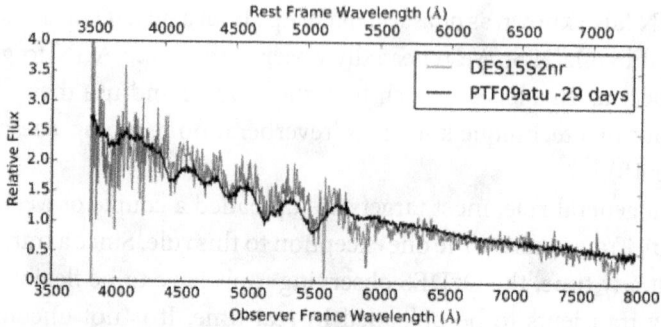

Fig. 14.4 The spectrum of a super-luminous supernova at $z = 0.22$. While this particular example was at a relatively low redshift, Dark Energy Survey (DES) can detect these extremely bright events up to $z = 4$.

Telescope (SALT) and the Very Large Telescope (VLT). In particular, spectra of Type Ia supernovae are obtained using all the follow-up telescopes by carefully coordinating DES Type Ia supernovae discoveries with telescope availability and capability. These spectra from OzDES and other telescopes play a key role in DES science, allowing all teams to better calibrate their photometric redshifts, the supernova team to precisely measure spectroscopic redshifts, and the AGN team to watch how AGN change over time. With these redshifts, we add the third dimension to DECam's pictures of the sky, mapping the distances to galaxies to complete our knowledge of their 3D positions.

CHAPTER 15

Photometric Redshifts

Chris Bonnett, Francisco Castander, Tamara Davis, Ben Hoyle, Ofer Lahav, Huan Lin & Carles Sánchez

The Dark Energy Survey (DES) is a photometric survey, which means that incoming light from stars and galaxies passes through one of a set of five filters (g, r, I, z, and Y), each sensitive to a different wavelength of light, before an image is recorded on the charge-coupled device (CCD) array. The aim is for every part of the 5000-square degree survey area to be imaged several times through every filter (see Chapter 8 for more details). As this chapter explains, these filters allow us to estimate the redshifts of distant objects, upon which all of our science analysis relies. Accurate and efficient measurement and calibration of photometric redshifts is, therefore, at the heart of DES. Many people in different science working groups in the collaboration are working to understand as well as possible how the photometric redshifts differ from full spectroscopic redshifts, because this knowledge is central to getting the best results possible from the science.

15.1 Measuring photometric redshifts

As introduced in Chapter 14, the redshifts of galaxies are a consequence of the expansion of the universe. As such, redshift information about the galaxies in the Dark Energy Survey (DES) is essential to achieving the primary goal of DES to constrain cosmological parameters. Another way to appreciate the importance of redshift information comes from the fact that although the redshifts of galaxies are not the same as the distances to those galaxies, redshifts serve as the most readily measurable proxy for those distances. Galaxy redshifts thus provide vital information needed to make a fully three-dimensional (3D) map of the universe as observed by DES, beyond the 'flat' images obtained directly from the Dark Energy Camera (DECam).

Redshifts may be measured most precisely using spectroscopy. Alternatively, redshifts may be measured less precisely via photometric redshifts derived from imaging data taken through multiple filters (or bands), such as the *g, r, i, z* and *Y* filters used by DES. Each filter blocks out all the light in the spectrum except for that within a narrow wavelength band, from optical to near-infrared. The primary advantage of photometric redshifts (or photo-*z*s for short) is that the required data is already there in the images, without the need to obtain follow-up spectroscopic measurements, which would be prohibitively expensive in terms of observing time given the hundreds of millions of galaxies in the DES sample. While not as precise as a spectroscopic redshift measurement, photometric redshifts are sufficiently accurate to obtain constraints on dark energy using three of the four DES cosmological probes (weak lensing, galaxy clusters and large-scale structure) if the uncertainties are well enough understood.

The key to unlock that third dimension of redshift information from the 2D images, which we have from DES, lies in the multi-band nature of the DES images, which we can think of as very low-resolution spectroscopy. For example, for the red galaxies that make up the bulk of the galaxy population in galaxy clusters, there is a relatively sharp drop in the spectrum at a wavelength of 400 nm. This 4000 Å break feature provides a clear signal that can be readily seen in multi-band data. In particular, as the redshift of the galaxy increases, this feature progressively shifts to longer wavelength filters (from *g* to *r* to *i*) resulting in changes in the measurements of the galaxy in the DES filters that we can translate into the galaxy's photo-*z*. This can be seen in Figure 15.1, in which the blue line shows the light spectrum for a typical elliptical galaxy at redshift $z = 0$ (near Earth). As a galaxy becomes increasingly distant and recedes at a faster rate due to the expansion of the universe, the break feature is observed at a higher and higher wavelength. This can be seen in the green and red lines on the graph, which are spectra of galaxies at $z = 0.5$ and $z = 1.0$, respectively. Four of the DES filters (labeled *g, r, i, z*) are also plotted on the graph, shown by the light green, red, purple and black peaked lines.

By measuring the amount of light visible in each of the DES filters from each of the millions of galaxies, we can estimate the redshift for each galaxy. For luminous red galaxies (LRGs), we are able to measure quite precise photometric redshifts, with uncertainties of a few percent, using an algorithm called 'redMaGiC'. However, the galaxy population also includes large populations of bluer spiral galaxies with weaker spectral features that make their photo-z measurements much less precise. Nonetheless, the overall galaxy population is characterized by distributions of spectral features and colours that change with redshift in such a way that reasonably precise photo-z measurements of a DES galaxy are still possible (with uncertainty of about 10%–15%). Although many photo-z measurement methods have been applied to multi-band data, from DES and other imaging surveys, such methods fall into two main categories: (1) template-based methods and (2) machine learning (or training-set-based) methods. Template-based methods use a library of galaxy spectral models, which span a range of redshifts and galaxy types. The multi-band data for a given

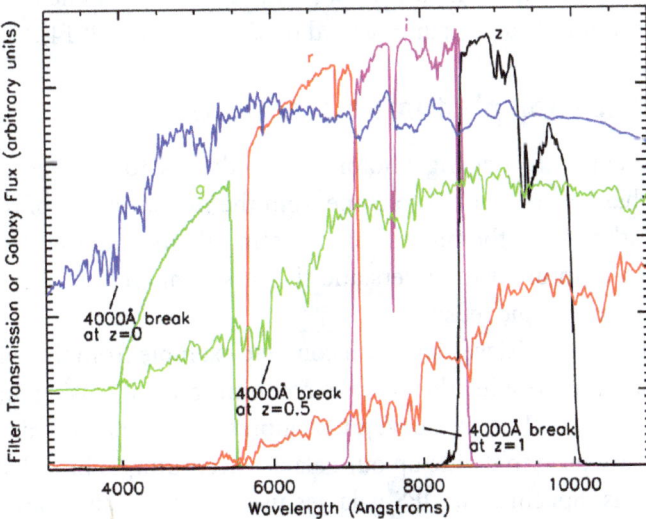

Fig. 15.1 The response functions of the DES *griz* filters overplotted on the spectral energy distributions of a typical elliptical galaxy at $z = 0$, $z = 0.5$ *and* $z = 1.0$. The passage of the 4000 Å break through the different filters with increasing redshift and the relative uniformity of the shapes of elliptical galaxy spectra permit accurate photometric redshifts to be measured for individual cluster galaxies. Credit: Huan Lin.

galaxy are compared against those from the library of models in order to find the best-matching model galaxy, whose redshifts (and type) are then assigned to the real galaxy. The 'redMaGiC' technique is an example of a template-based method, where there is only one template being considered (in this case a red galaxy model). Machine learning methods don't rely on galaxy models. They use a 'training set' of galaxies with known spectroscopic redshifts. These are then used to derive an empirical, mathematical relation between galaxy magnitudes (and/or colours) and true redshifts. This same relation is then applied to the much larger sample of galaxies that do not have true redshifts in order to derive photo-zs from the multi-band imaging data. In DES, we use training set data that comes from many existing, publicly available spectroscopic survey samples, as well as from new spectroscopic data that the collaboration obtains, such as the OzDES spectroscopic survey described in Chapter 14.

Early exploratory work on photo-z for DES was carried out using simulated data. In particular, we assessed template-based methods (e.g. Bayesian photo-z [BPZ]) versus supervised machine learning approaches (e.g. ANNz). Later, a comparison of photo-z methods applied to DES science verification data was summarized in Sánchez et al. (2014).

15.2 Validating photometric redshifts

The challenge of applying photometric redshifts to DES cosmological analysis has turned out not to come from the measurement of the photo-zs, but rather from the process of validating them. We need to calibrate our photo-zs in order to understand the biases inherent in them, relative to true spectroscopic redshifts.

To illustrate this, we can take a concrete example from the DES Year 1 (Y1) cosmology results, the analysis combining weak lensing and galaxy clustering (Dark Energy Survey Collaboration, 2018). In this analysis, photo-zs are used to divide up our galaxies according to how distant they are from us. Specifically, DES galaxies used in the analysis are assigned to a small number of so-called 'tomographic'[1] redshift bins, based on the photo-z of the galaxies. In a given tomographic bin, there is a spread in

---

[1]Tomography is a technique whereby a series of images are made in sections, in order to reconstruct a 3D map.

the true redshifts of the galaxies, because of the approximate nature of the photo-z. The cosmology analysis requires us to estimate the average true redshift of the galaxies belonging to a tomographic bin. We do have estimates of the average true redshifts, based on the BPZ method (a template-based method) or the 'Directional Neighborhood Fitting' (DNF) method (a machine learning method). The question is, how can we validate these estimates and check how accurate they might be? The most straightforward means of validation would have been to use the many large galaxy spectroscopic redshift samples that overlap with the DES footprint. The problem is that these spectroscopic samples are significantly incomplete. They don't obtain a successful redshift for every targeted galaxy, particularly at the faintest magnitudes covered by the DES galaxy sample. We have methods to attempt to correct for this incompleteness, but have not demonstrated that the corrections achieve the level of accuracy required by the cosmology analysis. As a result, we have resorted to two other methods for validation: (1) use of high-precision 30-filter photo-z data from the Cosmic Evolution Survey (COSMOS) and (2) use of the so-called 'cross-correlation' method.

15.2.1 Validation from COSMOS

Our first validation method uses data from the COSMOS field, a small, 2-square degree area of the sky that has been thoroughly observed by a wide variety of astronomical instruments. In particular, there exists a recently published catalogue of high-precision ($\sim$1% uncertainty) photo-zs, based on data from thirty filters spanning ultraviolet, optical and infrared wavelengths. These thirty-band COSMOS photo-zs suffice to serve as a complete sample of (nearly) true redshifts, and enable us to validate our BPZ-based estimates of the average true redshifts in tomographic bins. On the other hand, because of the small area of the COSMOS field, our confidence in this validation procedure is limited by 'cosmic variance', that is, uncertainties due to fluctuations from large-scale structure in the universe.

15.2.2 Validation from cross-correlation

The cross-correlation technique uses the fact that, because galaxies cluster via gravity, two galaxies with small angular separation are more likely to be spatially correlated and thus at similar redshift. Given a sample of galaxies with unknown redshifts, plus another sample of galaxies with known true redshifts, we can estimate the true redshift distribution of the first sample by cross-correlating the two galaxy samples, that is, by measuring how correlated the two sets of galaxies are in their positions on the sky. A key advantage of this method is that the sample with true redshifts does not need to be complete. The main requirement is that the true redshift sample covers the same sky area as the unknown redshift sample. In particular, the 'redMaGiC' galaxies, which cover the full DES footprint, serve as the true redshift sample and have provided us with a second validation method for our BPZ photo-zs.

15.3 Future work

With much hard work and effort by many in DES, the Y1 photo-zs have been validated successfully using both the COSMOS and cross-correlation methods. We have had six publications on making and validating the Y1 photo-zs, which demonstrate just how much new work had to be done. As DES moves on to our Y3 and Y5 data sets, we continue to work on improvements to our photo-z validation methods. The improved statistical power of those data sets will likely demand that we understand the biases in our photo-zs better than we have for the Y1 analysis. Additional high-precision many-band photo-z samples will be available, from both published data and potentially from the deep imaging observations in the DES supernova fields. Efforts are also ongoing to improve completeness corrections for spectroscopic redshift samples, so that we can make better use of these data in directly validating our photo-zs. Finally, we are exploring a number of new methods and approaches that will hopefully help us improve the understanding of our photo-zs and facilitate their use in future DES cosmology analyses.

References

Dark Energy Survey Collaboration (2018). Dark Energy Survey Year 1 results: Cosmological constraints from galaxy clustering and weak lensing, Physical Review D **98** , 4, p. 043526, doi:10.1103/PhysRevD.98.043526.

Sánchez, C., Carrasco Kind, M., Lin, H., Miquel, R., Abdalla, F. B., et al. (2014). Photometric redshift analysis in the Dark Energy Survey science verification data, *Monthly Notices of the Royal Astronomical Society* **445**, 2, pp. 1482–1506, doi:10.1093/mnras/stu1836.

CHAPTER 16

Simulating the Dark Energy Survey

Gus Evrard, Pablo Fosalba, Katrin Heitmann,
Andrey Kravtsov & Risa Wechsler

In this chapter, we look briefly at the role that cosmological simulations play in the study of the universe and how the Dark Energy Survey (DES) uses them to create synthetic galaxy catalogues that the science working groups use to test methodology and analysis.

16.1 Introduction to simulations

Cosmological simulations allow you to play God. Mix together the ingredients for your own universe, stand well back and watch it grow. They are the most recent addition to a 'triangle' of disciplines that allow us to study cosmology; the other two vertices are labelled 'observation' and 'theory'.

Humans have been observing the universe since we first looked up at the sky. We have been developing theories about the universe since the ancient civilizations made predictions based on those observations. With the invention of the telescope in the early seventeenth century and the advent of the scientific method during the Renaissance, this relationship developed at an increasing pace. Observe something; theorize about how and why it happens to make predictions. If the theory fits the observations, then it becomes a stronger theory to be further refined.

For the next 350 years, observation and theory were the only disciplines in town. However since the 1960s, thanks to the accelerating power of computer processing, huge progress has been made in developing codes that simulate the evolving contents of the universe across cosmic time. This has led not only to insights that were previously impossible, but also to the establishment of a new research discipline.

In a chemistry experiment, for example, we might want to see what happens if we add different amounts of blue liquid A to pink liquid B. Pour, mix, observe, theorize – and repeat. Whoops – accidentally added green liquid C? Never mind, clear the mess and start over.

Studying the universe is different. There is only one universe[1] to observe and we are inside the experiment, viewing it at this current time and this current position. We can't start over and repeat. Unless, that is, we can build our own simulated model.

Computer simulations not only allow us to 'mix' different versions of the universe together, but also to view it from whenever and whatever vantage point we wish. We can adjust the cosmological model accordingly. What would a non-flat universe look like? Or one without any dark energy?

16.2 N-body

Since the advent of computer simulations in the 1960s, the most widely used method has been the 'N-body' concept – where dark matter particles evolve inside a 'box'.

N-body takes as its starting point the picture we have of the very early universe from the cosmic microwave background (CMB). This gives us a statistical idea of what the universe looked like in its very early form.

We then take the observed fluctuations in the CMB radiation and convert them into density perturbations. This generates a dark matter particle distribution inside a cubic volume of a size relevant to the cosmic volume we are simulating.

To the computer, the dark matter particles (a practically collision-less fluid) are characterized by just seven numbers, which relate to their individual positions and velocities in three-dimensional space and their mass. We place the particles in initial positions determined by the CMB

---

[1] Theories of the 'multiverse' are beyond even the scope of DES and will be left to other books.

information and then leave them to evolve according to gravity, under the assumption of a given cosmological model.

Using only those ingredients (and a very large supercomputer) we can create the dark matter density maps that we observe in the universe today. We can then use other techniques to populate this dark matter 'scaffolding' with observable structure, the galaxies themselves.

In the past decade, advances in simulation software and computer processing have allowed us to resolve sub-halo populations of dark matter in detail, thus increasing the resolution of our synthetic cosmos. The first N-body simulations in the early 1960s used up to hundred particles. In order to simulate the universe at the resolution that Dark Energy Survey (DES) requires, we use $\sim 10^{10}$ particles in a box-size of the order of a gigaparsec ($\sim$Gpc).

It is important to ask ourselves whether our simulations are scientifically viable and the way we answer that is by comparing what we see in the simulations to what we see in the sky.

Observe. Theorize. Simulate. Repeat.

16.3 Simulations in DES – Of MICE and Buzzards

DES is the first survey to attempt to systematically integrate cosmological simulations into science analysis across the three principal lines of investigation – weak lensing, galaxy clustering and clusters of galaxies – outlined in earlier chapters. Two teams of researchers – one based in the United States and another in Spain – developed independent methods to build synthetic, also known as 'mock', expectations for survey outcomes under different cosmological models.

The simulation requirements are stringent; locations, sizes and fluxes in the five optical filters must be determined for hundreds of millions of galaxies within the full survey volume. A physics-based approach, one that involves modelling the coupled behaviours of dark matter and various phases of baryonic matter (stars, black hole, multi-phase plasma), is out of the question because even the largest such cosmological hydrodynamics simulations are capable of modelling only a small fraction of the DES survey volume. Instead, we produce large N-body simulations and statistically assign galaxies to locations based on local dark matter properties.

The DES simulations support cosmological analysis for the dark energy probes that are sensitive to the growth of cosmic structure: galaxy clustering including baryon acoustic oscillations (BAO); gravitational lensing; and population statistics of galaxy clusters. The synthetic sky surveys include the observed properties of galaxies, their large-scale distribution in space and time, the impact of lensing shear on the shapes and magnitudes of galaxies and as many relevant observational effects as possible. Multiple such simulations, each spanning the full area and depth of DES, are required for statistical analysis, and the simulations should explore a range of underlying cosmological models.

For the DES Year 1 analysis, two suites of simulations, known as BUZZARD and MICE, were produced.

16.3.1 BUZZARD

The production of mock catalogues are an integral part of the ongoing 'blind cosmology challenge' (BCC) within DES. The early BCC simulations (named 'Aardvark', 'Beluga' and 'Chinchilla') done around 2011 to 2013 were run on machines at the Extreme Science and Engineering Discovery Environment (XSEDE) (Erickson et al., 2012), the National Energy Research Scientific Computing Center (NERSC) and at the cluster run by the University of Chicago Research Computing Cluster. Some simulations were also run at the (Stanford Linear Accelerator Center) SLAC 'bullet' cluster. All of these were hugely important to the overall development of a suite of galaxy catalogues ('BUZZARD') (Busha et al., 2013; DeRose et al., 2019), summarized as follows:

1. Low resolution N-body simulations are run to map out the dark matter density field along the past light-cone of an artificial observer.
2. With the resultant dark matter distribution, galaxies are assigned to dark matter halos and particles using an empirical method called Adding Density Dependent GAlaxies to Light-cone Simulations (ADDGALS). The method attempts to match the correlations between local dark matter density and galaxy properties observed in previous surveys. Each galaxy is assigned the following observable quantities:

- Spectral energy distributions (SEDs) are assigned from a training set of galaxy spectroscopy from SDSS DR7 based on local environmental density. From these SEDs, the galaxy griz magnitudes are integrated.
- Galaxy sizes and ellipticities are drawn from distributions fit to the Subaru Prime Focus Camera (SuprimeCam) i-band data.
- Effects from the gravitational lensing of light from each galaxy are computed from the multiple-plane raytracing algorithm, Curved-sky grAvitational Lensing for Cosmological Light conE simulatioNS (CALCLENS). Galaxy positions, shapes and magnitudes are modified, or 'lensed, accordingly.
- Photometric errors are applied to the lensed magnitudes.
- The DES survey observing region is extracted.

3. The process is repeated six times, each time with a different cosmology, producing a small flock of BUZZARD simulations for DESY1 science.

Each set of BUZZARD catalogues is built from a combination of three separate N-body simulations. These have box lengths of 1.05, 2.6 and 4.0 h^{-1} Gpc, and 1400^3, 2048^3 and 2048^3 particles, giving mass resolutions of 2.7×10^{10}, 1.3×10^{11} and $4.8 \times 10^{11} h^{-1} M_\odot$.

16.3.2 MICE (Pablo Fosalba)

The Marenostrum Institut de Ciencies de l'Espai (MICE) simulations [http://maia.ice.cat/mice; (Fosalba et al., 2015a,b; Crocce et al., 2015)] is a project to develop large N-body simulations to model galaxy survey observables using one of the most powerful supercomputers in Europe, hosted at the Barcelona Supercomputing Center. The largest simulation evolves seventy billion (4096^3) dark matter particles in a volume of $3072\, h^{-1} \mathrm{Mpc}^3$ using the GADGET-2 gravity-solver code, which results in a particle mass of $2.93 \times 10^{10} h^{-1} M_\odot$.

Given its large volume and spatial resolution, spanning over five orders of magnitude in dynamic range, it allows for an accurate modelling of the growth of structure in the universe from the linear through to the highly non-linear regime of gravitational clustering. All-sky lightcone outputs are modelled using the 'Onion Universe' approach (Fosalba et al., 2008), which allows for a consistent modelling of galaxy clustering and weak-lensing observables within the Born approximation.

Galaxy assignment uses a Halo Occupation Distribution (HOD) approach combined with Halo Abundance Matching techniques, as described in (Carretero et al., 2015).

The MICE galaxy mocks, distributed to the DES collaboration through a dedicated web portal, CosmoHub, http://cosmohub.pic.es, have already been used for the Science Verification and Y1 analyses.

16.3.3 HALOGEN – BAO mocks

MICE and Buzzard provide excellent mock galaxy catalogues, but they are drawn from full N-Body simulations. This requires lots of computing CPU hours in supercomputers. For this reason, we can only build a few mock catalogues from these methods. An alternative, which is used to model BAO physics, is 'HALOGEN' (Avila et al., 2015; Avila et al., 2018). It is a fast, approximate method that can create simulations in a few hours in a local workstation. This allows us to create 1800 mock catalogues specific to the study of large-scale structure in DES. Using HALOGEN enables us to compute the covariance matrices (see Chapter 13 for more information), optimize the methodology in a statistical way and also understand how to deal with certain particularities of the data.

16.3.4 BALROG *(Brian Yanny and Yuanyuan Zhang)*

BALROG is a set of software modules that generate a hybrid between real DES data and simulated objects that are injected onto the real DES images. The fake stars and galaxies each have known location, size, shape and brightness. Then the images with the added fake objects are (re)processed through the standard DES software pipeline together with the real objects, and one tests how well their known properties have been recovered, and with what errors. Near the detection limit of a survey (between 24th and 25th magnitude in the case of DES), strong biases (offsets) in the measured properties are common, and can be quantified by this process for all DES survey objects in a statistically robust way. If left uncorrected, these biases can affect the derived science of the survey, for example, by throwing off the measurements of accurate photometric redshifts for faint galaxies.

Injecting fake objects into a large variety of real survey images allows DES scientists to study how the detection of objects can vary across the whole of the survey footprint. The BALROG program name, bestowed by an early DES software developer (Eric Suychta), was taken from that of the cave-dwelling monster in the 'Lord of the Rings' who wreaks havoc when disturbed. It serves as a reminder that scientists must be prepared to deal with the consequences of their investigations – however deeply they dig and whatever they may uncover.

16.4 Using the galaxy catalogue simulations

The simulated galaxy catalogues were then used as inputs into the Ultra Fast Image Generation, 'UFIG' (Bergé et al., 2013) pipeline to create simulated image catalogues. Generating simulated image catalogues allows us to understand how to connect observed properties (including complex systematics) with true properties.

A schematic diagram of the simulated galaxy and image catalogues built from the DES BCC is shown in Figure 16.1.

Each of the DES science projects makes use of the galaxy and/or image catalogues to test their specific methods and results. For example, the 'RedMaPPer' (red-sequence matched-filter Probabilistic Percolation

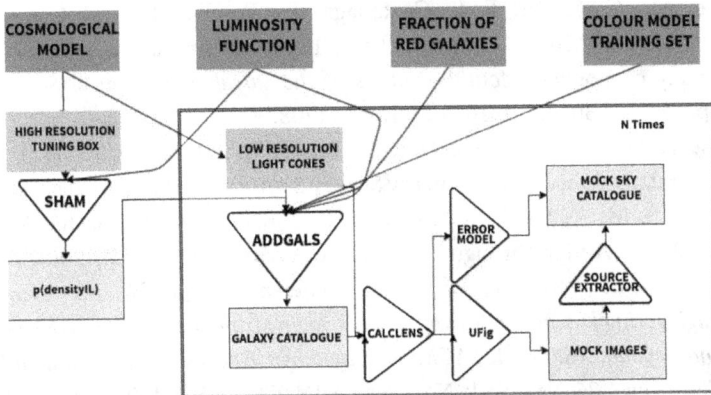

Fig. 16.1 A schematic diagram showing the Dark Energy Survey (DES) Blind Cosmology Challenge (BCC) pipeline, which generates simulated galaxy and image catalogues.

cluster finder) algorithm is applied to the galaxy catalogue to detect and create samples of simulated clusters.

References

Avila, S., Crocce, M., Ross, A. J., García-Bellido, J., Percival, W. J., et al. (2018). Dark Energy Survey Year-1 results: Galaxy mock catalogues for BAO, *Monthly Notices of the Royal Astronomical Society* **479**, pp. 94–110, doi:10.1093/mnras/sty1389.

Avila, S., Murray, S. G., Knebe, A., Power, C., Robotham, A. S. G., et al. (2015). HALOGEN: A tool for fast generation of mock halo catalogues, *Monthly Notices of the Royal Astronomical Society* **450**, pp. 1856–1867, doi:10.1093/mnras/stv711.

Bergé, J., Gamper, L., Réfrégier, A., and Amara, A. (2013). An Ultra Fast Image Generator (UFIG) for wide-field astronomy, *Astronomy and Computing* **1**, pp. 23–32, doi:10.1016/j.ascom.2013.01.001.

Busha, M. T., Wechsler, R. H., Becker, M. R., Erickson, B., and Evrard, A. E. (2013). Catalogue production for the DES blind cosmology challenge, in *American Astronomical Society Meeting Abstracts #221*, American Astronomical Society Meeting Abstracts, Vol. 221, p. 341.07.

Carretero, J., Castander, F. J., Gaztañaga, E., Crocce, M., and Fosalba, P. (2015). An algorithm to build mock galaxy catalogues using MICE simulations, *Monthly Notices of the Royal Astronomical Society* **447**, pp. 646–670, doi:10.1093/mnras/stu2402.

Crocce, M., Castander, F. J., Gaztañaga, E., Fosalba, P., and Carretero, J. (2015). The MICE Grand Challenge lightcone simulation – II. Halo and galaxy catalogues, *Monthly Notices of the Royal Astronomical Society* **453**, pp. 1513–1530, doi:10.1093/mnras/stv1708.

DeRose, J., Wechsler, R. H., Becker, R. M., Busha, M. T., Rykoff, E.., S., et al. (2019). The Buzzard Flock: Dark Energy Survey Synthetic Sky Catalogues Online.

Erickson, B. M. S., Singh, R., Evrard, A. E., Becker, M. R., Busha, M. T., et al. (2012). A high throughput workflow environment for cosmological simulations, in *Proceedings of the 1st Conference of the Extreme Science and Engineering Discovery Environment: Bridging from the eXtreme to the Campus and Beyond, XSEDE '12, ACM formatted by CE-CONV now formatted by CE-CONV on 6 26*, New York, NY, USA, ISBN 978-1-4503-1602-6, pp. 34:1–34:8, doi:10.1145/2335755.2335830.

Fosalba, P., Crocce, M., Gaztañaga, E., and Castander, F. J. (2015a). The MICE Grand Challenge light-cone simulation – I. Dark-Matter Clustering, *Monthly Notices of the Royal Astronomical Society* **448**, pp. 2987–3000; doi:10.1093/mnras/stv138.

Fosalba, P., Gaztañaga, E., Castander, F. J., and Crocce, M. (2015b). The MICE Grand Challenge light-cone simulation – III. Galaxy lensing mocks from all-sky lensing maps, *Monthly Notices of the Royal Astronomical Society* **447**, pp. 1319–1332, doi:10.1093/mnras/stu2464.

Fosalba, P., Gaztañaga, E., Castander, F. J., and Manera, M. (2008). The onion universe: All sky lightcone simulations in spherical shells, *Monthly Notices of the Royal Astronomical Society* **391**, pp. 435–446, doi:10.1111/j.1365-2966.2008.13910.x.

Non-Dark Energy Science

This section focuses attention on the wide range of research areas that the Dark Energy Survey scientists are working on, aside from dark energy, using the new information flooding in about the universe. This research includes Milky Way science, galaxy evolution, strong gravitational lensing, quasars, transients and moving objects and optical follow-ups to gravitational wave events.

Non-Dark Energy Science

This section focuses attention on the wide range of research areas that the Dark Energy Survey scientists are working on, aside from dark energy, using the new imaging data. It delights those that enjoy various research projects: Milky Way science, galaxy evolution studies, machine learning, transients and moving objects and more. It allows users to explore what was seen.

CHAPTER 17

Galaxy Evolution

Manda Banerji, Will Hartley, Daniel Thomas & Risa Wechsler

How do galaxies form? Why do they come in different shapes, colours and luminosity? And what can these features tell us about their population, evolution and environment? What do they have to do with the mysterious dark matter? In this chapter, we look at how the extensive catalogue of galaxies assembled by the Dark Energy Survey (DES) can be used to study galaxy evolution.

17.1 The Dark Energy Survey – A goldmine for galaxy evolution studies

For a few hundred million years after the Big Bang, the universe was a broadly homogeneous mix of matter and radiation. Had we been around to study it, astronomy would have been a much simpler science – if a little dull. Today, over thirteen billion years later, if we point our telescopes in any direction, we don't see a monotonous nothing – we see galaxies. Billions of them. Like any huge population, variety is the spice of life, in morphology, colours, luminosity and dynamics. Some galaxies are thousands of times brighter than our own, others just a fraction. There are galaxies that adhere to a regular shape, others which prefer to express their irregularity. Galaxies alive with star formation. Others – more sedate – harking back to their own star-forming youth in the distant past. Galaxies with 'Catherine wheel' – type rotating arms. Others with little interest in rotation. How can we account for this diverse range of properties and what clues do they hold about the evolutionary life of a galaxy? How did the universe evolve from a place devoid of interest to one teeming with galactic life? These are some of the questions that galaxy studies attempt to address.

The Dark Energy Survey (DES) is a goldmine for galaxy studies. Although the goal of DES is to answer questions about the expansion of the universe, its data-rich set of potentially 300 million galaxies makes it perfect for investigating galaxy evolution. A broad range of science is possible across the DES redshift space, from the nearby universe to around $z \sim 6$, which is close to epoch of re-ionization. The volume sampled by DES is several orders of magnitude larger than existing surveys, though shallower by several orders of magnitude. How the properties of galaxies are determined is described in detail in Chapter 7. DES photometry covers a wide range of wavelengths and so provides incredibly useful information beyond simply the derivation of galaxy distance through the photometric redshift. Optical observations from the Dark Energy Camera's (DECam) five filters can be combined with near-infrared data from the ESO VISTA Hemisphere Survey (VHS), which overlaps the DES survey area, providing a rest-frame wavelength range of between 2220 and 10,000 Å. Even at a moderate z band depth of 23 mag, this provides an excellent opportunity to analyze stellar population properties of galaxies.

17.2 Luminosity and stellar mass function

The general consensus is that structure forms from gravitational collapse following dark matter clustering. Galaxies assemble their mass, including both dark matter and luminous matter (gas and stars) within this 'bottom-up' hierarchical model. However, the growth of galaxies is also the result of the interplay of other processes, which include star formation, active galactic nucleus (AGN) feedback, supernova, mergers and galaxy interactions.

Consequently, there may be a less significant role for the halo-scale environment in influencing galaxy formation – especially at high mass – than a simple bottom-up model predicts. In fact, while theory suggests bottom-up, observations hint at top-down. This is one of the burning questions that the DES galaxy evolution group aims to help solve by studying the evolution of the galaxy luminosity and stellar mass functions (GLF and GSMF) with cosmic time (hence redshift). These can be thought of as a histogram of the stellar luminosity or mass of the galaxies within a large sample survey, with the number of galaxies in each luminosity (or mass) bin divided by the survey volume. DES can help shed some light on how galaxies build up their mass over time by studying the evolution of the galaxy luminosity and mass functions.

The key quantity that needs to be derived is the stellar mass of a galaxy, which is its mass without counting dark matter. DES photometry is ideal for this. Galaxy masses can be derived by simulating DES data with a model that predicts the energy output of a stellar population. Daniel Thomas and Claudia Maraston at the University of Portsmouth (United Kingdom) started to address this question with their research teams early on in the planning of DES. Together with PhD student Janine Pforr, they first used simulated DES data to develop their techniques in deriving galaxy masses. This technique was then later applied to real DES data by postdoctoral researcher Diego Capozzi when he joined the Portsmouth team, with the aim of building a catalogue of DES galaxies that could be used to study the galaxy mass and luminosity functions.

The team used the DES science verification (SV) data to generate a sample of four million galaxies at $0 \leq z \leq 0.3$ over an area of ~ 155 square degrees. Key challenges in this process included the separation of galaxies from stars in the data set, the fact that the calculation of galaxy masses and their reliability as galaxy distances are only known from photometric redshifts and the need to understand the so-called completeness of the data. Completeness is a measure of whether all galaxies of a certain mass and luminosity have actually been observed, or whether some particularly faint objects have been missed. An understanding of the latter is obviously key to studying the galaxy mass function, that is, the distribution of galaxy masses.

The team succeeded in mastering those challenges and was able to produce mass and luminosity functions of galaxies at very large distances (Capozzi et al., 2017). This meant they could see what the galaxy population looked like eight billion years in the past, when the universe was less than half its current age. This was a great success. Very importantly, the DES measurements agree well with results found before by others based on more accurate, so-called spectroscopic distances, but for much smaller samples, as shown in the left-hand panel of Figure 17.1. This allowed the team to calculate how the number density of galaxies was evolving over the past eight billion years. The door was now wide open to study galaxy populations at early times in the universe for very large galaxy samples.

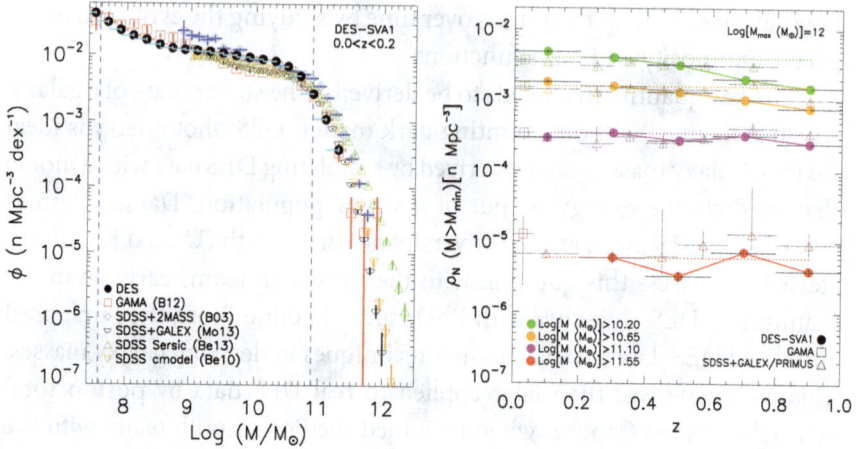

Fig. 17.1 Left-hand panel: The mass function as measured in the Dark Energy Survey (DES) in comparison with other measurements in the literature. Good agreement is found. Right-hand panel: The evolution of the number density of galaxies with lookback time for galaxies of different masses. It can be seen that the number of the most massive galaxies in the universe has not changed over the past eight billion years. From Capozzi et al. (2017).

It turns out that the number of the most massive galaxies in the universe has not changed over the past eight billion years (right-hand panel in Figure 17.1). This kind of evolution of the galaxy population has been seen before, but only for much smaller samples from spectroscopic surveys. The result is exciting for scientists as it cannot easily be explained within our current theoretical framework of galaxy formation.

17.3 Galaxy environment

The Portsmouth team was now ready to make the final step in their analysis. The major reason why they generated large galaxy samples from DES data was to study how the galaxy mass function evolved with time in different galaxy environments. Such different galaxy environments could not be addressed in the previously existing, smaller spectroscopic samples.

To put this in context: primordial fluctuations in the early universe sowed the seeds of the structure we see today. After inflation, in some regions dark matter accumulated, driven by gravity. These regions eventually collapsed to form dark matter haloes that grew by accreting and merging with other haloes. Baryonic matter fell into the potential wells of the haloes. Here, galaxies formed and evolved. It, therefore, seems plausible that there should be a measurable correlation between the properties of a galaxy and its 'environment'. And that a galaxy's environment plays an important role in hierarchical structure formation. However, the precise role has been strongly debated as it is not fully understood.

We can define environment as the density surrounding a galaxy. Two ways to quantify this are the fixed aperture and Nth nearest neighbour methods. In the fixed aperture method, we simply count the numbers of galaxies within a fixed volume centred on a target galaxy and divide by the comoving volume. In the second method, we compute the comoving distance to Nth nearest neighbour of a target galaxy (carefully choosing the value of N) and divide by volume to get a number density. With either method, however, it is very important that redshift information is as accurate as possible. If not, there is a possibility that galaxies in the same spatial vicinity on the sky are misclassified as neighbours due to redshift error.

PhD student James Etherington joined the Portsmouth team to study exactly this problem. Etherington and Thomas investigated the impact of the precision of photometric redshift using Sloan Digital Sky Survey (SDSS) spectroscopic data as well as simulations (Etherington and Thomas, 2015). They concluded that if the photometric redshift has a 10% uncertainty, then the galaxy sample must be between six and ten times larger than a spectroscopic survey to detect equivalent environment correlations – and this would be easily met by DES! Figure 17.2 shows an example from their work using SDSS data to show how simple number counts of galaxies are translated into environmental densities following the methods described earlier.

Using this method, the Portsmouth team was now in the position to investigate whether galaxy populations evolve differently with cosmic time in different galaxy environments. And it turns out they do! The result is shown in Figure 17.3, where we present the same kind of mass function as in Figure 17.1, but this time at different lookback times (left- and right-hand panels) *and* for different galaxy environments (blue and

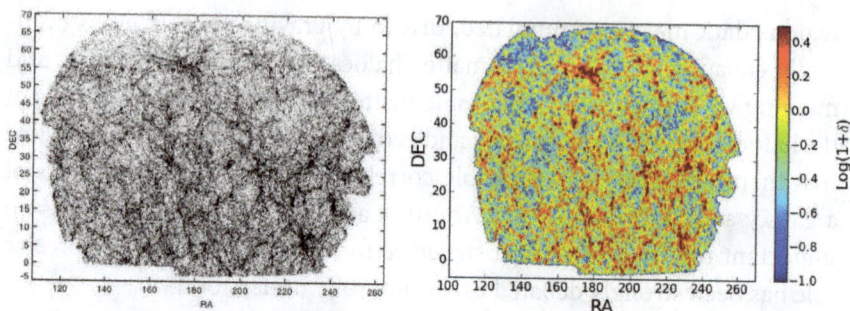

Fig. 17.2 Galaxy distribution for the Sloan Digital Sky Survey (SDSS) main footprint (left) and the projected galaxy environment (right). Dense environments are shown in red and sparse environments are in blue. From Etherington and Thomas (2015).

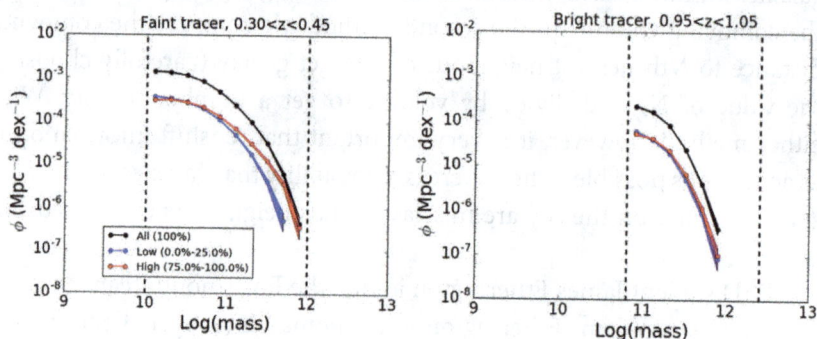

Fig. 17.3 Galaxy stellar mass functions for the lowest (blue), highest (red) and all (black) environment bins for two redshift bins. The vertical dashed lines mark the bounds of the common mass range. For clarity, error bars are not shown in this figure. From Etherington et al. (2017).

red curves). It turns out that, while massive galaxies tend to live in dense environments in today's universe (left-hand panel), they were much more evenly distributed in the past (right-hand panel). This is an exciting result. We are witnessing how high-density structures form around the massive galaxies as cosmic time goes by!

17.4 Evolution of galaxy properties in clusters

The most extreme and densest environment a galaxy can reside in is a group or cluster of galaxies. These structures can contain thousands of galaxies.

Again, DES is ideal for studying these extreme environments because of its large survey area.

Galaxies residing at the centres of such clusters or groups tend to be the brightest cluster galaxies (BCGs). Their centrality and large size reflect their special nature and set them apart from the general galaxy population. BCGs are surrounded by a subsidiary population of 'satellite' galaxies. As a consequence of their privileged position, BCGs tend to obey different relations between their mass, size and luminosity compared to satellite cluster galaxies. DES's rich sample of galaxies in clusters in DES data enables us to study BCG properties significantly further in redshift than before. For example, Zhang et al. (2016) use a sample of 106 X-ray selected clusters and groups, from the DES SV data, to study the stellar mass growth of bright central galaxies (BCGs) since redshift $z \sim 1$.

17.5 Galaxy biasing

The galaxy density maps discussed earlier are not only useful for characterizing the evolution and formation of galaxies, they are also a key measurement to study cosmology and the evolution of the universe itself. We can use galaxies and their distribution to map out the matter density field of the universe. While we are actually interested in the distribution of matter overall, including dark matter, we need to resort to measuring the distribution of galaxies, as dark matter by its nature cannot be detected directly. This carries a problem: the distribution of galaxies may not necessarily reflect the distribution of dark matter.

In practice, we observe the galaxies and assume that the relative density of light is, in some way, mirroring the clustering of the bulk of the matter. A simplistic correlation would be that there is a 1:1 relation. In this case, there is no bias, b, to that tracer, which is defined as $b = 1$. If this was the case, then galaxies would be a fair tracer of the underlying mass. However, b could be different. If galaxies are clumped more strongly than mass, then $b > 1$ and galaxies would be a biased tracer of (dark matter) mass, that is, by measuring the distribution of galaxies we are not measuring the actual distribution of dark matter. This, indeed, seems to be the case – galaxies are a biased tracer of the underlying mass. This is a consequence of the different physics involved in galaxy formation, which causes the spatial distribution of baryons to differ from that of dark matter. Galaxy biasing also increases with redshift and is dependent on scale.

In regard to the DES aim to derive cosmological parameters, biasing can be a nuisance parameter. However, it is very useful in its own right as a probe of galaxy formation and evolution. By cross-correlating mass maps derived from the cosmic microwave background (CMB) (measured by Planck and the South Pole Telescope) with DES galaxies, DES should have significant power to constrain galaxy bias.

17.6 Galaxy structure and morphology

So far, we have considered mass and environment and their evolution with cosmic time as the most basic galaxy properties describing a galaxy. Other obvious properties that characterize a galaxy are its colour and its shape. The latter is probably the most apparent when looking at a galaxy. Based broadly on Edwin Hubble's simplistic 'tuning fork' diagram, we can classify the general morphology of a galaxy as being spiral, elliptical or irregular. This gives an indication of the formation state of a galaxy. The morphological properties of the DES galaxies represent a powerful constraint on possible galaxy formation scenarios. When combined with other physical quantities – mass, luminosity, disc size and the so-called CAS parameters (Concentration, Asymmetry and clumpinesS), shape can provide added insights into the processes at play, and help us to develop new ideas about the mechanisms of galaxy evolution.

Using data from the first year data of DES, and covering over 1800 square degrees, Federica Tarsitano (at the ETH Zurich) and Will Hartley (at UCL) have catalogued structural and morphological information on forty-five million objects (Tarsitano and Hartley, 2018). This is the largest structural and morphological galaxy catalogue to date, which is a significant asset to future studies of galaxy evolution.

Although a morphological catalogue for the whole DES survey is not yet available, DES data was extremely useful to test whether the features learned by a convolutional neural network are meaningful for different data. In other words, how much of the knowledge acquired by a machine learning algorithm from an existing survey (e.g. SDSS) can be exported to a new data set (e.g. DES). We demonstrated that machines can quickly adapt to new instrument characteristics (e.g. PSF, seeing, depth), reducing by almost one order of magnitude the necessary training sample for morphological classification (excluding redshift evolution effects) (Domínguez Sánchez et al., 2019).

17.7 Galaxy intrinsic alignments

The catalogue of galaxy shapes also plays an important role for one of the main probes central to the goals of DES: weak gravitational lensing (see Chapter 11), in which the apparent shapes of galaxies are sheared by foreground large-scale structure. It is thought, however, that even without this lensing effect, there is some intrinsic (i.e. unlensed) alignment (IA) due to the galactic neighbourhood. This IA becomes a nuisance parameter for restraining cosmological parameters through weak lensing studies. It needs to be accounted for as it may contribute between 10% and 50% of the overall apparent shear at different scales. However, aside from being an 'unwanted guest' in the study of cosmology, the IA of galaxies can tell us much about the galaxy formation and evolution. It can also shed light on the relation between the baryonic content and that of its host dark matter halo. In general, different physical alignment mechanisms are believed to be the source of IAs in red and blue galaxies, while IAs may also vary as a function of redshift, luminosity and other properties. The DES data set is extremely useful for joint studies of IAs in combination with data from other surveys.

17.8 Galaxies in the early universe

We discussed earlier how DES gives us insight into how the galaxy population has been evolving over the past eight billion years. DES also allows an exciting glimpse into very early epochs, much closer to the Big Bang, when the universe, and the galaxies within it, was still in their infancy.

DES is designed for the study of galaxies at distances covering lookback times of about half the age of the universe, with the aim of constraining cosmological dark energy parameters. But DES data is also very powerful for the identification of rare and massive ($> 10^{12} M_{\odot}$) galaxies at very large distances, implying much greater lookback times into the early universe.

These distances imply redshifts of about $z \sim 4 - 6$. Studying galaxies at such high redshifts will greatly improve our understanding of how galaxies form and evolve. Such galaxies will have low number densities and will be very faint. Observations need to be designed with long exposure times, which makes the experiment time-consuming and costly. As a consequence, surveys aimed at studying galaxies at such distances generally only cover small areas on the sky. This means that rare objects, such as the most massive galaxies, tend to be missed. We do not really know whether massive galaxies already existed early on in the universe. And most importantly, our theoretical understanding of galaxy formation suggests they should not yet have existed at such early cosmic epochs – imagine the excitement if they are found!

DES can help to make key progress in this regard. Claudia Maraston and Daniel Thomas at the University of Portsmouth, together with postdoctoral researcher Luke Davies, looked into DES's potential to discover such objects. In Davies et al. (2013), they show that the unique combination of sky coverage and survey depth that DES provides will indeed allow the identification of massive, high redshift star-forming galaxies. This is illustrated in Figure 17.4, where survey area is plotted as a function of survey depth (in terms of the detection limit in the z band). DES places itself at a sweet spot: other surveys are either wide enough but too shallow (top-left of the diagram) or deep enough but too narrow (bottom-right of the digram) to detect massive galaxies at very high redshifts.

The method to detect such objects is called the *Lyman-break technique*. This relies on the fact that stars (and hence galaxies) do not emit much radiation beyond the Lyman-break wavelength of ~ 91 nm. This limit is directly connected to the energy required to remove an electron from a hydrogen atom in its ground state. This results in a sharp cut-off in the spectrum of the galaxy at this wavelength. In galaxies at $z > 3$, this Lyman-break is shifted into the visible part of the spectrum and so visible light from the galaxy will not appear in optical telescopes. At $3.5 \leq z \leq 4.5$, the galaxy would not show in the blue g band filter of DES. At higher redshifts, $z > 6$, Lyman-limited photons are redshifted such that the DES r band and i band would not pass any light, but the redder z band filter will. This dropout behaviour in one or more filters allows the detection of high redshift galaxies. Davies et al. (2013) studied simulated DES data and

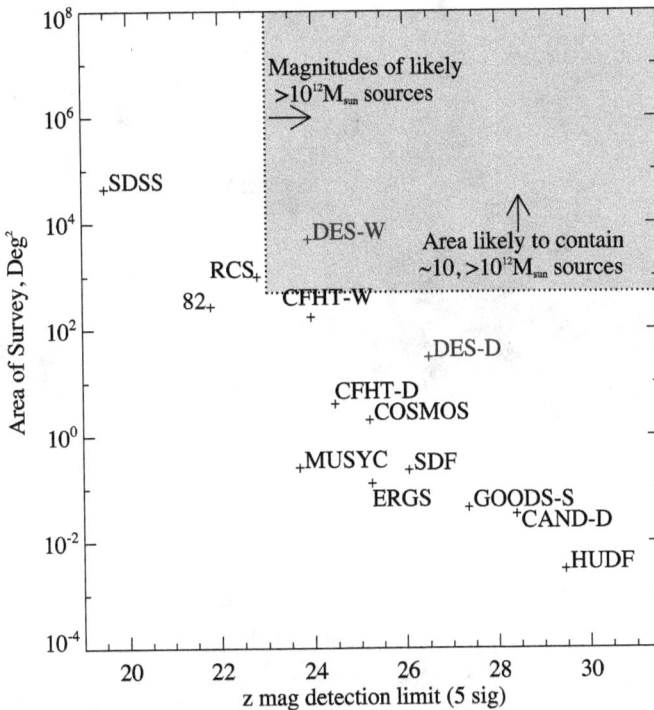

Fig. 17.4 A comparison of z band magnitude limit against area for a sample of surveys. The grey region represents the area of parameter space, which is likely to detect massive galaxies at $z > 3$. The Dark Energy Survey (DES) falls within this region as it covers a large area and is relatively deep in redshift. From Davies et al. (2013).

worked out the best combination of photometric filters that would allow the detection of massive, high-z galaxies in DES.

PhD student Pierandrea Guarnieri and undergraduate student Joakim Carlsen later joined the Portsmouth team and finally applied these selection criteria to real DES data. The result is very exciting. In a painstaking exercise analyzing several million galaxies, the team managed to isolate a handful of candidates – galaxies that may be very massive and very distant (Guarnieri et al., 2019). Figure 17.5 shows an example of such a candidate. It can be seen clearly that this object emits no light in the bluest, g band filter, but is detected in all other DES and also redder filters (JHK) from observations with the VISTA telescope.

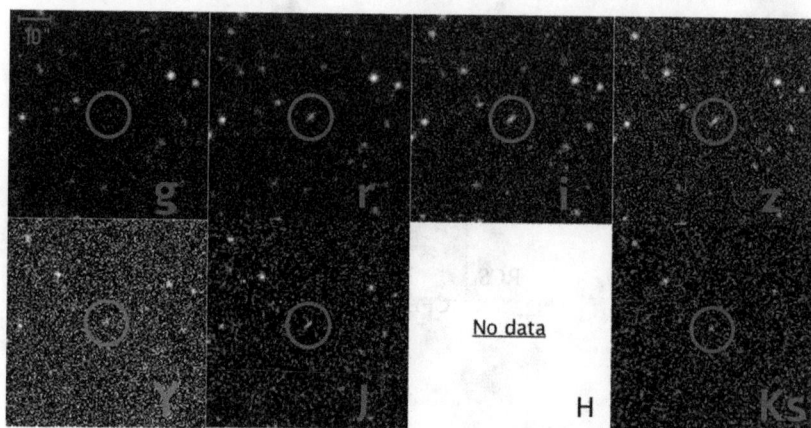

Fig. 17.5 Images of a candidate for a massive galaxy at high redshift in different filters. This object is detected in all bands except the bluest (*g* band), hence it is a candidate for a massive galaxy at redshift $z \sim 4$. From Guarnieri et al. (2019).

They show that all their candidates could well be very massive galaxies, ten times more massive than the Milky Way, already formed at an epoch when the universe was only a few billion years old (compare this to the current age of the universe, which is fourteen billion years!). If true, this discovery would challenge much of what we know about the formation and evolution of galaxies. The next step is now to go to very large telescopes and take spectra of these candidates, to measure their distances more accurately and confirm (or falsify) the discovery. Either way, this work is exciting as it paves the way to the continuing search for more candidates from DES and other data from future photometric, large-scale surveys.

Another approach is followed by Manda Banerji and her team at University College London (UCL). They studied a section of the DES data that is deeper, but shallower. This allows them to detect galaxies at even larger distances than the Portsmouth team. Their approach is complementary as these galaxies will be somewhat less massive than the ones discovered in Portsmouth. And they succeeded!

Figure 17.6 shows the filters for their candidate. Compare this to Figure 17.5, and you will see that Banerji and collaborators found an object that only emits light in the reddest DES band, the *z* band. This galaxy is therefore even more distant and has emitted its light at an even earlier epoch, when the universe was less than one billion years old! Very

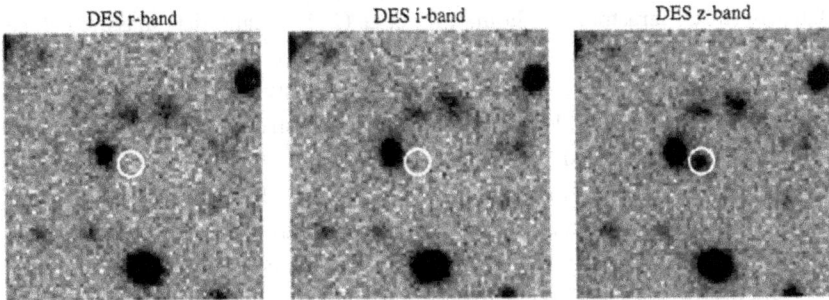

Fig. 17.6 The Dark Energy Survey (DES) r, i and z band images in the DES deep fields for a spectroscopically confirmed galaxy at $z = 6.07$. The galaxy position is indicated by the circle in each of the three images but is detected only in the redder z band image and undetected in the other two. This 'dropout' behaviour demonstrates the Lyman-break technique of detecting high-z galaxies.

excitingly, this galaxy is more than a candidate – its redshift has already been confirmed spectroscopically. DES is definitely on the right track to make new and exciting discoveries now and in the years to come!

References

Capozzi, D., Etherington, J., Thomas, D., Maraston, C., Rykoff, E. S., et al. (2017). Evolution of galaxy luminosity and stellar-mass functions since $z=1$ with the Dark Energy Survey science verification data, *ArXiv e-prints*.

Davies, L. J. M., Maraston, C., Thomas, D., Capozzi, D., Wechsler, R. H., et al. (2013). Detecting massive galaxies at high redshift using the Dark Energy Survey, *Monthly Notices of the Royal Astronomical Society* **434**, pp. 296–312, doi:10.1093/mnras/stt1018.

Domínguez Sánchez, H., Huertas-Company, M., Bernardi, M., Kaviraj, S., Fischer, J. L., et al. (2019). Transfer learning for galaxy morphology from one survey to another, *Monthly Notices of the Royal Astronomical Society* **484**, pp. 93–100, doi:10.1093/mnras/sty3497.

Etherington, J. and Thomas, D. (2015). Measuring galaxy environments in large-scale photometric surveys, *Monthly Notices of the Royal Astronomical Society* **451**, pp. 660–679, doi:10.1093/mnras/stv999.

Etherington, J., Thomas, D., Maraston, C., Sevilla-Noarbe, I., Bechtol, K., et al. (2017). Environmental dependence of the galaxy stellar mass function in the Dark Energy Survey science verification data, *Monthly Notices of the Royal Astronomical Society* **466**, pp. 228–247, doi:10.1093/mnras/stw3069.

Guarnieri, P., Maraston, C., Thomas, D., et al. (2019). Candidate massive galaxies at z ~ 4 in the Dark Energy Survey, *Monthly Notices of the Royal Astronomical Society*, **483**, pp. 3060–381, doi:10.1093/mnras/sty3305.

Tarsitano, F. and Hartley, W. (2018). A catalogue of structural and morphological measurements for DES Y1, *In preparation*.

Zhang, Y., Miller, C., McKay, T., Rooney, P., Evrard, A. E., et al. (2016). Galaxies in X-ray selected clusters and groups in Dark Energy Survey data. I. Stellar mass growth of bright central galaxies since z~1.2, *The Astrophysical Journal* **816**, **98**, doi:10.3847/0004-637X/816/2/98.

CHAPTER 18

Quasars

Xin Liu, Paul Martini & Richard McMahon

In this chapter, we discuss work being done in the Dark Energy Survey (DES) to identify and study quasars, which are among the oldest, most powerful and distant objects in the universe. Quasars were first detected in the 1940s and named quasi-stellar radio sources due to their star-like appearance and strong radio emission. They are one of a class of objects called 'active galactic nuclei' or AGN. The quasar group in DES work with OzDES (based at the Anglo-Australian Telescope) to estimate the masses of the black holes at the centre of galaxies, using a technique called reverberation mapping.

18.1 Inescapable

Dark Matter. Dark Energy. Cosmologists use the term dark to describe something that does not emit light, or that simply escapes understanding. Quasars are ultimately powered by another type of 'darkness:' black holes. Black holes are not just dark but literally black, as nothing can escape the event horizon. Yet unlike dark matter or dark energy, some black holes are incredibly easy to find. This is because some black holes are accreting enormous quantities of gas, dust and even shredded stars from their surroundings, and in the process this accreted material is heated to extraordinary temperatures and emits extraordinary amounts of light as thermal radiation. In these cases, the central region outshines the rest of the combined luminosity of the stars in the galaxy with emission across all wavelengths.

Quasars, also known as active galactic nuclei (AGN),[1] are the most extreme, most luminous and most massive of the accreting black holes in

---

[1] Although the terms are generally interchangeable, AGN typically refers to all types of Active Galaxies, of which quasars are the most distant and luminous.

the universe. The black holes at the centers of quasars can range in mass from a million to a billion times the mass of our Sun, or one solar mass, and are appropriately enough termed 'supermassive' black holes (SMBH). To emit so much light they may accrete many solar masses of material each year.

18.2 Evolution of quasars

Quasars are fascinating because of their extreme luminosities and their extreme masses, but the quasar population has also changed in remarkable ways over cosmic time. Quasars are quite rare in the local universe, where they are associated with only a very small fraction of all galaxies. But long ago, when the universe was about a quarter of its current age of 13.8 billion years, quasars were at least a thousand times more common. And even further back, when the universe was within only a few billion years of the Big Bang, quasars are again observed to be rare. What drives this remarkable evolution? Most likely the answer is connected to how galaxies form, evolve and interact with one another. An important clue to this puzzle is that there is very good evidence that all massive galaxies in the local universe have an SMBH at their centres, although usually a quiescent one that is only detectable due to its gravitational influence on the surrounding galaxy.

A lot of our detailed knowledge of quasar evolution is from the Sloan Digital Sky Survey (SDSS), which discovered hundreds of thousands of quasars in the course of an imaging and spectroscopic survey across much of the northern sky. One particularly exciting discovery from SDSS was quasars at the enormous redshifts of six and above summarized in Fan et al. (2006). These discoveries showed that SMBHs with masses as large as billions of solar masses already existed less than a billion years after the Big Bang. In addition, the spectra of these quasars show a pronounced increase in the amount of intervening absorption by the intergalactic medium (IGM) at redshifts above six. Such IGM absorption is seen at all redshifts, and is due to clouds of neutral Hydrogen gas that absorb flux at specific wavelengths that correspond to electron transitions in the Hydrogen atom. The strongest is the Lyman $-\alpha$ transition at 121.6 nm. While there are clouds of neutral Hydrogen gas scattered throughout the IGM, they are quite rare, as most of the Hydrogen in the IGM is ionized and thus does not absorb. The striking thing about redshifts of six and above is that there is a very pronounced increase in the amount of absorption, which is tantalizing evidence that we are seeing the beginning of the transition from a mostly ionized universe to a mostly neutral universe. These striking results from SDSS have raised lots of new questions that only a deeper survey like the Dark Energy Survey (DES) could answer.

18.3 Enter DES

Paul Martini and Richard McMahon joined DES in 2007. Paul had been working on trying to figure out the typical lifetime of quasars. This was a big challenge because all of the evidence suggested they remained active for a million years and Paul was not optimistic he'd be able to stick around long enough to watch the whole process. He was therefore working on various indirect ways to make estimates with demographic studies of quasars, and the unprecedented area and depth of DES was very exciting. He was also intrigued by the existence of billion-solar mass black holes in quasars above $z > 6$, which might have required continual growth for hundreds of millions of years to grow so large. Richard had recently initiated the VISTA Hemisphere Survey (VHS), an ambitious program to image a large area of the southern sky at near-infrared wavelengths. Richard's focus was very high-redshift quasars, particularly quasars at even higher redshifts than the SDSS record of $z > 6.4$.

DES was the perfect complement to VHS, as it would provide shorter-wavelength data for the same region of the sky. This was important because the main way to detect very high-redshift quasars was to take advantage of all of the neutral absorption in the IGM, absorption that erased essentially all of a quasar's flux below the observed wavelength of the Lyman $-\alpha$ line at 121.6 nm. As observed wavelength increases by a factor of $(1 + z)$, a quasar at $z = 6$ will emit essentially no light below about 850 nm. Consequently it will be invisible in all but the reddest (z and Y) DES filters. Almost no other objects in the universe exhibit such a stark change in brightness at these wavelengths, so this is a great way to find high-redshift quasars. The only problem is that these quasars are still amazingly rare, with only one or fewer expected per ten square degrees, even at the extraordinary sensitivity limit of DES. In contrast, there are tens of thousands of other objects in the same area of the sky, such that even a small amount of contamination from very red stars (or asteroids or detector defects) could look enough like a quasar that we could waste precious hours of telescope time trying to get a confirmation spectrum.

18.4 The plan

We (Richard and Paul) were asked to activate the 'DES Quasar Working Group' in early 2008. In doing so, we considered what quasar science could be done with DES as well as organizing the efforts of other scientists in the collaboration who were similarly interested in quasars. The ultimate purpose of this exercise was to try to maximize the scientific return of DES. While the driving purpose of DES was dark energy, it was clear to everyone that this survey would enable much more. We consequently did a lot of brainstorming over the next few years and organized a lot of meetings with other DES scientists. We ultimately reaffirmed the unique strength of DES to help identify new quasars at redshift six, seven and even higher, and that this population would be invaluable to study the IGM, the masses of the earliest SMBHs and perhaps tell us something about the lifetime of quasar activity.

The next step was to plan spectroscopic observations. All of the science we had identified required at least a spectrum to confirm that a faint blip in our images was actually a quasar, and higher quality spectra were needed for studies of absorption in the IGM and for estimates of black hole masses. Thanks to extensive quantitative predictions by Manda Banerji, who at the time was a post-doctoral researcher working with Richard, we had a good estimate of how many quasars we should find and could begin to plan proposals to many (very competitive) large telescopes. Fortunately, our large collaboration helped a lot in this regard, as among the group we had at least the 'right to propose' for telescope time on most of the large telescopes in the world. Paul, as a faculty member at Ohio State, had special access to the twin 8.4-m Large Binocular Telescope (LBT), although unfortunately the LBT is in the wrong hemisphere to observe all but the most northern fields in DES. However, he also has national access to the United States share of the 8-m Gemini Observatory, which has telescopes in both the northern and southern hemispheres. In fact, Gemini South is just a short drive from the Blanco Telescope. Richard has national access to all telescopes that are part of the European Southern Observatory, in particular the four 8-m VLT telescopes in Chile. Other members of the quasar working group had special access to other large telescopes in Chile and Hawaii.

In addition, the quasar working group was not the only group interested in getting as much access to spectroscopy as possible. A very exciting new opportunity arose when Chris Lidman of the Australian Astronomical Observatory expressed interested in proposing for a major spectroscopic survey of the DES supernova fields, using Australia's premier 4-m telescope. The proposal was to monitor all ten DES supernova fields about once a month during the DES observation season to measure redshifts for as many supernovae as possible, or their host galaxies after the supernovae had faded. These observations would use the amazing two-degree fibre positioner, which could take spectra of 392 objects at a time in its two-degree field of view. The exciting opportunity for the quasar working group was that many of these fibres would not be needed for supernova observations, as supernovae are somewhat rare. Paul realized that this proposal was also an excellent opportunity to conduct an unprecedented monitoring campaign of hundreds of the quasars in these fields to measure their black hole masses via reverberation mapping.

18.5 Reverberation mapping

Reverberation mapping is a novel method to measure the masses of SMBHs in quasars through time-domain observations. The method relies on the variability of the quasar accretion rate, which manifests as variability in the amount of high-energy radiation emitted by the accretion disc in the immediate environment of the black hole. The accretion disc is the disc of hot gas around a black hole. The continuum (continuous spectrum) emission of this disc in turn excites the atoms in clouds of gas on somewhat larger scales, and produces emission in discrete lines such as the Balmer series of Hydrogen, as well as lines further in the ultraviolet such as singly ionized Magnesium and triply ionized Carbon. These clouds of gas reverberate in response to variations in the continuum emission. The clouds of line-emitting gas are moving at thousands of kilometers per second and they are consequently called the broad line region (BLR) because these Doppler motions substantially broaden the lines. The observed velocity for the BLR around a given black hole is ultimately determined by the black hole mass and the average distance of the clouds via a relationship known as the virial theorem:

$$M = \frac{f R_{BLR} dv^2}{G} \tag{18.1}$$

The quantity f is a geometric parameter, R_{BLR} is the radius of the BLR, G is the gravitational constant and dv is the characteristic velocity. The key to reverberation mapping is that the timescale for the BLR to reverberate only depends on the light travel time R_{BLR}/c between the BLR and the source of the continuum emission, which is near the inner edge of the accretion disc and therefore extremely close to the black hole.

Paul was excited by the potential of these new observations. He had participated in a number of previous reverberation mapping projects led by Bradley Peterson at Ohio State, who pioneered the method and led many reverberation studies over a number of decades. The most exciting aspect was that the project could potentially monitor hundreds of luminous quasars at high-redshift and thus measure black holes at the 'quasar epoch' when most black hole growth occurred. This stood in contrast to most previous reverberation studies, which measured fairly low-luminosity quasars in the very local universe. Those studies concentrated on observations of one object at a time, and over the years had measured nearly seventy black hole masses in the nearby universe. The new DES program had the promise to dramatically expand the number of measurements and focus on the relatively unexplored high-redshift universe. The significant challenge was that no one had successfully conducted a reverberation study with a fibre spectrograph up to that point. Extensive simulations by Anthea King, then a PhD student of Tamara Davis at the University of Queensland, predicted we should have a useful yield of lag measurements and we pressed ahead.

18.6 Using OzDES

In Chapter 14, we introduced OzDES, a spectroscopic redshift follow up survey of DES targets, based at the Anglo-Australian Telescope (AAT).

Although the main science motivation of OzDES is to obtain host galaxy spectroscopic redshifts for a sample of 2500 Type Ia supernovae discovered by DES, another primary goal is the repeated reverberation mapping of 771 AGN.

The emission from the accretion disc, or continuum emission, is measured using the DES photometry, which is taken on an approximately weekly basis. Some of this emission comes directly to us but some of it ionizes the gas forming the broad line emission. Since the gas has been ionized, we observe emission lines, in particular Carbon-IV, Magnesium-II and Hydrogen-β. The radiation with emission lines will be observed after the continuum emission because it has to travel a longer distance as shown schematically in Figure 18.1. The changes in these emission lines are observed approximately once a month in the OzDES spectra. By comparing the continuum light curves obtained by DES and the emission-line light curves measured with OzDES, it is possible to determine the time delay between the two signals and this tells us how far away the BLR clouds are orbiting the black hole. As the emission lines are emitted while the gas is orbiting the black hole, the lines are broadened due to relativistic effects and so we can measure the velocity of the clouds by looking at the widths

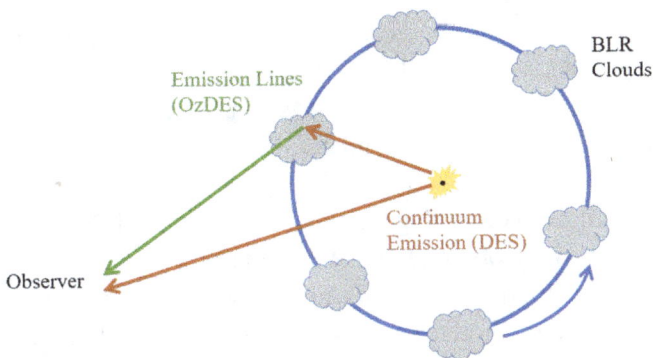

Fig. 18.1 A cartoon picture of an active galactic nuclei (AGN) with a black hole at its centre. A flash of light from the accretion disc can either travel directly to the observer (orange line) or less directly via the clouds in the broad line region (BLR) (orange and green lines). The difference in the path lengths between the two routes gives an idea of the size of the BLR. The size, together with an estimate of the velocity at which the clouds move around the black hole, can be used to estimate the mass of the black hole.

of these lines. Together, the velocity of the clouds and the radius at which they orbit can then be used to measure the mass of the SMBH.

Simulations show (King et al., 2015) that we can expect to recover time lags and in turn measure black hole masses for around 30% to 40% of the 771 AGN targets (most $z > 0.8$), which will result in around 250 new mass measurements! With this significantly improved data set, it will be possible to address a number of pressing questions, such as how SMBHs got to be so massive in the first place. OzDES is not only more than quadrupling the number of black holes for which we have masses, it is also pushing these masses to previously unmeasured redshifts. This will allow us to study how SMBHs have evolved over the past twelve billion years.

18.7 Discovery

DES began Year 1 observations in the second half of 2013, and after heroic effort the images and catalogues became available to the collaboration in 2014. Richard and his PhD student Sophie Reed immediately began combing through these data, identified many artifacts in the data processing, but also found a very promising $z > 6$ quasar candidate. Richard contacted Michael Rauch, a colleague who happened to be observing at the Magellan Clay telescope, who quickly confirmed that the candidate was really a quasar $z > 6.1$ (see Figure 18.2). Sophie wrote a paper on this discovery shortly thereafter (Reed et al., 2015), and proceeded to carefully comb the data and identify more candidates. Richard obtained observing time with the 3.6-m New Technology Telescope to observe these new targets, while Paul obtained observing time with the 8-m Gemini South telescope. The success of these observations led Sophie to write a second paper that announced the discovery of eight more $z > 6$ quasars, as well as describe her sophisticated method to identify candidates with multi-wavelength fits of the data to quasar models.

Sophie also found some interesting results on the size of the quasar 'near zone', which is the region immediately surrounding the quasar that it has completely ionized. This region will be larger if the IGM around the quasar is already mostly ionized, as it is then easier for the quasar to ionize the remaining material. Two of the highest redshift quasars in Sophie's

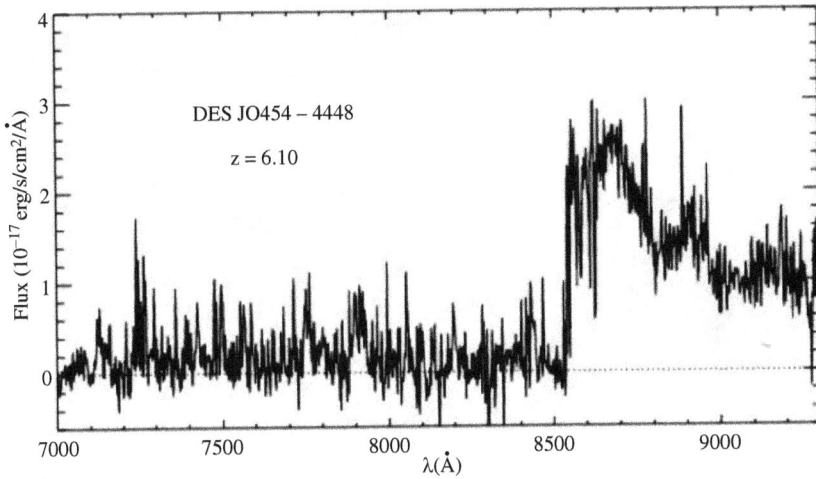

Fig. 18.2 First redshift $z > 6$ quasar discovered by the Dark Energy Survey (DES) and published as Reed et al. (2015).

larger sample have unusually small near zones, which may be evidence that the universe is more neutral at those redshifts, although the small near zones could also be because the quasars are relatively young. This topic will be an important part of our analysis as we identify higher redshift quasars in larger numbers.

The first two years of DES observations also led to a flurry of activity to set up the reverberation mapping project. Manda Banerji and Paul used the DES data to identify quasar candidates in the supernova fields and request spectroscopy from the OzDES collaboration. Over this time, we built up a sample of nearly 1500 quasars in these fields and ultimately selected nearly 800 for regular spectroscopic observations. One big surprise at this time was the serendipitous discovery of a 'Broad Absorption Line' (or BAL) quasar with a post-starburst spectrum (shown in Figure 18.3). These objects simultaneously show evidence for a quasar as well as a luminous and massive moderately-aged galaxy, which typically had intense star formation that ended 10^8 to 10^9 years ago.

We had planned to monitor a few BALs to look for changes in their BALs, but the one-of-a-kind post-starburst aspect was only later noticed by Chris Lidman. This object is extremely interesting because the age of the post-starburst population is relatively straightforward to calculate, and

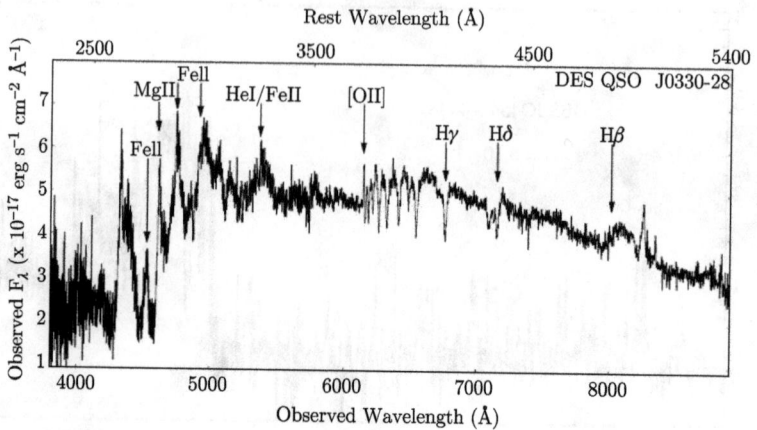

Fig. 18.3 Stacked spectrum of the Dark Energy Survey (DES) quasi-stellar object (QSO) J0330-28 at $z = 0.65$. The LoBAL features are prominent at wavelengths shorter than the MgII line at rest-frame 2798 Å. The absorption features around rest-frame 3900 Å are from host galaxy stars. *Source:* From Mudd et al. (2017).

provides a potential age for the quasar. Paul's PhD student Dale Mudd wrote up a paper to announce this discovery (Mudd et al., 2017).

Chris Lidman and others in Australia also put a substantial amount of development work into improvements to the OzDES pipeline, which have dramatically improved the calibration of the data, while Anthea King and Dale Mudd joined an effort lead by Gisella de Rosa at the Space Telescope Science Institute to produce a code for automatic measurement of the broad emission lines. By the end of year four (second half of 2016) of DES and OzDES observations, we had obtained between fifteen and twenty spectroscopic observations of most quasars, as well as nearly a hundred photometric observations, and anticipate we now have enough data to measure reverberation lags for a subset of our data. In the meantime, Natalia Somer and Janie Hoormann have joined the analysis effort in Australia. As we have gradually accumulated enough observations to measure emission-line reverberation lags, we have also calculated the relative lags of the continuum emission as a function of wavelength. These calculations can provide measurements of the sizes of quasar accretion discs because longer-wavelength continuum emission generally originates from larger distances from the central black hole. Dale Mudd has led this work and measured accretion disc sizes for about a dozen quasars, and the results are very interesting. He has found good evidence that quasar accretion discs are up to several times larger than expected, a result that should motivate further work on accretion disc theory.

18.8 Future

The boundary between quasars and their host galaxies is not particularly well defined, and this partly motivated the merger of the quasar and galaxy evolution working groups in 2016. This has helped to move a lot of related research together into one group. Some examples of this work include a study of active galactic nuclei (AGN) in cluster galaxies led by Tesla Jeltema and her students Erica Bufanda and Devon Hollowood. Another is work led by Manda Banerji and her PhD student Clare Wethers at Cambridge on the host galaxies of heavily obscured quasars.

As the next seasons of DES data are processed, we will continue to look for more $z > 6$ and the first $z > 7$ quasars, look forward to the first black hole mass measurements from reverberation mapping, and to lots of additional, novel uses of the DES data, such as recent work by Nick Rumbaugh, Yue Shen and others on quasars that exhibit extreme variability.

References

Fan, X., Strauss, M. A., Richards, G. T., Hennawi, J. F., Becker, R. H., et al. (2006). A Survey of z>5.7 Quasars in the Sloan Digital Sky Survey. IV. Discovery of Seven Additional Quasars, *The Astronomical Journal* **131**, pp. 1203–1209, doi:10.1086/500296.

King, A. L., Martini, P., Davis, T. M., Denney, K. D., Kochanek, C. S., et al. (2015). Simulations of the OzDES AGN reverberation mapping project, *Monthly Notices of the Royal Astronomical Society* **453**, 2, pp. 1701–1726, doi:10.1093/mnras/stv1718.

Mudd, D., Martini, P., Tie, S. S., Lidman, C., McMahon, R., et al. (2017). Discovery of a z = 0.65 post-starburst BAL quasar in the DES supernova fields, *Monthly Notices of the Royal Astronomical Society* **468**, pp. 3682–3688, doi:10.1093/mnras/stx708.

Reed, S. L., McMahon, R. G., Banerji, M., Becker, G. D., Gonzalez-Solares, E., et al. (2015). DES J0454-4448: Discovery of the first luminous z ≥ 6 quasar from the Dark Energy Survey, *Monthly Notices of the Royal Astronomical Society* **454**, pp. 3952–3961, doi:10.1093/mnras/stv2031.

CHAPTER 19

Strong Gravitational Lensing

*Adam Amara, Elizabeth Buckley-Geer, Martin Makler &
Brian Nord*

Strong gravitational lensing is an effect that gives rise to extraordinary images. It is also a powerful tool with which to study objects in the universe and measure cosmological parameters. In this chapter, we describe how strong lensing works and what we can learn from it. We discuss how we find strong gravitational lenses in the Dark Energy Survey (DES), the lensing systems we have already discovered and the advances we are making in strong lensing science.

19.1 Strong gravitational lensing

Light is deflected by gravity when it passes close to a massive object. Einstein made the first accurate prediction of this effect using his general theory of relativity, which was published in 1915. A source of gravity that can distort a background image is called a gravitational lens, by analogy with an optical lens. Chapter 11 discusses 'weak' gravitational lensing, where the bending of light is hardly perceptible, but we can also learn a lot from studying systems where the effect of the lensing is strong enough to be observed directly on an astronomical image. This only happens when the intervening object, such as a galaxy or a cluster of galaxies, is massive enough to produce a very strong gravitational field, and is lined up almost perfectly between Earth and distant galaxy. This configuration is quite rare.

See Figure 19.1 for a schematic of the situation. Light from a distant galaxy curves around the gravitational lens and travels along more than one path to Earth. There are three effects characteristic of strong

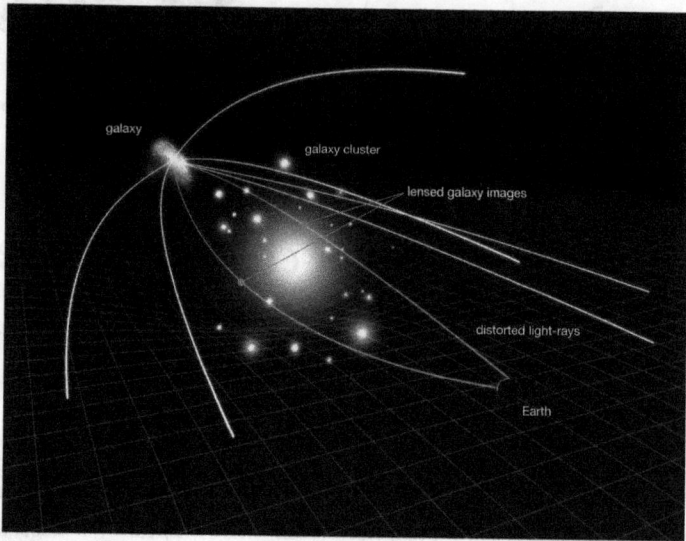

Fig. 19.1 Light travelling from a distant galaxy is lensed by a massive foreground galaxy or galaxy cluster which is closer to Earth. Credit: NASA, ESA & L. Calçada.

lensing: multiple images, strong distortions and magnifications and time delays between images. The number of images observed and the amount they are distorted depend both on the mass distribution of the lens and on the relative positions of the source and the gravitational lens and the observer. Typically, a giant cluster of red galaxies acts to distort a distant blue galaxy into a number of blue arcs. Researchers found the first giant arc in 1986, in the Abell 370 galaxy cluster. If the source, lens and observer are exactly aligned and the mass distribution of the lens is axially symmetric,[1] light from the distant galaxy is deformed into a ring shape called an Einstein ring, with a radius called the Einstein radius. The first complete Einstein ring was discovered in 1988 in the radio wave part of the electromagnetic spectrum (Hewitt et al., 1988). The first complete Einstein ring in the infrared was discovered in 1998 (King et al., 1998).

---

[1] For finite sources, the requirement is not so strict. As long as the source is large enough to encompass the caustic it will produce a ring that is pretty much circular (even if the lens is elliptical and source–lens–observer are not perfectly aligned).

Figure 19.2 shows an image of a nearly complete Einstein ring found in the Dark Energy Survey (DES) footprint, called the Elliot arc (further described later). If the distant object is a quasar, then usually two or four images of this light source appear around the foreground lensing galaxy – the latter configuration of four is known as an Einstein cross. The current picture of a quasar is a supermassive black hole, which is pulling in gas at the centre of a galaxy. The in-falling gas heats up and emits huge amounts of radiation, making quasars incredibly luminous and visible from extreme distances.

General relativity describes gravity entirely in terms of the geometry of space and time, or spacetime, so the bending of light does not depend on the energy (i.e. the wavelength of the photons). This means that, unlike refraction of light, gravitational lensing of light is an achromatic, or colour-independent, effect: blue light is bent the same as red light, for example.

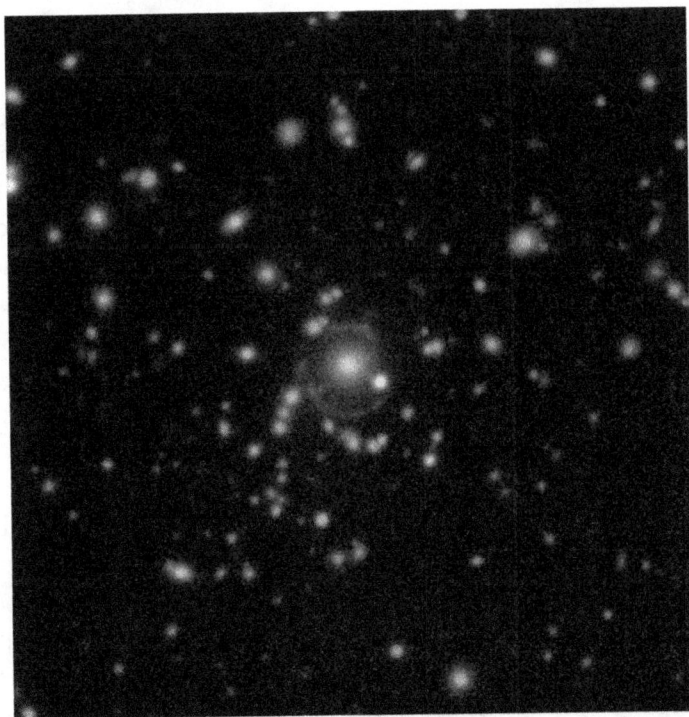

Fig. 19.2 Image taken in the *g, r* and *i* band filters of the Elliot arc and its cluster environment. The image size is 2 × 2 arcminutes.

Strong lensing typically happens if the lens has a surface mass density, or mass per square centimetre, larger than some critical value. Once the surface mass density falls below this critical value, it is usually no longer possible to produce the multiple images characteristic of strong lensing. At surface mass densities well below the critical value, we can no longer measure the precise shape of an individual object, so we make a statistical measurement of an ensemble of images of background galaxies, each of which has been slightly stretched, or sheared. This is the regime of weak lensing (see Chapter 11), but there is no precise dividing line between weak lensing and strong lensing. Strong lensing effects can sometimes still occur for densities below the critical, due to the effects of the shear.

19.2 What can we learn from strong gravitational lensing?

The amount of distortion of the image of a source object is directly related to how much mass is in the gravitational lens and how this mass is distributed, so we can use strong lensing to measure the amount of matter in large galaxies or clusters of galaxies. Since most of this matter is invisible dark matter, clumped around and between the luminous matter, strong lensing gives scientists a way to directly measure the amount of dark matter in a galaxy or cluster. Strong gravitational lenses are selected entirely on the basis of how much mass they contain (both luminous and dark), not on how bright they are or whether they are emitting any X-rays (which involves making assumptions). Therefore, strong lensing doesn't have some of the biases that other selection techniques for measuring mass have.

Once we have measured the distribution of matter, or 'density profile', in the main bulk of a galaxy cluster using strong lensing, weak lensing analyses can be used to study the distribution of matter in the cluster at greater distances from the centre. Combining weak lensing with strong lensing gives us a much more complete picture of the density profiles of clusters. We can use this information to test models of the growth of structure over time, which can tell us more about dark energy.

The light rays from the lensed images of the source take longer to reach the observer than they would if there were no lens present, causing a time delay in the arrival of the light from different images. There is an additional time delay caused by the passage of the light through the gravitational potential of the lens. If the light emitted by the source is varying over time, then we can measure this time delay, which can vary from days to months. The length of the time delay depends on the relative positions (angular and in redshift) of source, lens and observer, and on the lens mass distribution. Quasars are one such light source. Many quasars are variable in their light output, so there is a time delay in the appearance of the images. Studies of this time delay can be used to measure the Hubble constant H_0, because the time delay between different light paths is proportional to H_0. Strongly lensed supernovae (incredibly bright exploding stars) can also be used to accurately estimate the time delays between images. The first strongly lensed supernova was detected in 2014 by Quimby et al. (2014). Since then eight others have been found.

In rare cases, a strong lensing system contains a single large galaxy lens and multiple sources, each at a different redshift. These 'double source plane' lens systems can be used to determine the dark energy density Ω_Λ, dark matter density Ω_M and dark energy equation of state w (the ratio of the dark energy pressure to its density) by measuring the ratio of the angular diameter distances of the different sources. You only want one lens object because that is the simplest potential. If the lens is a galaxy cluster, the model becomes much more complicated and it is harder to make the measurements.

Beyond the measurement of cosmological parameters, such as the dark energy equation of state and the Hubble constant, strong gravitational lensing gives scientists the opportunity to learn more about galaxy evolution, the substructure within dark matter and the properties of the interstellar medium. Gravitational lensing magnifies light from the source, allowing us to see images of high redshift galaxies that would otherwise be too faint to observe. The surface brightness of the source is conserved when an object is lensed, even though the image is distorted, so the same amount of flux is spread over a different area. This means the source becomes brighter. Some objects can be brightened up tens to hundreds of times, depending on where they sit in relation to the lens. If you study galaxy evolution, it's a great way to investigate high redshift sources that would otherwise be too faint to observe unless you had access to a large 10-m telescope. And since lensing increases the apparent angular size of the source, it enables the study of source galaxy substructure with greater effective spatial resolution than for an unlensed source of the same physical size at comparable distance.

19.3 Strong lensing in DES

19.3.1 Identifying strong lenses

Before we can use strong lensing systems for science, we first have to find them, and this turns out to be the hardest part of the process. In DES, we use a combination of automated arc finders, catalogue searches and crowdsourced citizen science efforts. All of these searches involve some amount of visual scanning as the final step in the process, meaning someone has to physically look at each image. A trained human can identify the morphological and photometric features that imply a lensing configuration better than any computer algorithm so far, but it takes a long time to search through the images by eye.

Most of the gravitational lenses we have discovered are 'red and dead' (no longer forming stars) elliptical galaxies at a redshift of about 0.4 to 0.6. The source can be any object emitting light behind the lens. Since greater redshift means further back in time and there are more star-forming galaxies further back in time, a lot of the source objects are bright blue, star-forming galaxies. This results in striking images of blue, stretched-out arcs around bright red lenses.

Before DES, the total number of strongly lensed systems known worldwide was about 1000. Based on simulations (Collett, 2015), we predict that by the end of DES we will have found of order 2000 new strongly lensed systems, thus doubling the pre-DES number. However, there is only a certain amount of image resolution in DES and many of the lensing systems have small Einstein radii, which makes them harder to find. The 'blue around red' search involves finding blue peaks on the images, but if two objects are too close together, they become blended in the data and do not appear as separate peaks. This means finding them by searching through the catalogues is difficult.

Gravitational lenses are relatively rare so we look for them in particular samples of images, not in random pictures of a piece of sky. Discovering a gravitational lens simply by looking at the pictures as they come in off the telescope is very unlikely, but it can happen. Before DES began, some of the same people now in the strong lensing working group were involved with a small (80 square degrees) precursor survey called the Blanco Cosmology Survey. It imaged a couple of patches of sky within the DES footprint, using the predecessor camera to the Dark Energy Camera (DEcam), the Mosaic II. Liz Buckley-Geer (Fermilab) travelled to Chile to do some observing for this survey at the Blanco Telescope and she recalls:

At 4 a.m. I was sitting there looking at the pictures coming off the telescope. Huan Lin was with me and I was saying, 'Look, there's a nice little spiral galaxy', and he was clearly getting a bit bored because he's a real astronomer and he's seen all these things before. But then I looked at this one picture and said, 'There's an interesting little fuzzy blob down there'. So he loaded the image onto his laptop and blew it up a few times and we looked at it and he said, 'I think you just found a gravitational lens'! It was nearly a complete Einstein ring. We observed it again in DES and took follow-up spectra and confirmed it was a lens. Finding one this way was beginner's luck. I named it after my nephew, Elliot.

See Figure 19.2 for a picture of the Elliot arc. The full report on the serendipitous discovery is in Buckley-Geer et al. (2011).

One good way to find lenses is to look at all the galaxy clusters found in DES using redMaPPer (see Chapter 12). Clusters make good lenses because they are very massive, so you can typically find giant arcs around them. Since the number of lensed arcs is a function of the cluster mass, cluster lensing can be used to explore the mass distribution in clusters. However, complicated systems with multiple galaxies can be challenging to model. It requires a fair amount of fine-tuning to understand where you are in the parameter space and how the system is behaving, so lens modelling is something of an art.

RedMaPPer found about 7500 galaxy clusters in the Year 1 data and scanning all of these images by eye is a time-consuming process. To help us, we work with students from the Illinois Mathematics and Science Academy (IMSA) in Aurora, west of Chicago, which takes academically bright Illinois students for the last three years of high school. Figure 19.3 shows an example of one of the lenses selected from this search. It was confirmed using spectra taken by the Gemini South telescope in Chile.

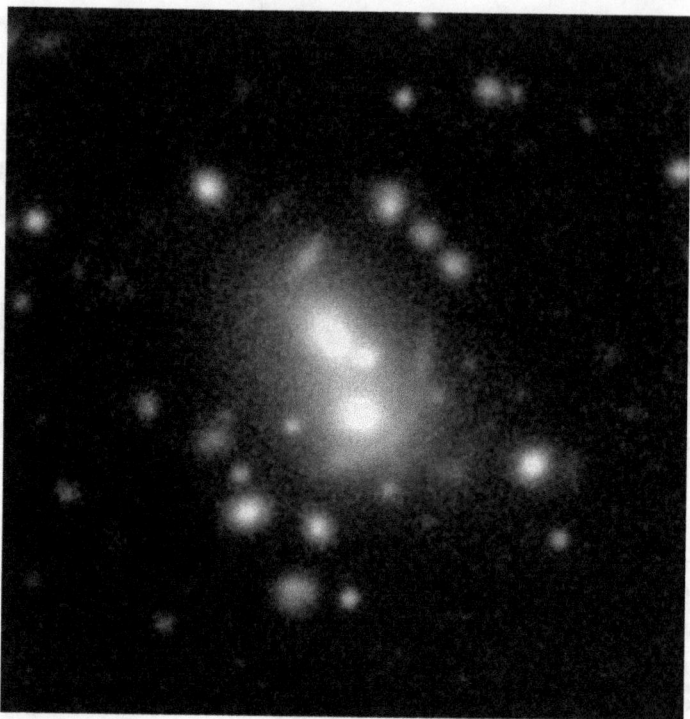

Fig. 19.3 This is an excellent example of gravitational lensing by a cluster of galaxies. The image size is 1 × 1 arcminute. Credit: DES collaboration.

We are also collaborating with the Space Warps project,[2] which is part of Zooniverse and uses crowdsourcing to do the visual inspection. Volunteers join online and classify images via a web-based interface. For example, Tom Collett (University of Portsmouth) is interested in asking for help on Space Warps to find double source plane lenses, where there are two distant sources being lensed. If you take a galaxy image and make a model of that galaxy and then subtract the flux of the image, any residual flux left in the image might be a lensed galaxy. He wants to look for blue residuals around red lensing galaxies, then put the images into Space Warps and ask people to identify them. Citizen science is increasingly useful now that imaging surveys are getting larger. As well as enabling lens discovery,

---

[2]spacewarps.org

it provides large samples of training sets for testing automated lens-finding algorithms. Engaging non-specialist volunteers in science projects gives members of the public a chance to interact with scientists, helps build public understanding of how science works and encourages young students to choose science as a career.

Scientists are also developing automated systems to select the images, based on neural networks and machine learning techniques. Neural networks are made of artificial neurons (filters) that process input image data and can 'learn' important features in the data over time, such as rings around galaxies or arc-like features. Someone still has to visually inspect the selected images, but these kinds of techniques will be critical for future surveys like Large Synoptic Survey Telescope (LSST), which is even larger than DES.

People can make mistakes, of course. An arc caused by lensing can be mistaken for a simple un-lensed galaxy, or vice versa. The only way to really know whether the object you're looking at is a lens is to take the spectra of light from the system, to prove that the source galaxy is more distant than the lensing galaxy. The redshifts of the objects are measured from the spectra and from these we can work out the distances with high accuracy. Once we have visually scanned huge numbers of DES images and identified candidates, we use other telescopes to measure the redshifts of the lensing and source galaxies.

Using data from the science verification season (which covered 250 square degrees of the sky), we identified fifty-three candidates. We then obtained spectra for twenty-one of these candidates using the Gemini Multi-Object Spectrograph (GMOS) at the Gemini South telescope and the Inamori-Magellan Areal Camera and Spectrograph (IMACS) at the Magellan/Baade telescope. Six of the systems were confirmed to be strong gravitational lenses (see Nord et al., 2016). The process is very time consuming – confirming these six systems took more than ten hours of telescope time – and there isn't currently enough telescope time in the world to follow up and confirm thousands of lenses.

The Year 1 search program covered 1500 square degrees of sky and H. Thomas Diehl (Fermilab) did a catalogue search of this data, looking for blue objects near red objects. Then a group of high school students from IMSA, Phillips Academy (in Andover, MA) and other local Illinois high schools inspected them all, yielding a list of about 400 medium- to high-quality candidate lensing systems, with typical Einstein radii of a couple of arcseconds to about ten arcseconds (Diehl et al., 2017). From these we have so far found nine spectroscopically confirmed systems (Nord et al., 2019, submitted). We are trying to find the systems that are most useful for cosmology (i.e. measuring large scale properties of the universe), such as the quasars and the double source plane systems. Most of the objects in our list of 400 don't fall into that category, but they will be useful to someone who wants to study dark matter, or early galaxies. When we publish the list of candidates, we are essentially saying to the astronomy community: 'Here are 400 interesting objects, go play. Pick your favourite ones and follow them up'.

19.3.2 Science!

We can use our sample of strongly lensed systems to study the spatial mass distribution of dark matter on scales from individual galaxies all the way to massive clusters. Cluster-scale lenses are particularly useful because, as explained earlier, they allow us to study the effects of strong lensing in the core of the cluster and weak lensing in the outer regions. Strong lensing provides constraints on the mass contained within the Einstein radius of the lensed arcs, whereas weak lensing provides information on the mass profiles in the outer reaches of the cluster. Combining the two measurements allows us to make tighter constraints on the mass and concentration of a common model of the cluster mass density profile, over a wider range of radii than would be possible with either method alone. In addition, if one has spectroscopic redshifts for the member galaxies, one can determine the cluster velocity dispersion, assuming the cluster is virialized, and hence obtain an independent estimate for the cluster mass. These different methods, strong plus weak lensing and cluster velocity dispersion, provide independent estimates of the cluster mass and can then be combined to obtain improved constraints on the mass and concentration.

DES Year 1 observations covered only about a third of the full survey. When we analyze the full six years of data, we will have an even larger sample of sources at varying redshifts, which will be very valuable for studies of galaxy evolution. The excellent red sensitivity of the DECam charge-coupled devices (CCDs), along with the *grizY* filter set, provides sensitivity to high redshift 'Lyman-break' galaxies.

We expect to find about ten double source plane systems in DES, where one galaxy is lensing two sources in the background. It's not easy to find these systems because the source objects have fairly small Einstein radii, so we are probably not going to see them as two separate sources in DES; they will be blended together. The systems that are currently known to exist are all from the Sloan Digital Sky Survey and they were only identified as double source planes by the Advanced Camera for Surveys (ACS) aboard the Hubble Space Telescope. The Sloan Lens ACS (SLACS) took all the Sloan spectra and looked for additional emission lines in the spectra of red galaxies that might indicate there was something else in the image that you couldn't necessarily resolve but could see in the spectra. That enabled them to find a lot of lenses where there is an elliptical galaxy with something in the background, and out of about a hundred systems, there were three that had a second source (Gavazzi et al., 2008).

Liz Buckley-Geer (Fermilab) and Martin Makler (Brazilian Center for Physics Research, CBPF) were the first co-conveners of the Strong Lensing science working group (SLWG). The working group was created in 2009, from a study group formed in 2008. In 2014, Martin was succeeded by Adam Amara (ETH Zurich). In 2016, Liz stepped down after co-leading the group for six years and Brian Nord (Fermilab) took her place. There are about ten active members of the strong lensing group at any one time. Quite a few of the group, including also H. Thomas Diehl and Huan Lin, work at Fermilab (the Fermi National Accelerator Laboratory, near Chicago) and became involved with DES early on, as the Sloan Digital Sky Survey was coming to an end. Along with the Fermilab participants, members of the CBPF group in Brazil have also made a substantial contribution to the SLWG, working on forecasts, simulations, automated arc finding and the eyeballing effort, among other areas. We also have an external collaboration led by Tommaso Treu (University of California, Los Angeles), called 'STRIDES' (STRong-lensing Insights into the Dark Energy Survey). This collaboration, which consists of experts in the field of measuring the Hubble constant from lensed quasars, aims to find approximately hundred strong lensed quasars and determine their time delays in order to measure the H_0 and other cosmological parameters.

Lensed quasars can be identified by their variable nature, and by using colour and morphology information from the catalogues. As of summer 2017, we have confirmed five quadruply lensed quasars and about twenty doubles in DES. Figure 19.4 shows the first quadruply lensed quasar discovered in DES (Lin et al., 2017). A measurement of the Hubble parameter using these systems is completely independent of the measurement made using studies of the cosmological background radiation by the Planck space observatory (Planck collaboration 2016), and it is also independent of measurements from supernovae and Cepheid variable stars (Riess et al., 2016). At the moment, there is a tension between these latter methods (about 7.5%). The measurements from the three lensed quasar systems that are currently being studied by STRIDES are much more consistent with the supernovae measurement. However, there could be unknown uncertainties in all of these measurements, so we want to find more of these lensed quasar systems to make our estimation more precise. The more different ways you have of measuring the Hubble parameter with

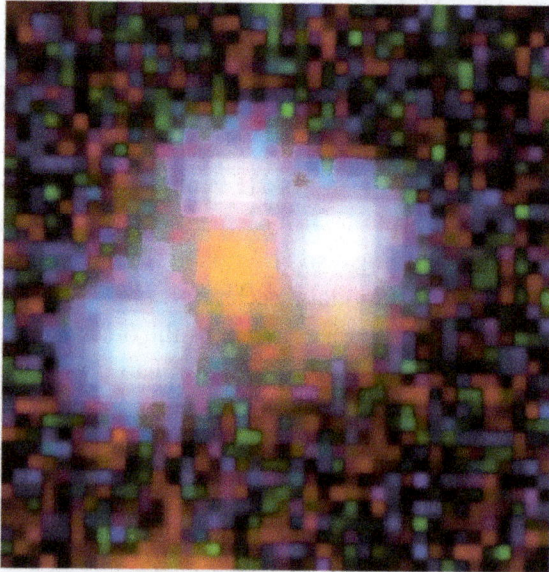

Fig. 19.4 The first quadruply lensed quasar system discovered in the Dark Energy Survey (DES). The bright bluish objects are the lensed quasar, the fourth object at the bottom right is blended with another galaxy that causes its colour to be different. The central lensing galaxy is in the centre of the system in orange. The image is 10 × 10 arcseconds.

precision, the more chance you have of disentangling new physics from unknown systematic uncertainties.

Considerable additional science can also be carried out using these systems. For example, stars in the lens galaxy that cross in front of the source cause microlensing (about one microarcsecond) effects on the multiple quasar images, which can be used to study the stellar contents of the lens galaxy. The same microlensing can also provide constraints on the inner structure of the lensed quasar, such as the size of the accretion disc. In addition, millilensing (about 1 milliarcsecond) caused by flux ratio anomalies probes the presence of substructure in the lens. High-resolution imaging of the lens galaxy allows us to reconstruct the source and can give us a fascinating direct view of how quasars have co-evolved with their host galaxies, looking back ten billion years into the past.

Finally, there is always the chance to discover some weird system we didn't even know existed. Just paging through images, you see the most amazing galaxies out there. Those of us who study strong lensing are fortunate enough to spend a lot of time looking at pictures of the universe. It is an extraordinary place.

References

Buckley-Geer, E. J., Lin, H., Drabek, E. R., Allam, S. S., Tucker, D. L., et al. (2011). The serendipitous observation of a gravitationally lensed galaxy at z = 0.9057 from the Blanco cosmology survey: The Elliot arc, *The Astrophysical Journal* **742**, 48, doi:10.1088/0004-637X/742/1/48.

Collett, T. E. (2015). *The Astrophysical Journal* **811**, 20, The population of galaxy-galaxy strong lenses in forthcoming optical imaging surveys, doi:10.1088/0004-637X/811/1/20.

Diehl, H. T., Buckley-Geer, E. J., Lindgren, K. A., Nord, B., Gaitsch, H., et al. (2017). VizieR Online Data Catalog: Candidate strong lens systems from DES obs. (Diehl+, 2017), *VizieR Online Data Catalog* **223**.

Gavazzi, R., Treu, T., Koopmans, L. V. E., Bolton, A. S., Moustakas, L. A., et al. (2008). The Sloan Lens ACS Survey. VI. Discovery and analysis of a double Einstein ring, *The Astrophysical Journal* **677**, 1046–1059, doi:10.1086/529541.

Hewitt, J. N., Turner, E. L., Schneider, D. P., Burke, B. F., andLangston, G. I. (1988). Unusual radio source MG1131+0456 - A possible Einstein ring, *Nature* **333**, pp. 537–540, doi:10.1038/333537a0.

King, L. J., Jackson, N., Blandford, R. D., Bremer, M. N., Browne, I. W. A., et al. (1998). A complete infrared Einstein ring in the gravitational lens system B1938 + 666, *Monthly Notices of the Royal Astronomical Society* **259**, p. L41, doi:10.1046/j.1365-8711.1998.295241.x.

Lin, H., Buckley-Geer, E., Agnello, A., Ostrovski, F., McMahon, R. G., et al. (2017). Discovery of the lensed quasar system des j0408-5354, *The Astrophysical Journal Letters* **838**, 2, p. L15.

Nord, B., Buckley-Geer, E., Lin, H., Diehl, H. T., Helsby, J., et al. (2016). Observation and confirmation of six strong-lensing systems in the dark energy survey science verification data, *The Astrophysical Journal* **827**, 1, p. 51.

Planck Collaboration (2016). Planck 2015 results. XIII. Cosmological parameters, *Astron-omy and Astrophysics* **594**, A13, doi:10.1051/0004-6361/201525830.

Quimby, R. M., Oguri, M., More, A., More, S., Moriya, T. J., et al. (2014). Detection of the gravitational lens magnifying a Type Ia supernova, *Science* **6182**, pp. 396–399, doi:10.1126/science.1250903.

Riess, A. G., Macri, L. M., Hoffmann, S. L., Scolnic, D., Casertano, S., et al. (2016). A 2.4% determination of the local value of the Hubble constant, *The Astrophysical Journal* **826**, 56, doi:10.3847/0004-637X/826/1/56.

CHAPTER 20

Stellar, Milky Way and Local Group Science

Keith Bechtol, Alex Drlica-Wagner, Ting Li, Jennifer Marshall, Basilio Santiago & Brian Yanny

The Dark Energy Survey (DES) was established to study the properties of hundreds of millions of distant galaxies, in order to learn more about why the expansion of the universe is speeding up. However, it quickly became obvious that the survey could also be used to study objects in and around our home Galaxy, the Milky Way. The Dark Energy Camera (DECam) was the first super-sensitive, wide-field digital camera to operate in the southern hemisphere and this represented an amazing opportunity. Within the first season of observations, scientists using DES data discovered eight new satellite dwarf galaxies of the Milky Way, and by the end of the second year, we had found a total of seventeen dwarf galaxies. In this chapter, we explain how new data is enabling us to characterize the structure, mass assembly and star formation history of our Milky Way Galaxy, and to improve the census of low-mass and low-luminosity stars and sub-stellar objects. DES Milky Way working group attempts to address fundamental questions in Galactic archaeology and near-field cosmology such as what can we learn about the nature of dark matter from its distribution in and around the Milky Way? How was the Milky Way formed? and How do environmental effects influence the formation and evolution of galaxies?

20.1 The Milky Way

The Earth orbits our local star, the Sun, at an average distance defined as one astronomical unit (A.U.) – a useful distance measure for objects within the solar system. At greater distances, astronomers talk in terms of light-years – the distance light travels through empty space in one year – or in terms of parsecs, which are even grander: one parsec (pc) is equal to 3.26 light-years. Most of the stars in the Milky Way occupy a disc-shaped volume that is about 50 kiloparsecs (kpc) in diameter and about 0.6 kpc thick. This is so huge compared to the size of the Earth that until about a hundred years ago most people thought our Galaxy was the entire universe. The first evidence that there were galaxies beyond our own finally came from observations of Andromeda by Vesto Slipher and Edwin Hubble in the early twentieth century. Since we live inside the Milky Way, if we look at the night sky towards the plane of the disc, the stars appear as a hazy (milky) band of light stretching across the sky. Interstellar dust obscures the view in this direction by scattering light at visible wavelengths, but if we observe starlight at near-infrared and radio wavelengths we can see much further, because light at these longer wavelengths can penetrate the dust.

Observations have established that the Milky Way is a spiral Galaxy with four major spiral arms, several short arm segments and a central bar-shaped bulge. The Sun is located about 8.2 kpc from the Galactic centre, just outside a relatively short arm segment called the Orion arm. Perpendicular to the plane of the Galaxy there are relatively fewer stars, which means we have an almost clear view out to the rest of the universe. There are estimated to be up to 400 billion stars in the Milky Way, with the disc dominated by radiation from hot main-sequence stars of spectral class O and B. Stars are grouped according to similarities in their spectra, from hottest to coolest in the sequence OBAFGKMLT, and this sequence is further refined into 'spectral types' by attaching a number from zero (hottest) to nine (coolest). The halo around the flat galactic disc is populated by ancient, post-main-sequence stars, indicating that star formation has ceased in the halo. These stars are said to be 'metal poor', that is, they contain very few elements heavier than hydrogen and helium. Some of them are packed into large, bright 'globular clusters' that are fairly evenly distributed around the halo.

The Milky Way disc can be further differentiated into a 'thin disc', containing 85% of the stars in the Galactic plane, and a 'thick disc', stretching up to 5 kpc into the halo from the plane and composed mainly of older stars. 'Open clusters' are found throughout both parts of the disc. They contain fewer stars than globular clusters and the stars are typically young O and B class stars, with older open clusters generally found in the thick disc. The Pleiades is an example of a young open cluster dominated by hot blue stars that are visible to the naked eye. It is becoming apparent that there is a complex substructure in the Galactic halo, and data from the Dark Energy Survey (DES) is expanding our knowledge of this local environment.

Stars, gas and dust all orbit around the Galactic centre, which is a source of intense radio waves called Sagittarius A* ('A star') since it is located in the direction of the constellation Sagittarius. Hundreds of stars are concentrated in this area and their motions indicate that Sagittarius A* is a supermassive black hole, with an estimated mass 4.1 to 4.5 million times that of the Sun! Gravity holds matter in orbit around the Galactic centre but in the far outer regions of the Galaxy there appears to be much less matter, so we would expect that the orbital speed of the gas and stars would decrease with increasing distance from the centre. Instead, observations show that the speed of Galactic rotation remains uniform well beyond the edge of the most of the Galaxy's visible mass. The implication is that the visible matter is embedded within an enormous halo of unseen material, or 'dark matter', which accounts for about 90% of the Galaxy's total mass and may extend to a distance of 150 kpc from the Galactic centre.

20.1.1 Brown dwarfs in the galactic disc

A sub-stellar object with a mass several times greater than the mass of Jupiter is called a brown dwarf. Like stars, brown dwarfs are thought to form by gravitational contraction of gas, and they have essentially the same chemical composition: three-quarters hydrogen, one-quarter helium and 1% or less metals (elements other than hydrogen or helium). Brown dwarf spectra are defined by the spectral classes L and T (which were created as an extension to the OBAFGKM system), along with Y to classify infrared spectra of cool stars. A large number of sub-stellar objects are necessary to infer the fundamental properties of brown dwarf populations and, unlike normal main-sequence stars, determining the masses of a brown dwarf requires both their luminosity and age to be known.

DES is predicted to observe about 30,000 objects with masses smaller than or equal to 8% of the mass of the Sun. This number is about an order of magnitude larger than the previous census of known objects of sub-stellar mass. Given the extremely low luminosities of these objects, observation of them is largely restricted to within 300 to 600 pc distance from the plane of the Milky Way. It has been challenging to extract the sample of L and T dwarfs from DES data, partly due to these objects being detectable only in the redder filters.

DES records images using a photometric system of five filters, spanning 400 nm to 1080 nm. Each filter is sensitive to a different wavelength band of light: g (green), r (red), i, z and Y (all near-infrared wavelengths). Photometry is the measurement of the flux, or intensity, of electromagnetic radiation. The difference in brightness between two wavelength bands is referred to as 'colour' and on colour–colour diagrams, the colour defined by one brightness difference is plotted on the horizontal axis and then the colour defined by a second brightness difference is plotted on the vertical axis. Since observed stellar colours (unlike apparent magnitudes) are independent of the distance to the star, these diagrams can be used to directly calibrate or test colours and magnitudes in the data. The near-infrared Y-band filter is helpful in distinguishing brown dwarfs from high-redshift galaxies and quasi-stellar object (QSO) populations, specifically based on their positions in $(z - Y)$ versus $(i - z)$ colour–colour diagrams. Data from the VISTA Hemisphere Survey (VHS), a ground-based, wide-field infrared sky survey run by the European Southern Observatory in Chile, can be combined with DES data to enable more accurate identification.

A complementary approach to finding nearby sub-stellar objects is to use proper motion measurements. Proper motion is the observed movement of a star in space relative to the Sun, against the background of the distant 'fixed' objects. A local disc star with a transverse velocity of twenty kilometres per second relative to the local standard of rest, and a distance of fifty parsecs, has a proper motion of eighty milli-arcseconds per year. Given the six-year baseline duration of the DES survey, the resulting displacement of 0.4 arcseconds should be measurable with DES data.

The accuracy of the measurements is limited by atmospheric distortions, but the large number of L and T dwarf candidates provide excellent targets for spectroscopic follow up, which may help to constrain atmospheric models of these objects.

20.1.2 Red dwarfs

Over 70% of the stars observed by DES belong to the thick disc and halo of the Milky Way. Low-mass main-sequence stars dominate the stellar sample and by the end of the survey about seventy-five million of these red dwarfs are expected to be imaged. Red dwarfs are small stars of either K or M spectral type, which is relatively cool. They range in mass from about 0.075 to about 0.5 times the mass of the Sun. The relatively straightforward colour selection for low-mass stars coupled with the large survey footprint of DES is allowing us to study the structure of the Galactic disc with unprecedented detail.

20.1.3 White dwarfs

White dwarfs are the endpoint of stellar evolution for the majority of stars. They are stars that are no longer undergoing nuclear fusion and have shrunk to about the size of Earth. Sufficient compression of a mass comparable to that of the Sun causes matter to become degenerate, that is, electrons are stripped from their atoms and packed very closely together. The pressure of these densely packed electrons prevents the stellar remnant from collapsing completely. White dwarfs are classified as spectral class D (for Degenerate), further divided into spectral types DA, DB, DC, DO, DQ, DX and DZ, depending on the composition of the atmosphere. The faint end of the white dwarf luminosity function (WDLF), that is, the number of white dwarf stars with a given luminosity, is occupied by the oldest members of a given stellar population, which are the furthest along the white dwarf cooling sequence. Therefore, the WDLF of the disc and the halo provides a method of dating the early star formation in each of these Galactic components. Additionally, modelling of the WDLF can be used to constrain the stellar initial mass function and star formation history. White dwarfs are also an important class of objects for photometric calibration. Both DES and other future large-area surveys will benefit from the discovery and characterization of white dwarfs in DES footprint.

20.1.4 Galactic open clusters

Galactic open clusters are unique astrophysical laboratories for understanding stellar evolution because they typically contain stars with a range of different masses, all of which began to form at around the same time from the same parent nebula (a cold, dense interstellar cloud of gas and dust). The age of a cluster can be fairly accurately determined by measuring the apparent magnitude and colour ratio of many of the stars in the cluster and then plotting a colour–magnitude diagram. The age can be found from this diagram by finding the point where stars have exhausted their supplies of hydrogen and have left the main sequence. Their main-sequence lifetimes, estimated from their masses at the turn-off point, are equal to the age of the cluster. Seventeen open clusters are located in DES footprint but only three of these clusters have well-determined physical properties, tending to be relatively nearby (much closer than one kpc), young (much younger than four billion years) and either thin- or thick-disc (redshift smaller than 500 pc) systems. The well-studied Blanco One open cluster lies on the boundary of the footprint, and was surveyed in the (*griz*) filters during the science verification period.

It is relatively straightforward to establish which stars are members of a typical open cluster, such as the rich, nearby (120 pc) and young (125 million years) Pleiades cluster, by using colour–magnitude and colour–colour diagrams. However, based on science verification data, all higher-mass stars cause saturation of the pixels in the DECam CCDs in low-extinction (not much absorption or scattering of the starlight by dust and gas) DES fields out to 3 kpc. At that distance, all cluster G, K and M-type dwarfs are detected, whereas all sub-stellar objects (brown dwarfs) and degenerate compact objects (white dwarfs) are far below the detection threshold.

20.1.5 Variable stars

In addition to a large spatial footprint, DES spans a six-year temporal period. Each of the wide-field pointings (the area of the sky the telescope points at) is observed ten times in each band over the duration of the survey. At the same time, each of the ten DES supernova fields (covering a total of twenty-seven square degrees) is observed regularly over intervals of seven days. This large 'multi-epoch' data set has presented us with an exciting opportunity for discovering temporally variable stars. Variable stars were discovered as long ago as 1595, by the Dutch amateur astronomer David Fabricus, and are known to be post-main-sequence stars that undergo substantial changes in size, alternately growing and diminishing. As these giants pulsate, they also vary dramatically in brightness. Cepheid variables are probably the best–known class of pulsating stars because they can be used to estimate distances to remote galaxies. This is possible for two reasons. First, Cepheids are very luminous which means we can detect them and, second, there is a direct relationship between the time it takes for the Cepheid to pulsate and its average luminosity. DES has discovered a large number of RR Lyrae variables, which have lower mass than Cepheids but are also important 'standard candles' that can be used to estimate astronomical distances. Even the limited number of observations available in the wide-field survey have been sufficient to discover new pulsating variable stars.

20.1.6 Galactic stellar streams

Stellar streams are loose groups or associations of stars originating in globular clusters and small 'dwarf' galaxies that were torn apart by 'tidal' stripping due to the gravitational pull of the Milky Way and stretched along their orbit. The resulting streams of stars can span tens of degrees on the sky. Models of cold stellar streams can tell us more about the total mass distribution, including the substructure of the Milky Way dark matter halo, and also allow us to infer Galactic parameters and the properties of the progenitor system.

Several known tidal streams, including one associated with the Sagittarius dwarf galaxy, extend into the DES footprint. Often, these halo substructures are found by inspecting the spatial distribution of halo tracers, such as horizontal branch, red giant, or halo main-sequence turn-off stars. Data from the first three years of operation have allowed DES to detect more than a dozen new stellar stream candidates. Figure 20.1 shows some of the stellar streams discovered by DES.

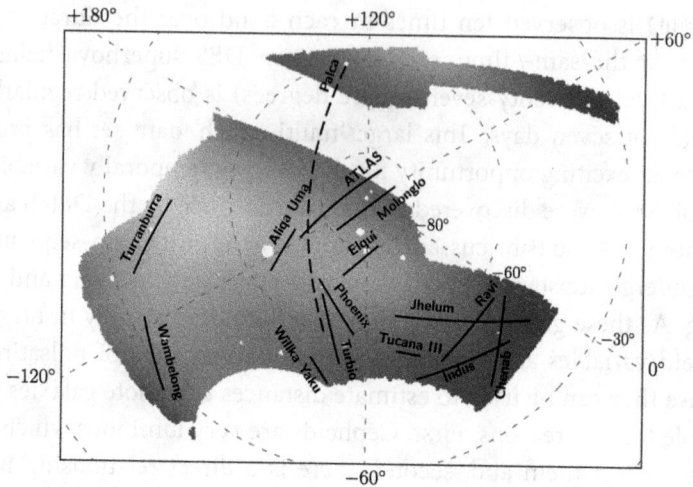

Fig. 20.1 Image showing the full area of sky mapped by the Dark Energy Survey (DES) with some newly discovered stellar streams. Two of these – ATLAS and Molonglo – were previously known. The other thirteen streams were discovered in DES. From Shipp et al. (2018).

DES is also sensitive to low-density extra-tidal features around globular clusters and such features have been detected around several globular clusters.

20.2 The Local Group

Our Milky Way Galaxy and its nearest neighbours are gravitationally bound into a vast cluster of galaxies called the Local Group. The Local Group is ten million light-years across and consists of three large spiral galaxies – Andromeda, the Milky Way and Triangulum – and more than fifty known dwarf galaxies. Some of these dwarf galaxies are satellites of Andromeda and the Milky Way, but we don't know their exact number because many of them are too dim to be easily observed. The largest of the Milky Way's dwarf satellite galaxies is the Large Magellanic Cloud (LMC), which is about a tenth the size of the Milky Way and contains several billion stars.

20.2.1 Local Group globular clusters

Local Group globular clusters have provided a rich population of resolved (i.e. seen as distinct) objects for study with DES. These objects trace some of the oldest and closest stellar populations. They allow us to probe the gravitational field of the Local Group both through the bulk orbital motion and through the tidal disruption of their member stars. Additionally, the dense resolved stellar populations of globular clusters provide a testbed for star-galaxy classification, determinations of proper motion and precision astrometric (measurement of the positions and movements of the stars and galaxies) calibrations. Globular clusters can also be used to calibrate a photometric metallicity scale for DES filters, enabling us to determine what fraction of the mass of an astronomical object is not hydrogen or helium. Two globular clusters of particular interest to the DES survey are Omega Centauri and Reticulum.

Omega Centauri is a dense Milky Way globular cluster in orbit in the opposite direction to the Milky Way as a whole. The retrograde orbit suggests that this globular cluster may have originated as a dwarf galaxy that was tidally stripped by the Milky Way. Recent searches for tidal tails of Omega Centauri show no current tidal disruption; however, the wide coverage of DES and its depth at red wavelengths should provide unprecedented sensitivity to tidal features in the old stellar population of Omega Centauri. In addition, the relative proximity and high stellar density of Omega Centauri makes it an excellent candidate for studying stellar proper motion.

The Reticulum globular cluster is at a distance comparable to that of the LMC and is likely gravitationally bound to that galaxy. Its age and metallicity have been determined using Hubble Space Telescope data and are typical of the LMC old cluster population. DES coverage around this cluster will allow the detection of extra-tidal features associated with this system, which in turn will help constrain the gravitational field in the outer regions of the LMC. The combination of spatial coverage and photometric depth will allow a more thorough investigation of the cluster's structure, with improved determinations of its current mass function, mass segregation and the fraction of its stars that are orbiting in binary systems. These characteristics will allow us to diagnose the role played by internal dynamical effects. The Reticulum cluster serves as a distance reference, since it is known to be associated with the LMC and to contain many RR Lyrae stars, which are stars whose brightness fluctuates periodically.

Besides Omega Centauri and Reticulum, Harris (2010) lists eight additional Milky Way globular clusters that are covered by DES and several other globular clusters that lie close enough to the survey border to allow for searches for extra-tidal features.

20.2.2 Stellar structure at the outskirts of the LMC

The largest contiguous area that falls within the footprint of DES Science Verification data is the South Pole Telescope-E (SPT-E) field (the telescope is located at the Amundsen-Scott South Pole station in Antarctica). The southern border of the SPT-E region is located approximately four degrees from the centre of the LMC. Therefore, SPT-E provides a uniform and deep sample of LMC stars ranging from 5 to 22 kpc in projected distance from the LMC centre. The contiguous coverage of the SPT-E field has already allowed us to constrain the structure and density distribution of the LMC disc out to approximately 20 kpc.

The existence of a spherical stellar halo associated with the LMC remains an open question. Determining the extent out to which LMC stars are to be found and whether this galaxy has its own system of satellite galaxies and other substructure also requires a large coverage in area and depth. When jointly analyzed, data from DES and from the Survey of the Magellanic Stellar History should provide a much more detailed picture of the stellar content of the LMC, the Small Magellanic Cloud (SMC) and the entire Magellanic system.

A more complete census of RR Lyrae stars belonging to the LMC and SMC is expected to result from these data, especially in the outermost regions. This will, in turn, provide an independent assessment of the structure and geometry of the external components of these galaxies. Finally, new star clusters and stellar substructure associated with the LMC have been identified and characterized with science verification data.

20.2.3 Milky Way satellite dwarf galaxies

The LMC and its neighbour, the SMC, are visible to the naked eye in the southern hemisphere, but most dwarf galaxies contain very few bright stars and this makes them hard to find. They are thought to be made predominantly of dark matter, which interacts gravitationally but cannot be seen. The Sloan Digital Sky Survey (SDSS) discovered fifteen new dwarf satellites in the northern part of the sky, bringing the total known to twenty-six. The Dark Energy Camera (DECam) can see objects ten times fainter than SDSS and is imaging a whole new area of sky. The promise of exceptional data from DES motivated the development of several algorithms to search for dwarf galaxies.

Very sensitive observations are required to distinguish groups of stars that form a dwarf galaxy from the light of background galaxies and from foreground stars in the Milky Way. For example, the smallest confirmed dwarf galaxy neighbour discovered by SDSS, Segue I, contains only a few hundred stars and is 99.97% dark matter. To overcome this challenge, we combined two characteristics of dwarf galaxies and searched for these characteristics in the data: firstly, dwarf galaxies tend to have more stars near their centres than on their peripheries and, secondly, stars in dwarf galaxies tend to follow a distinct pattern of colour and brightness. Analysis of the first season of DES data revealed nine new Milky Way satellite galaxy candidates that had sizes and luminosities consistent with ultra-faint dwarf spheroidal galaxies. Analysis of the second year of data added eight more systems to this tally. The dwarf candidates ranged in physical size from a slight 120 light-years to a more substantial 1300 light-years across. Some are orbiting just beyond the edge of the Milky Way disc while others are twenty times farther away, almost half a million light-years distant. See Figure 20.2 for a map of the newly discovered dwarf galaxies within the DES footprint.

It is possible that some of these candidates might actually be globular clusters within the Milky Way Galaxy, so for final confirmation we need to obtain spectroscopic information from other telescopes. We have been able to use data from the Magellan telescope, the Very Large Telescope (VLT) and the Gemini telescope to study the nearest dwarf candidate, Reticulum II, which is almost 100,000 light-years away. Magellan measured the velocity of the stars, while VLT and Gemini measured their metallicity. The latter measurements told us that Reticulum II contains very few heavy elements, implying that it is an extremely old stellar system with an age of more than twelve billion years. The measured velocities of the stars, together with Einstein's theory of gravitation, told us that this system contains hundreds of times more dark matter than baryonic matter (visible as stars). This was clear confirmation that Reticulum II is a dwarf galaxy.

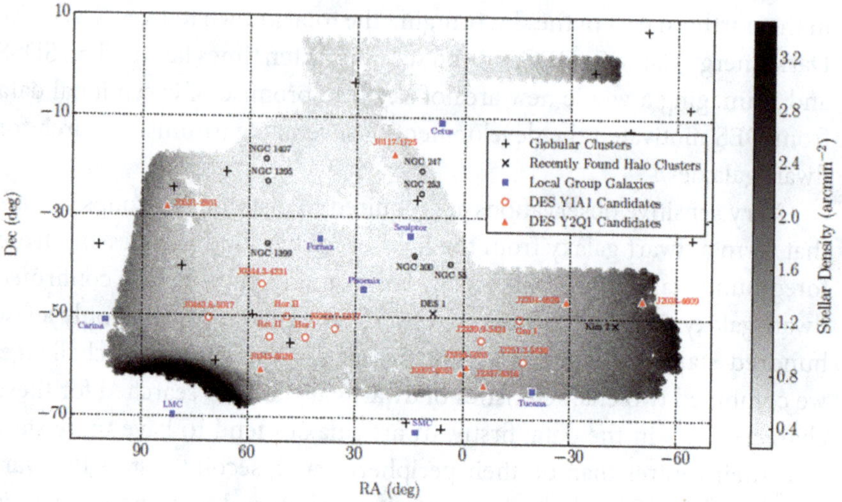

Fig. 20.2 Seventeen new Milky Way companions from the Dark Energy Survey (DES) (in red). Credit: Drlica-Wagner et al. (2015).

Dark matter has not yet been directly detected and its exact nature remains a mystery. One possibility is it is made up of weakly interacting massive particles (WIMPs) that could collide with one another to produce gamma rays. The abundance of dark matter in dwarf galaxies makes them ideal targets for the indirect detection and characterization of dark matter particles through this annihilation. Not only are dwarf galaxies the most dark matter dominated objects we know of, but since there is less activity within them compared with a large galaxy, there is less chance of confusing dark matter gamma-ray signals with those due to known astrophysical processes.

In 2015, we used the Large-Area Telescope, the principal instrument on NASA's Fermi Gamma Ray Space Telescope spacecraft, to look for a gamma-ray signal from eight of the recently discovered Milky Way satellites. There is a useful synergy between DES, which is an optical survey of the sky, and the Fermi experiment, which is searching for gamma rays. This is a good example of the way in which DES data can be combined with data from other experiments to maximize its effectiveness. We found no significant excess of gamma rays coming from any of the candidate galaxies. However, this lack of signal was important since it meant we could set a limit on the rate at which dark matter annihilates. A confirmed dark matter signal would be very exciting news, but it is important that the evidence holds up to scrutiny. For example, we need to look for a signal in observations from a large sample of dwarf galaxies that can be combined statistically. We can then use these observations as a cross-check on gamma-ray signals coming from the centre of the Milky Way.

Finally, dwarf galaxies are also important for what they can tell us about the evolution of the universe. Current models suggest that galaxies began as small structures that merged together to form larger ones, and that this process is still ongoing. Simulations indicate that the Milky Way began as a collection of dwarf galaxies and that in four to five billion years it will merge with the Andromeda Galaxy. Most dwarf galaxies formed more than ten billion years ago, so they can give us an insight into the composition of the first generation of stars and potentially tell us more about what happened in the early phases of structure formation.

References

Drlica-Wagner, A., Bechtol, K., Rykoff, E. S., Luque, E., Queiroz, A., et al. (2015). Eight ultra-faint galaxy candidates discovered in year two of the Dark Energy Survey, *The Astrophysical Journal* **813**, p. 109, doi:10.1088/0004-637X/813/2/109.

Harris, W. E. (2010). A New Catalog of Globular Clusters in the Milky Way, https://arxiv.org/abs/1012.3224.

Shipp, N, Drlica-Wagner, A., Balbinot, E., et al. (2018). Stellar Streams Discovered in the Dark Energy Survey, *The Astrophysical Journal* **862**, 114.

CHAPTER 21

Solar System Science

David Gerdes

The Solar System wasn't anyone's focus during the construction of the Dark Energy Survey (DES). Hardly surprising. After all, DES is at its core a cosmological survey, with the intended purpose of observing hundreds of millions of distant galaxies and helping cosmologists seek answers to the greatest mysteries of the Universe. However, while it wasn't built to be used by mere planetary scientists, its design did also make it an incredibly helpful tool for observing (and discovering) objects far closer to home. From near-Earth asteroids on potentially fatal orbits, to distant dwarf planets with bizarre orbital parameters, DES has been used to probe all reaches of the Solar System, with some of its most exciting discoveries to be found not in the furthest reaches of the Universe but inside our own cosmic backyard.

21.1 Introduction

Most Dark Energy Survey (DES) scientists don't think about the solar system very much. So when the automated data-processing programs examined a sequence of exposures from the night of December 26, 2014 and had the computer equivalent of a hissy fit, the data management team was perplexed. A previously respectable region of the sky was suddenly blasting photons into the camera and saturating the images, making them unusable for cosmology. Was it an airplane, or, more worryingly, a hardware failure? The interloper turned out to be Comet Lovejoy, which thousands of amateur astronomers were then photographing from backyards around the world. We hadn't realized it was there, and the sophisticated software that plans each Dark Energy Camera (DECam) exposure was never instructed to avoid it. This happy accident resulted in the iconic image in Figure 21.1 that's also a kind of astronomical Rorschach test. If your first thought upon seeing this image is, 'annoying foreground object!' you are probably a cosmologist. Indeed, cosmologists consider the entire Milky Way Galaxy to be a nuisance that's merely blocking the view of the really good stuff billions of light-years away.

Fig. 21.1 Dark Energy Camera (DECam) image of Comet Lovejoy, taken by accident.

But cosmology, originally, was the study of the solar system. The quest to understand the mysterious motions of the planets – the word comes from the Greek 'planetes', or 'wanderer' – gave birth to the science of astronomy and nurtured its growth over the centuries. With his invention of the telescope, Galileo discovered the moons of Jupiter, the rings of Saturn and the phases of Venus. Isaac Newton put forth his theory of gravity, and invented calculus, to solve problems of planetary motion. Einstein used his new General Theory of Relativity to correctly calculate the precession of Mercury's perihelion. Dramatic close-up images of planets, moons and comets continue to thrill the public and stimulate new science. The branches of astronomy now stretch beyond the Galaxy and across the universe to the Big Bang itself, but its roots remain here in our own cosmic backyard.

Yet the solar system wasn't on anyone's mind when the Dark Energy Survey started out. In all of the studies, white papers, proposals and project reviews leading up to the survey, the idea of using DES to study the solar system never came up. The cosmologists who planned the survey simply had other fish to fry.

21.2 Near-Earth asteroid of DOOM!

I was one of those cosmologists back in early 2014, as DES Year 1 was wrapping up. I had just agreed to take on a couple of talented undergraduates for the summer. But what project should I give them? A good summer student project is self-contained, feasible on the timescale of a few months, and a little off the beaten path. I'd worked on techniques for measuring the redshifts of galaxies and was interested in using them to do cosmology with galaxy clusters. Many people were working on redshift estimation, though, and I worried that novice students would be overrun. I wanted the students to see the fruits of their work, but there was not yet enough data to do cluster cosmology. Every project I could think of seemed wrong. I went down to Chile for a week to observe for DES and decided to consider it some more. The quiet darkness of Cerro Tololo is my favorite place to think.

On my last day on the mountain, we received an email from NASA's Jet Propulsion Laboratory with the subject line 'Urgent request for DDT observations of a near-Earth object candidate'. That got my attention. DDT is short for 'Director's Discretionary Time' and it refers to an observatory director's godlike powers to override regularly scheduled observations for anything he or she deems sufficiently important and time sensitive. The message requested immediate DECam observations of 'a critical near-Earth asteroid … on a potentially hazardous orbit'. Other telescopes had not been able to observe the object and very soon it would become unobservable, lost in the Sun's glare. It was up to DECam to hunt down a killer asteroid!

That night, the observing team quickly got to work and captured a sequence of five exposures that showed the object streaking through our field of view. These measurements showed that this football-field-sized rock would miss Earth by a full twenty million km – thus, in the words of observer Steve Kent, 'posing no threat to DES operations'.

This gave me an idea. Apart from one-off requests like this, no one was using our data to search for moving objects. The distant galaxies that DES was designed to measure are stationary – they appear in the same place night after night. Well, they are not completely stationary, because the universe is expanding, but their position does not change measurably on human timescales. And we already had a perfect place to look for such things: DES supernova fields. The ten supernova search fields are specific locations on the sky that we visit approximately weekly throughout each six-month observing season. The supernova analysis team, led by Bob Nichol, Masao Sako and Rick Kessler, had developed a sophisticated set of image-processing tools designed to detect anything that had changed in each new set of images – what astronomers refer to as 'transients'. The supernova people were looking for a particular kind of transient: a new object that appears in a distant galaxy, rapidly rises in brightness, then slowly fades over the next few months. But their software found all kinds of other transients as well, including objects that appeared in a given place in only one exposure. Which of these single-exposure transients were actually the same object appearing in different places on different nights? Could we find new members of the solar system this way? Now I knew what my students and I would do that summer.

When I mentioned the idea of using DES data to search for transients within the solar system, the response ranged from 'Great idea!' to 'Really? Good luck with that!'

21.3 The working group begins

One issue was that the funding agency supporting DES is the US Department of Energy and their view is that they will support only ideas which tie into the science objectives of DES, that is, those tightly circumscribed to dark energy and cosmology. Perhaps it was a little naïve of me to imagine they would think, 'Wow cool – a huge new data set, what an opportunity to explore the solar system with. It would be a shame if someone else did it instead'. That was not at all the reaction I got, which was, 'We support DOE science with DOE facilities – we will not support planet hunting or solar system science.' Funding agencies are nervous about crossing boundaries. This forced me to move away from a funding model I'd had for twenty years and become more entrepreneurial. I sought and eventually received a funding grant from the National Science Foundation (NSF) and NASA, and the solar system group was launched.

Collaboration conversations would start in the hallway in Michigan with theorist Fred Adams and his student Juliet. 'Hey Fred, I've found something cool in our data set – how do you get an orbit like this?' We were able to hire a postdoc using the grant from NASA, as well as several undergraduates. Other folks in the collaboration included Masao Sako and Gary Bernstein at the University of Pennsylvania, William Webster and Jim Annis at Fermilab, as well as a group in Brazil interested in known solar system objects.

21.4 A treasure trove of transients

DES data set offers a huge treasure trove of data with which to explore the solar system, and a serendipitous opportunity to do science that no one else was going to do. To date there have been around 90,000 exposures, in which there appear about 150,000 known solar system objects. The vast majority of these are asteroids in the main belt between Mars and Jupiter – not so interesting to me – and some are objects further out. For every known object, there are probably three unknown objects because we go fainter than other surveys.

Trans-Neptunian objects (or TNOs), objects that orbit the Sun beyond Neptune's own orbit, appear in DES images in different places at different times, while the stars and galaxies don't move over five years. If you see something in an exposure that is there one time but every time you go back it isn't there, that's a transient.

Although we started out just looking in supernova fields, the big difference that our collaborators in Pennsylvania made was to take the supernova pipeline and make it work with the wide survey images. The ideas are the same but it is harder. In the supernova fields you visit the same spot over five or six nights, whereas in the wide survey it might be twice on one night and then maybe once more two weeks later, and the next observation isn't for another year. As a result the analysis is more complicated.

We start with a list of transients – exposure by exposure – but we don't know which are the same things appearing on different nights and which are different objects completely. We need a way to connect the dots in order to confirm that these several detections are actually the same thing – and that's what the student and I spent that summer doing.

We also needed to take into account that these objects are in orbit around the Sun, and so have a particular movement across the sky. However most of the apparent movement from one image to the next isn't due to the motion of the object but to our motion on Earth – we are observing these things from a moving platform! Since we are closer to the Sun than the objects are, we are moving much faster than them, so our line-of-sight projects the object onto a different fixed background from night to night.

Taking all this into account, we worked out that it takes three (ideally four or five) observations of a transient object on different nights to uniquely define its elliptical path around the Sun. To my delight we discovered five new objects. In early August 2014 – on DECam engineering nights when the telescope was effectively schedule free – I was able to ask for the telescope to point to locations where we predicted these objects would be. Lo-and-behold, they were there!

We had demonstrated that predicting the orbit of a transient with DES is possible, but measuring its size and other physical properties is more difficult. Even in DECam, they appear like pixelated points. We can, however, make an educated estimate of the size by measuring the reflected sunlight or 'albedo'. You don't know what fraction of light the object is reflecting – so can't tell if it is from something small and shiny or big and dark – but a reasonable value is somewhere between 5% and 25%.

We can also measure the current distance of the object from the Sun. Fifty astronomical units (AU) is where many of these things live (Neptune is 30 AU from the Sun). With the help of the refined pipeline code from the University of Pennsylvania operating over DES wide-field, we began to run jobs on lots of candidates.

21.5 Finding 'DeeDee'

One day in the summer of 2016, I was working with another summer student. Together we were cranking through data and browsing through a list of candidates. We saw the usual expected objects at, say, 42 AU, 47 AU, 49 AU and then we came across one at 92 AU! At the time there was only one other known object in our solar system observed beyond 90 AU – the dwarf planet, Eris.

Using the two years' worth of data that had been collected by then, we built the best orbit we could and confirmed that, sure enough, it is at 92 AU; it had not been seen before and takes 1100 years to go around the Sun. At its closest to the Sun, it is 38 AU away. That we had discovered this object at 92 AU, where the illumination from the Sun is very low, meant it had to be big! For comparison, this thing was about as bright as a candle situated half way to the Moon. (This gives an idea of the power of the camera). The students and I were really excited about this discovery, but we wanted to know how big it was. The heat an object emits is in proportion to its surface area. So it was worth asking for time on another instrument, one which can measure the heat from a 1000 tonne rock thirteen billion kilometres from the Sun. And there is one instrument on Earth that would allow us to measure this – the Atacama Large Millimeter/submillimeter Array (ALMA) of microwave dishes in the desert of Northern Chile.

Time on ALMA is scheduled a year in advance but, like all other instruments, there is a small amount of director's discretionary time. So we wrote a proposal and they thought about it for a week – and accepted! (I had to convince my undergraduate student that he had got a completely wrong idea about how quick and easy it is to get telescope time.)

Even then, there were complications. Although ALMA gave us the time, they didn't know the precise time when they could take the data. It could be 'any time from now to two months from now,' so could we provide them with an ephemeris (a table of this object's positions) at two minute intervals between now and two months from now. And make sure we take into account that ALMA is at this particular point on Earth and Earth is rotating. I carefully made the ephemeris, but I was scared to death I had made a mistake – worrying that they would take our data and point the ALMA at a blank sky. Fortunately, a few weeks later, we got the image back and there was the dim glow from our object! In June 2016, this 'thing' was just a couple of pixels of a DECam image on a laptop. Six weeks later ALMA pointed their billion dollar array at it and a few weeks after that we had our data.

We worked out that it reflects about 13% of light, which in turn told us that it was about 630 km in diameter – big, but not Pluto sized – and therefore a dwarf planet candidate[1] (Gerdes et al., 2017). Until then we had referred to it as 'distant object' so thought we needed a better name. My student Stephanie Hamilton proposed 'DeeDee' for 'distant dwarf' and that stuck. The International Astronomical Union is the body that bestows the official astronomical name but the discoverer can propose a name to the committee. The rules state that for objects in the Kuiper belt and beyond, the orbit has to be well measured and named after a creation deity. When the time comes, I am interested in reaching into the culture of the native people of the Great Lakes region as a nod to where our team was located, but also to bring those cultures into the international astronomy nomenclature.

Finding DeeDee also confirmed that our code worked really well, including on transients that moved more slowly than a typical TNO. DeeDee was moving about one arcsecond (1/3600 of a degree) an hour and goes almost nowhere over the five years of DES observations.

---

[1]The International Astronomical Union define a dwarf planet as one which, unlike a planet, has not cleared its orbit of other, smaller objects and so is considered too small to be a fully fledged planet.

21.6 Planet Nine from outer space

What has motivated us greatly in our work has been new interest in a hypothetical 'Planet Nine'. If it exists, DES could help find this enormous scientific prize. When we first started this work, the idea of an undiscovered planet well beyond the orbit of Neptune was not generally considered. However, two of our discoveries may have strengthened 'smoking gun' evidence towards its existence.

One of the very first objects we found in the fall of 2014, in the supernova field, was an unusual object – 2013 RF98. Its semi-major axis of 350 AU indicated that it takes about 5000 years to go around the Sun – and as such has one of the longest TNO orbits.

Objects with a semi-major axis of more than 250 AU and a perihelion orbit of more than 30 AU beyond Neptune have been dubbed extreme-TNOs (or ETNOs). When people started looking at them, a funny pattern emerged. 'Normal' TNOs follow elliptical orbits at fairly random individual orientations. However the most extreme objects, the ETNOs, align in a particular direction and cross the ecliptic plane in a certain way. Taken together they all lie in a plane, as though some 'distant perturber' might be lining them up.

In the winter of 2016, Konstantin Batygin and Mike Brown,[2] both at Caltech, wrote a paper (Batygin and Brown, 2016) about Planet Nine. This presented a mechanism whereby a distant planet of around ten Earth masses could, over the age of the solar system, make the orbits of these ETNO align. One of the original objects they used to motivate this was 2013 RF98.

As a consequence of their analysis, they gave a general description of the shape, size and orientation of Planet Nine and estimated its orbit to be around 20,000 years (this would make it even slower moving than DeeDee, at a few arc-seconds a day). However, they couldn't tell where it is 'now'. If it exists, it is 'somewhere out there' with a path that probably crosses DES survey area.

---

[2]Mike Brown is famous in the world of outer solar system objects. While he discovered many dwarf planets, like Eris, he is considered to be the individual who got Pluto 'demoted' from the ranks of planets. His twitter handle is @plutokiller.

This was enough to provide us with the motivation, not only to find more of these ETNOs, but also Planet Nine itself! Although we haven't found it (yet?), we can insert fake Planet Nines into our pipeline and test if our search codes can find them – usually they do.

We have now found four ETNOs in DES (Hamilton and Gerdes, 2017) and other surveys have found more – but it's uncertain what this means for the existence of a Planet Nine. We need to understand better the detection bias of the surveys. Maybe ETNOs all line up because we're better at finding things there and our surveys are not equally sensitive to all objects. Also, ETNOs are more likely to be observed when close to perihelion. These are some of the biases we need to consider very carefully.

21.7 Going nuts about Caju

The other question to ask is, if Planet Nine exists, could it be producing objects with interesting dynamics that would be hard to explain without a Planet Nine?

One of the most interesting objects we've discovered so far is 2015 BP519 (Becker et al., 2018), or 'Caju'[3] as we've dubbed it. This TNO is angled at fifty-four degrees to the plane of the solar system. There is nothing like this beyond Neptune – other extreme TNOs are angled less than thirty degrees. Caju takes ~9000 years to go around the Sun. Models of solar system formation have a hard time explaining how an object beyond Neptune can have an orbit like that. Simulations we have run that include just Neptune and the other planets result in a stable orbit for Caju over ~four billion years. However, we find that including a Planet Nine in the simulation naturally excites this object to a high-inclination orbit.

The original paper of Batygin and Brown made a few predictions about Planet Nine. One of these was that it would produce a population of extreme high-inclination objects crossing inside Neptune's orbit. The inclination would evolve over time, even into retrograde. At the time of the paper, there were no known objects beyond Neptune like that, but this new object, Caju, may be the first example of an object undergoing inclination oscillation. Although not proof of Planet Nine, finding Caju

---

[3] Portuguese for 'cashew'.

was an important discovery for us as nothing else has this combination of high inclination and perihelion. And Caju would probably not have been discovered in a survey that was only looking in the ecliptic plane. That is another advantage of using DES – we are mostly looking out of the plane of the Galaxy, to where objects have a high inclination and a complicated dynamical history.

21.8 Looking back

In the space of a few years, we have discovered some really fascinating objects within our solar system and written some significant papers. All of this with a cosmology survey that is probing the universe far far beyond our backyard.

References

Batygin, K., and Brown, M. E. (2016). Evidence for a distant giant planet in the solar system, *The Astronomical Journal* **151**, p. 22, doi:10.3847/0004-6256/151/2/22.

Becker, J. C., Khain, T., Hamilton, S. J., Adams, F., Gerdes, D. W., et al. (2018). Discovery and dynamical analysis of an extreme trans-neptunian object with a high orbital inclination, *The Astronomical Journal* **156**, p. 81.

Gerdes, D. W., Sako, M., Hamilton, S., Zhang, K., Khain, T., et al. (2017). Discovery and physical characterization of a large scattered disk object at 92 au, *The Astrophysical Journal Letters* **839**, L15, doi:10.3847/2041-8213/aa64d8.

Hamilton, S. and Gerdes, D. W. (2017). Detection bias for trans-neptunian objects on highly elliptical orbits with the Dark Energy Survey, in *AAS/Division for Planetary Sciences Meeting Abstracts #49, AAS/Division for Planetary Sciences Meeting Abstracts*, Vol. 49, p. 405.04.

CHAPTER 22

Optical Follow-ups to Gravitational Wave Events

Marcelle Soares-Santos

The joint detection of electromagnetic (EM) and gravitational wave (GW) emission from astrophysical sources is one of the holy grails of present day astronomy. These multi-messenger observations allow for novel measurements, including standard siren measurements of the Hubble constant. In this chapter, we describe how the Dark Energy Survey (DES) GW working group was set up and how the first optical confirmation of a GW event was made.

22.1 A long time ago...

'...in a galaxy far far away...' is as good a beginning to any science fiction story. But this one is science fact.

A long time ago, in a galaxy far far away, two black holes collided. All the energy from that cosmic collision, which took just a tenth of a second, made ripples in the very fabric of space-time itself. About 1.3 billion years later, on Monday September 14, 2015, that gravitational wave (GW) passed through Earth and was detected by the Laser Interferometer Gravitational Wave Observatory (LIGO). It was the first detection of a GW and the first time that we were able not just to 'see' the universe, but to 'feel' the vibrations of space-time.

22.2 Gravitational waves 101

Einstein's 1916 General Theory of Relativity predicts that when massive accelerating objects, such as black holes and neutron stars, rapidly orbit each other and merge, space-time itself becomes distorted and sends a 'gravitational-wave' that propagates outwards from the source. The analogy often used is that of throwing a pebble into a pond and watching the ripples spread out in a wave-like fashion. Like electromagnetic (EM) waves, GWs travel at the speed of light and carry information about the astronomical event that caused them. However GWs are not part of the EM spectrum and cannot be detected by traditional observatories. Instead, GWs are detected by giant laser interferometers, which 'see' nothing but 'sense' the distortion of the space-time wave as it passes through Earth.

LIGO is the world's largest and most sensitive GW detector. Conceived and built over forty years on the work of Kip Thorne, Rainer Weiss and Barry Barish,[1] LIGO comprises two detectors – one in Hanford, Washington and the other in Livingston, Louisiana. Having two detectors separated by 3000 km means that any local vibration is not falsely detected as a GW. Each detector has two 'arms' at right angles to each other, each more than 4 km long.

As a GW passes through the detector, causing space to stretch and squeeze, the length of the arms changes by a minuscule amount (a supernova exploding in our own galaxy would result in a change of less than one thousandth the diameter of an atomic nucleus). These tiny changes in distance are registered by the detector as an interference pattern made by laser light returning from the two arms. GWs produced by different cataclysmic cosmic events are detected as unique patterns by LIGO. The signal from colliding black holes is different to that of coalescing neutron stars.

The original LIGO, in operation between 2002 and 2010, was not sensitive enough to detect any GW events. However, a huge amount was learned during this period, which led to the Advanced LIGO detector, rebuilt between 2010 and 2014, which is ten times more sensitive – increasing the potential detection volume by a factor of 1000.

As well as LIGO, a network of other GW interferometers exists around the world. Virgo, a 3 km interferometer located in Pisa in Italy, collaborates closely with LIGO. Data from both detectors are combined and

---

[1] For which they were awarded the 2017 Nobel Prize in Physics.

analyzed jointly. Using three detectors allows us to triangulate the source position far more precisely than by using only the two LIGO detectors.

22.3 Putting the Dark Energy Survey GW team together

At the end of 2013, the Advanced LIGO collaboration sent an email to the astronomical community stating they would be beginning observations in 2015 sensitive enough to see GW events for the first time and inviting astronomers who were interested in searching for EM counterparts to partner up.

Marcelle Soares-Santos was working as a staff scientist at Fermilab in 2014.

'I was driving to the lab thinking about the email and initially dismissed it. My reservation was about whether they would be able to achieve the sensitivity to actually see far out enough in order that we would have enough events to follow-up. But after driving for forty-five minutes to get to the lab, I thought that they wouldn't make a call if they weren't thoroughly certain.'

The next step was to contact a few people in DES who might be interested in the idea – one of them was Jim Annis and the other one was Josh Frieman.

Jim was also initially sceptical when Marcelle suggested this might be a great opportunity, worried that the localization capability of LIGO was limited to an extent that they could not pinpoint which galaxy a GW came from. It would be no better than 'somewhere out there in that direction!'

But if DECam and DES team couldn't do it, no one could. After talking it over, Marcelle and Jim arrived at a realization: a wide field-of-view camera, such as DECam that can scan the sky in a short period of time, is exactly what you need! They had the right instrument and the right expertise. The type of searches were similar to what DES does every day when searching for supernovae (see Chapter 9) – essentially, compare images at the same location at a different time and look for something new.

Given the expertise available, a GW team was put together that included Josh Frieman, Masao Sako and Rick Kessler from the Supernova working group. Josh suggested they also reach out to Daniel Holz at the University of Chicago because, although he was not a DES member, he was in LIGO. Dillon Brout, a graduate student at the University of Pennsylvania, became another recruit to the GW team after listening to Marcelle give a talk at the 2014 DES collaboration meeting at Michigan. He asked Marcelle if he could get involved because 'it sounds cool!' Dillon became one of the central people involved in the strategy.

22.3.1 Developing the strategy

The Supernovae working group had already developed pipelines to find transient events, by returning the same location to compare DECam images and look for new objects. However, it became clear that the GW team needed to make significant adaptations to the strategy. The supernova strategy concentrated on ten pre-defined fields in the sky that are always revisited and their pipeline was designed with those ten fields in mind. A GW strategy would require the telescope to point to any random location in the sky, even where there may be no previous DES images. So the code was adapted accordingly.

Having Daniel Holz from LIGO on the team was hugely important when it came to simulating GWs. The LIGO group had put together a simulation of what they thought they'd see in the first two years of operation. Daniel, together with his then students Hsin-Yu Chen and Zoheyr Doctor, helped DES interpret those simulations, which were used for designing the optimal GW follow-up strategy. This includes determining the duration of each exposure as well as choosing which filters to use.

It would be too much to physically point the telescope to the location of a simulated event, but we can add fake sources to existing DECam images to see whether they can be detected. Using simulations allowed DES to create a wide range of models of what the emission might look like, and then to run the observing strategy code to determine the actual observations.

22.4 GW150914 – This is not a rehearsal

On September 14, 2015, the first GW event was detected by the LIGO detectors. The signal indicated that two black holes, about thirty times the mass of the Sun, had merged. At the time LIGO was still operating in 'engineering' mode, due to go into full 'research' mode only three days later! Virgo was off-line at the time so was unable to confirm the signal.

For DES, this sooner-than-expected detection meant an unexpected reality check on how ready we were to go live with our GW strategy.

The information DES received from the LIGO collaboration was that there was 'an event'. Although LIGO knew that it was a massive binary black hole merger, they were bound to secrecy as to its nature.

LIGO has two pipelines, one for mergers and one for bursts. If an event shows up as a merger, then it also very likely shows up as a burst. However, the converse is not always true – for example, a supernova would show in the burst pipeline but not in the merger one. This event told us that it had been detected in the burst pipeline but we were given no information about whether it was detected as a merger. Since only the two LIGO detectors had picked up the signal, the location of the source was imprecise at best – around 610 deg^2 with a probability of 90%. Nor was there any information about its distance.

Jim Annis described this as like 'searching for a needle in a fast moving haystack'!

With the limited information we had, we designed a search on the assumption that the event was a merger of two neutron stars, and we concentrated on an area that overlapped with the Large Magellanic Cloud (LMC) (Annis et al., 2016). In total we covered around 100 deg^2 of the sky but found nothing.

Weeks later, we received more information from LIGO that included the fact that the event was a binary black hole merger not located in the LMC. On both counts, it was therefore not surprising that we saw nothing (theoretically, merging black holes have no associated optical emission).

Even so, it was a big success that we were able to move quickly and observe the area that was available to us to search the 'haystack' (Soares-Santos et al. 2016). We were also able to demonstrate that our search was sensitive to 200 Mpc, a distance that LIGO is sensitive to.

Detecting GWs is analogous to detecting sound. With only two 'ears' we can more or less say that a sound is coming from the right or left but no better. With three, you can triangulate the localization much better. When Virgo joined in the second season, this provided a dramatic factor of eight to ten improvement in determining the localization area, which was essential when it came to the GW event of August 17, 2017.

22.5 GW170817 – Discovery

On August 17, 2017, at 12:41:04 UT (Universal Time) the Advanced LIGO/Virgo (ALV) observatories detected a binary neutron star merger, GW170817.

Dillon was in Philadelphia, Jim was in Tuscon and Marcelle had returned home from Fermilab. She recalls,

'I had just gone to bed two hours earlier since I was pursuing a black hole merger event from two days earlier (GW170814). I woke up to a text message from an automated listener system to say that LIGO had picked up an event, which I assumed was the same black hole merger. Suddenly my phone started to ring. It was Hsin-Yu Chen. She was a key person in developing the optimal search strategy and she had rung to say it was a new and different event – we had to wake everyone up!'

Jim, Dillon and Marcelle hooked up remotely to work out a strategy based on the outputs of Dillon's code and to update new sky maps (in order to point the telescope). They immediately contacted H. Thomas Diehl, the operations manager at DES, to flag that they were going to do something different to the regular plan that night. Over the day they began to put their plan into action. Jim recalled,

'We knew it was a neutron star merger and we knew we had a chance of viewing it but we had to get our act together to actually make the measurement. LIGO tells you that it is somewhere over there – so you have to work out where to point DECam to cover that entire area in order to find one faint fuzzy blob. So we had to evaluate all of that in order to make the maps and the files and get them to the telescope all before sunset.'

At the same time, Marcelle was sitting in an increasingly empty room. 'That day a truck was scheduled to come and pick up my stuff as I was moving from Fermilab to Brandeis University. Throughout the day, as we were chasing a neutron star merger in a far-off galaxy, my entire apartment was being taken apart. In the end there was only one folding chair and my laptop!'

At 23:12:59 UT (10.53 hours after the GW detection) we began to image a 70.4 deg^2 region of the sky that covered 93% of the localization probability in the map provided by the LIGO–Virgo collaboration.

We knew we were hunting for a neutron star merger, and a 3D sky map indicated a distance of around 40 ±8 Mpc, which is relatively close. There was concern that, being so close, DECam images would be saturated, so the exposure time was dialled back to thirty seconds. This had the advantage that we could cover more area in the same time. Further, the GW strategy meant that only images in the i and z bands are taken. Changing the filter takes time, which would mean too much time spent at each location. At 40 Mpc, there are not that many bright galaxies so we decided to 'eye ball' the images for speed, comparing DECam images with archive images from the PAN-Starrs1 survey to look for 'something new'. Douglas Tucker and Sahar Allam, both from Fermilab, also dropped everything and worked overnight on the observations from remote stations at Fermilab, where the fast image processing pipeline (developed by Ken Herner to run on grid environments) was also working overtime. In addition to DES and LIGO, we also joined forces with a group at Harvard, led by Edo Berger, where Ryan Chornock was the first in the collaboration to win this game of cosmic 'spot the difference'.

An email from Ryan read 'WOW – Check out NCG4933! Galaxy is at 40 Mpc'.

Part of the protocol is that once a detection has been made this is reported to the Gamma-ray Coordination Network (GCN), the network that includes LIGO and all the partners. This allows for confirmation from people who have telescopes with a small field-of-view (FOV). It also plants a flag that says, 'We got here first!'

Although we had found the source, we were not the number one. Another group, the One-Meter, Two-Hemisphere (1M2H) team, who had a smaller telescope, had also found it and submitted their report nine minutes before us.

22.5.1 Multi-messenger astrophysics

Now that the candidate had been found, it was monitored in more detail with all the band filters in order to confirm that it was what we were looking for, rather than a supernova. Both the colour and light-curve were analyzed and compared. A supernova's light-curve would decay over a month or so, whereas the model of a merging neutron star should decay over about ten days. The colour allowed us to discount the possibility that we were observing a supernova at the end of its light-curve. In this case, the analysis of colour and light-curve was very detailed and consistent with what we expected from the transient emission from two merging neutron stars – known as a kilonova.

We also needed to confirm that this was THE one, not ANOTHER event in another galaxy within the LIGO localization that could have been the true transient. Based on our models of kilonovae, we revisited the entire region when we expected it to have gone.

Once again, we processed all the images and compared them, noting how many candidates were astrophysical, were not moving objects and had light-curve decays consistent with a kilonova but not a supernova. When we did that there was only one candidate left, the same one we had found by visual inspection. Figure 22.1 shows, on the left, the observation of GW170817 about a day after the LIGO signal and, on the right, the same location more than fourteen days later. Now you see it; now you don't. We had our first confirmed optical follow-up of a GW event (Soares-Santos et al., 2017).

DES scientists at University College London (UCL) were part of a team that observed the aftermath of the event, recording images using DECam. Simultaneously, Samuel Emery, Alice Breeveld and Kuin at UCL's Mullard Space Science Laboratory (MSSL) were working on the Swift Mission. As part of a study led by the University of Leicester, they successfully detected

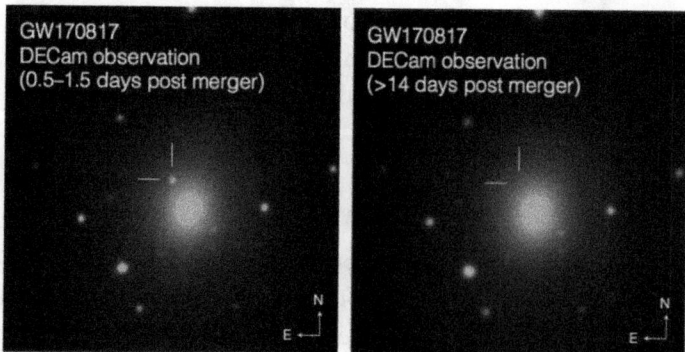

Fig. 22.1 NGC4993 composites (1.5'1.5'). Left: Composite of detection images (z taken on August 18, 2017 00:05:23 UT and the g and r images taken one day later). The optical counterpart of GW170817 is at right ascension (RA), Dec = 197.450374, 23.381495. Right: The same area two weeks later.

a brightly glowing and rapidly fading ultraviolet light source from the event location but were unable to detect any X-rays because they were too faint.

By analyzing data from DECam, UCL DES scientists Antonella Palmese, Will Hartley and Ofer Lahav were able to better understand how binary neutron star systems form by further characterizing the properties of the galaxy (NGC 4993) where the GW event originated (Palmese et al., 2017).

GW170817 was the first detection of an optical counterpart of a GW source (Abbott et al., 2017). It will not be the last!

22.6 Standard siren?

The follow-up study of GWs had originally been seen as a small side project in DES and it was not obvious how it would connect to the main goal of understanding dark energy. However, one of the exciting results that followed from the observation of GW170817 was an independent estimate of the Hubble constant H_0.

Up until now, our estimates of H_0 have come by way of a so-called 'distance ladder' and its dependence on standard candles. These are astronomical objects that have a known luminosity – usually Cepheid variable stars and Type Ia supernova – from which we estimate distance. But one of the niggling concerns is just how 'standard' they are.

By combining the distance to the source solely from the GW signal with the redshift (inferred in this case from DECam), the GW programme allows for an independent measurement of H_0, allowing us to bypass the distance ladder completely. The measurement of H_0 from the optical follow-up of GW170817 was found to be 70 kms^{-1} Mpc^{-1} (Abbott et al., 2017), which compares very well with existing methods, albeit with a greater uncertainty (of around 15%). With more events, this error bar will decrease to a level that could make an impact on the cosmology goal of DES, with the possibility of adding another probe to DES's existing four.

22.7 Looking forward – Reflecting back

The emergence of the DES GW effort and, independently, of the trans-Neptunian objects project (see Chapter 21) prompted the DES collaboration to form a new working group, the 'Transients and Moving Objects'

(TMO) working group. It was the last DES Working Group to be created, in 2014, co-coordinated by Marcelle and David Gerdes. Julio Camargo replaced Gerdes as co-convener of the TMO group in 2019. It is much smaller than traditional giants such as Supernova and Clusters, but despite its slim size, it is very active.

Just over a hundred years ago, Einstein brought new insight to our understanding of the universe with his General Theory of Relativity. From his genius came the concepts of dark energy and GWs, which are today at the cutting edge of research in cosmology. To be working in either of these fields is hugely exciting. To be working in *both* is a privilege.

References

Abbott, B. P., Abbott, R., Abbott, T. D., Acernese, F., Ackley, K., et al. (2017). A gravitational-wave standard siren measurement of the Hubble constant, *Nature* **551**, pp. 85–88, doi:10.1038/nature24471.

Abbott, B. P., Abbott, R., Abbott, T. D., Acernese, F., Ackley, K., et al. (2017). Abbott, Multi-messenger observations of a binary neutron star merger, *The Astrophysical Journal Letters* **848**, L12, doi:10.3847/2041-8213/aa91c9.

Annis, J., Soares-Santos, M., Berger, E., Brout, D., Chen, H., et al. (2016). A dark energy camera search for missing supergiants in the LMC after the advanced LIGO gravitational-wave event GW150914, *The Astrophysical Journal Letters* **823**, L34, doi:10.3847/2041-8205/823/2/L34.

Palmese, A., Hartley, W., Tarsitano, F., Conselice, C., Lahav, O., et al. (2017). Evidence for dynamically driven formation of the GW170817 neutron star binary in NGC 4993, *The Astrophysical Journal Letters* **849**, L34, doi:10.3847/2041-8213/aa9660.

Soares-Santos, M., Holz, D. E., Annis, J., Chornock, R., Herner, K., et al. (2017). The electromagnetic counterpart of the binary neutron star merger LIGO/Virgo GW170817. I. Discovery of the optical counterpart using the Dark Energy Camera, *The Astrophysical Journal Letters* **848**, L16, doi:10.3847/2041-8213/aa9059.

Soares-Santos, M., Kessler, R., Berger, E., Annis, J., Brout, D., et al. (2016). A dark energy camera search for missing supergiants in the LMC after the advanced LIGO gravitational-wave event GW150914, *The Astrophysical Journal Letters* **823**, L33, doi:10.3847/2041-8205/823/2/L33.

PART IV

Reflections and Outlook

This final part of the book broadens the focus to consider the Dark Energy Survey from the perspective of a philosopher, an anthropologist and artists. In doing so, it raises more questions than it answers. The section concludes with a chapter speculating on the future of dark energy research.

CHAPTER 23

An Anthropology Angle: Credit and Uncertainty in the Dark Energy Survey

Lucy Calder with Martin Holbraad

The Dark Energy Survey (DES) is a scientific investigation that aims to reduce our uncertainty about the universe. It is also a collaboration between hundreds of scientists from all over the world. This creates a remarkable opportunity to investigate how the idea of uncertainty in scientific knowledge relates to the uncertainty in the lives of scientists carrying out the research. The material in this chapter is based on seven months of non-intensive fieldwork carried out by astrophysicist and anthropologist Lucy Calder from January to July 2014, predominantly among the group of DES scientists working in the Cosmology Group at University College London (UCL). Her supervisor in the Department of Physics and Astronomy was Professor Ofer Lahav and her training in anthropology was guided by Professor Martin Holbraad.

23.1 Introduction

In line with customary anthropological practice, I aimed to participate as much as possible in the activities of the Dark Energy Survey (DES) scientists, in order to observe and record their daily life. I attended regular weekly meetings of the DES University College London (UCL) group and joined workshops, lectures and telecons. I also attended the biannual full collaboration meeting in Urbana-Champaign, Illinois, where I met a broad range of DES members from different institutions. During this period, the theme which emerged most persistently and pervasively was that of *uncertainty*. It was obviously present in the scientific knowledge ('truth' in science is probabilistic), but it was also hinted at in the majority of my conversations with DES scientists about their lives in academia. Working in science seemed fraught with unpredictability.

At the same time, I found it useful to refer to the idea of *credit* as described in Latour and Woolgar (1979). Their notion of credit came from analyzing the frequent use scientists made of the term and finding that it incorporated ideas of both reward (in terms of funding and awards) and credibility (essentially trustworthiness). They then used this concept to explain the apparently very different motivations of scientists to do science. This model was a convincing starting point for my own analysis, but I was persuaded by my fieldwork to acknowledge the wider connotations of the word in order to incorporate a scientist's sense of self-belief and self-worth. When I refer to credit in this essay, I am using it in its broadest sense, unless I indicate otherwise.

The specific ethnographic question I eventually explored became: How is an individual's uncertainty about accumulating credit connected to uncertainty in scientific knowledge (if at all)? Furthermore, how does this relate to ambivalence over what constitutes 'science' and to the pressure on scientists to associate certainty with their results?

23.2 Background literature

Latour and Woolgar's (1979) concept of credit includes both the notion of reward, in terms of citations or awards, and also the wider idea of credibility among a scientist's peers, resulting from a belief in his or her ability 'actually to do science' (Latour and Woolgar, 1979; Latour, 1987). 'Doing science', is understood to be anything ostensibly connected to the process, from data and papers to institutions and funding. Credit acts as the connective between these apparently different aspects, rather like a currency. The author acknowledges a debt to Bourdieu's idea of symbolic capital 'which in the form of the prestige and renown attached to a family and a name is readily convertible back into economic capital' (1972:179). However, the idea of scientists as investors looking to maximize their symbolic profit does not explain the source of their social power. Latour and Woolgar (1979) recognize that the *content* of science itself has value. Information is considered to be a commodity, except 'only rarely is information itself 'bought'. Rather, the object of 'purchase' is the scientist's ability to produce some sort of information in the 'future' (Latour and Woolgar, 1979: 207). This economic metaphor allows the authors to explain why scientists variously consider it to be of prime importance who they work with, where they work, the project they choose to work on, grants and awards and academic position. All of these are used as resources 'in the struggle for credible information and increased credibility' (Latour and Woolgar, 1979: 213). Money, data, prestige, credentials, papers, and so on can all be converted into credit and then 'reinvested' and transformed into another kind of credit in an endless cycle. The scientific 'community' is envisaged as a market, in which the individual success of a scientist depends on how quickly he or she can progress through the credit cycle.

Latour and Woolgar's model has been criticized for painting too agonistic a picture of scientists, in which they resemble aggressive, accumulating capitalists. I acknowledge these concerns, but I suggest that if the entire range of associations encompassed by the notion of credit is emphasized, the concept can still be used as a way to make connections between diverse and seemingly separate issues, without reducing scientists entirely to politicians. The Oxford English Dictionary lists sixteen definitions of credit, including 'a source of pride or honour', 'the quality of being generally believed', 'good name or standing', 'personal influence resulting from the confidence or trust of others, or from one's reputation'. There is a difference between external gains, like recognition and reward, and internal states of mind, such as self-worth and self-identity, but the concept of credit can incorporate both.

On the question of whether scientific knowledge can be thought of as a commodity, Biagioli (1998) points out that, according to intellectual property law, since scientific knowledge is considered to be 'truth' it cannot be created by anyone, only discovered. Therefore, it cannot be owned by anyone and cannot be exchanged for money, including scientific funding.[1]

---

[1] Data, however, can be owned. The majority of the funding for DES comes from the US DOE, which is public money: thus, in effect, the US public 'owns' the data.

Scientific knowledge is supposed to be entirely objective, although numerous science studies scholars have attempted to overturn this idea.

With regard to the problem of what constitutes science, later science studies came to regard economic metaphors as insufficiently complex. Actor-network theory (ANT) develops the idea of the cycle of credit into a much wider web of interrelationships spreading far beyond the laboratory. In *Science in Action* (1987), Latour claims that scientific facts are constructed over a period of time by a rhetorical process whereby people constantly either qualify or lend support to a particular statement. From being a conjecture connected to a particular historical situation, a statement about an object can become anonymous, taken for granted knowledge, until finally the object is seen as a real part of the world and the statement a subsequent description of it. According to ANT, scientific knowledge within DES has no meaning outside a complex network of social relationships. In effect, this is a nominalist philosophical position. 'Social constructionists' are not claiming that reality does not exist, but that the only structure we can conceive lies within our representations: facts are made, they are not 'out there', independent of observers. Hacking (1983), a philosopher of science, points out that the standard view of science, as discovery of facts that exist pre-structured in the universe, is also a philosophical position. So what we have are two opposing metaphysical pictures of the relationship between thought and the world, and there may be no way to decide between them.

Since I did not wish to become embroiled in a philosophical discussion about theories of truth and knowledge and my original intention was to side step the realism/constructionism debate, I decided to take Latour's earlier work as the methodological point of departure for my analysis. His fieldwork in Salk's laboratory in 1979 was key to the development of many of the core concepts in ANT, but at this stage constructionism[2] has not already been assumed. I submit that the cycle of credit, itself drawing on Bourdieu's idea of symbolic capital, continues to be a powerful explanation for what makes science happen, without necessitating fact construction. However, my ethnographic material reveals that the model needs some

---

[2] The term 'constructivism' often appears in place of 'constructionism', but I shall avoid it in this essay.

modification and, having made these refinements, I can then engage with the consequences this has for ANT itself.

23.3 Credit and uncertainty: Science and infrastructure

Over the years, the number of students going to university to do PhDs in physics or astronomy has steadily increased, but the number of permanent positions has remained relatively stable, with the result that the competition for jobs among post-doctoral research associates (postdocs) continues to intensify. The academic process is essentially the same worldwide: people progress from undergraduate degrees to PhD studies then post-doctoral research. Postdocs are paid positions, but the salary is less than someone less qualified academically can earn in a City job. Furthermore, a postdoc position usually only lasts from two to three years and it is considered good practice for the scientist to move to a different institution for each post, at least to another city and very often abroad. This leads to the 'two-body problem', which refers to the fact that their partner, if they have one, may not be willing or able to move abroad with them. In most of the world, a first permanent job in science would be as a junior lecturer in a university or research institution, while in North America postdocs are ultimately aiming for 'tenure', which means a university staff ('faculty') position protected by contractual rights.

The general consensus among those I spoke to was that if you want to stay in academia you have roughly three chances after your graduate studies, but if you can't find a job during your third postdoc, it is better to leave the field. This results in an ever present anxiety about time.[3] The PhD students are generally more sanguine about their chances, but Postdoc 4[4]

---

[3]Traweek (1988) identified time as one of the 'key symbols of the culture' (Traweek, 1988: 15) of particle physicists in the 1970s and 1980s. 'In the course of a career a physicist learns the insignificance of the past, the fear of having too little time in the present, and anxiety about obsolescence in the face of a too rapidly advancing future', in contrast to their 'cosmological vision that transcends change and mortality' (Traweek, 1988: 17). All of this is recognizable among scientists in DES.

[4]Specific individuals are referred to using letters or numbers. Following DES guidelines, all students, postdocs, research fellows and anyone on a short-term contract is termed an 'early career scientist'. In contrast, scientists with a 'permanent' job are referred to as 'senior scientists'.

described himself as 'long in the tooth' at the age of 31. Working hours are flexible but many early career scientists told me they worked long days of up to twelve hours, and then more time at the weekends. One postdoc said: 'It's not a reasonable career choice', while another remarked: 'Of course we take holidays because it's impossible to be efficient if you don't take some breaks, but holidays are time taken out of work, so then we get behind'. Nevertheless, the large communal offices in UCL have an informal, collegiate atmosphere. The scientists come to work in casual clothes of their choice and spend the majority of their days in front of computer screens, surrounded by a random assortment of plants, books, photographs, empty mugs and shelves full of colourful files.

There is a general feeling that DES is important, being (by chance) the only large scale astronomical survey of its kind taking place over the next few years, and the impression among the postdocs that time is running out is exacerbated by a general sense of urgency running through the entire collaboration. It was part of the agreement with the funding bodies that the raw data recorded by DECam is made publicly available a year after it was collected by the telescope. From 2017, there will be a regular release of data that has been cleaned and calibrated by Dark Energy Survey Data Management (DES DM). This essentially means that members of DES have only few years' advantage over their competitors in 'the field' (with its intimation of the battlefield) and are under pressure to analyze the data and write their papers as quickly as possible. The DES director told me his biggest pressure was 'keeping things working smoothly so we will get forefront science out in a timely way'.

All the early career scientists agreed that to have a successful career in astronomy and physics it is essential to publish research papers, preferably as first author. There was, however, no real agreement about exactly how many were needed, or who was checking. Publishing papers in DES is a big issue, in particular whose names will be on them and in what order. Within a collaboration, the need to write scientific papers must be balanced by other demands, primarily the pressure to do 'infrastructure work', which is work that is essential to the success of the survey, but is not classed as 'science' so cannot (usually) be used as material to be written up. More obvious examples of this include maintaining the camera, writing computer code to link the data in the camera to the National Center for Supercomputing Application (NCSA), participating in one of the committees, spending nights observing at the telescope in Chile. Broadly, it should be work that benefits a large number of people in the collaboration, while 'doing science' means learning something new about the universe, but the line between the two is not clear cut. The data must be subjected to many hours of analysis before it can be used to answer any of the questions about the universe and this data analysis, while it is fundamental and necessary, has a particularly liminal status with regard to being either infrastructure or science.

> Cosmology is really big and it's really hard now. It's really complicated. There's so much stuff to know just to do one little piece of science, especially weak lensing. It's so, so hard to measure [galaxy] shapes, and then to use the shapes to compare the models. We have to check that we understand the data, measure the shapes correctly, make proper catalogues. Even before the science. We won't do any proper science with this for years. [PhD 1]

Several postdocs told me that some of the code they wrote was no different to that needed in, for example, a commercial financial institution; the only difference was the end goal. Were they then, in that case, still 'doing science'? Professor B said, 'It's actually almost getting to the extreme where we have students coming up now who are very good at manipulating the data, analyzing the data, but who may not always appreciate that there's a larger theoretical structure that we're trying to fit this into. So I worry, because it's very easy to get kind of sucked down into the data. I want people to not forget the context'.

Professor M admitted that 'the boundaries are blurred', but because early career scientists feel under great pressure to publish papers in order to get a job, defining these boundaries becomes important. Postdoc 3 broadly defined doing science as 'doing work that is actively allowing you to get a paper written, that you could publish in a scientific journal'. Conversely, if a scientist can prove to the Management Committee that they have spent at least two years working full time on infrastructure, they are compensated for it by being awarded 'builder' status. This gives them the right to put their name on every paper that comes out of DES and also gives them access to the data even if they move to a non-DES institution. One year working full time on infrastructure gives someone personal data rights alone. Nevertheless, one postdoc commented, 'Being a builder is valuable if you've also at the same time got yourself a position in a university to give you time to exploit that builder status. But builder status on its own, which just allows you to use the data, isn't any good if you've had to leave astronomy because you can't get another job'.

One of the first DES papers to be submitted to astro-ph[5] lists the three key authors first, presumably in order of amount of work done; then twenty-nine less central authors, in alphabetical order; then forty-four builders, in alphabetical order. And the number of builders is only going to increase over time. When I asked Postdoc 8 if he thought having his name on lots of papers as a builder would count the same as having his name on a few as a principal author he sighed deeply, looked away from me and said, 'Does it count? That's a good question. I don't know'.

It became clear to me that 'doing science' is the main criterion for gaining respect and success, that is, credit, in this (self-styled) community. This was apparent in the pride with which a particle physicist showed me his business card with 'scientist' as the job title. Or the near despair of the postdoc who told me he was 'buried' in infrastructure work and just wanted to 'finish it quickly and move to real science'. In comparison, a scientist's other work can be overlooked. During an Early Career Scientists Question and Answer session at the Illinois meeting, one senior scientist on the panel said, 'I like teaching but it's a funny thing. It takes up a whole lot of time but it doesn't feel like it makes much difference in how you're evaluated, which is odd'.

When I asked scientists what motivated them, the answers were diverse and sometimes contradictory. The word 'interesting' was often used. Postdoc 5 told me what an 'interesting' problem the neutrino mass was because 'they're well described in the theory and that's something solid [he slapped the desk] and cosmology can bring an answer here'. Yet later on, when I specifically asked what motivated him, he replied: 'Sometimes coding is just fun – you're solving problems and building something on

---

[5] It is now standard practice for all astrophysics papers to be published on Cornell University's online archive: http://arxiv.org/list/astro-ph

your own. Seeing something that you've thought of reach completion. As for something physical like the mass of the neutrino. No, to be honest I just don't care'. Many people agreed that grappling with the problems of analyzing data was a fundamental incentive: 'the prosaic, the nitty gritty, the technical issues', since that is what they were primarily engaged in. 'It's that process – the praxis of doing that kind of stuff is really what motivates me more than the scientific questions' (Postdoc 4). There were others who claimed it was about working with interesting people, or looking for new astronomical objects.

In this situation, the currency-like nature of credit can be usefully invoked to connect these diverse and seemingly separated issues, without necessarily imagining scientists as Machiavellian. It may be that this latter interpretation of Latour and Woolgar's 1979 model was a result of his economic and war-like (Latour and Woolgar, 1979: 212) analogies, but I do not think it is mistaken to think of scientists to some extent as strategists. In light of my research, I suggest that the uncertainty in the boundary between what does and does not count as doing science, among scientists, leads to uncertainty in the amount of credit an individual scientist will receive. Unlike Latour, scientists do not collapse the boundaries between what is internal and what is external to science.

For example, Postdoc 3 told me: 'It's difficult to give credit to everyone because of the paper system. It's a system with down sides. It's quite rigid. The only thing that people outside DES will see are the publications. But I work well with this guy P and he will vouch for me even if I'm not on so many papers. If I apply to a non-DES institution I can ask P to write a letter to this person, which relies on two things: this new person trusting and believing in P'. He is hoping that the credit (credibility) of P will be enough to make up for any lack in his own, caused by choosing to concentrate on work that may not be counted as science.

When I spoke to Professor M about the boundary between infrastructure and science, he told me that to help the early career scientists he tries, with his students, 'to turn infrastructure work into papers' (i.e. into what is then perceived as science). For example, someone could write a paper comparing different methods, or showing what can be achieved by combining different data sets. It seems that if infrastructure work is judged to be creative and interesting by enough people, it becomes equivalent to doing science. Perceptions can also shift over time, depending on the way that science develops.

For senior scientists in permanent positions, uncertainty is no longer linked to finding a job, but to how successful they are at applying for funding, attracting students and postdocs and gaining access to new data. In a system where 'doing science' is given the most credit, those people who are primarily interested in solving technical problems through numerical computation are in the most precarious position, but scientists at all stages of their careers worry about how much science they are producing. Senior scientists can claim a share of the credit for work done by their students, but some of the professors I spoke to told me they felt frustrated about not having time to work on their own science because they have so much administration to do. Furthermore, since it is a challenge to keep up to date with computer software, they often rely on postdocs to do the initial stages of the data processing (infrastructure work), which can cause tension. Some people simply decide not to play the game at all and leave academia, but due to the strong connection between knowledge and self-worth, leaving the field has long been viewed negatively within physics and astronomy. Even scientists who do not seem to enjoy their work feel under an obligation to stay. Many people are working to try to shift this attitude, but leaving science still means relinquishing one's identity as a scientist.

23.4 Uncertainty and credit: Scientific knowledge and social relationships

The astronomical catalogue produced by the data management (DM) team contains a huge amount of information about the visible objects detected in the sky, but still more analysis needs to be done on the data before it will yield any new information about the universe. Different working groups extract only the data they need from DES DM and then use it to make their own 'value added catalogues'. For example, the Weak Lensing Group, in order to make a map of invisible dark matter, needs to measure the shapes of distant galaxies, because the light from these galaxies is distorted more or less depending on how much dark matter is between those galaxies and Earth. The group is thus working on making a shape catalogue.

Within each working group, numerous analysis groups work on a particular problem within their working group's area, and the idea is that each individual contributes a piece of the puzzle. But complications arise because many people end up working on similar problems and write computer codes to do essentially the same thing. Since there are potentially an endless number of different ways to solve the same problem, and people make different assumptions about the cosmology, the codes can end up giving slightly different results.[6] In particular, different codes are being used to estimate the photometric redshifts ('photo-zs') of objects in the images, resulting in slightly different photo-z catalogues. Since estimations of all the other cosmological parameters rely on the values found for the redshifts, this is a fundamental source of uncertainty.

During one of the UCL Monday meetings, Postdoc 1 told the others: 'We need to have some kind of guideline for the catalogues. We have to have some kind of party line that says this is the code or these are the two or three codes you can use. Otherwise it will lead to trouble'. Postdoc 5 protested that it was hard to come up with objective criteria to choose between codes because 'one approach can be better at one thing and less good on something else, and the other one can be opposite'. Postdoc 1 added, 'even if someone uses exactly the same code it's not going to give the same answer because it gets tweaked and refined by the person who wrote it'. This exchange gets right to the heart of what is happening in DES. It appears that in order to produce a unique set of DES results, the uncertainty among individuals about how much credit they will receive, and hence about their future career in science, is increased.

Hacking (1999) identifies contingency as one of the fundamental philosophical disagreements between scientists and social construction-ists. For example, Pickering, in *Constructing Quarks* (1984), claims that the development of high-energy physics was highly contingent such that, in different circumstances, divergent, incommensurable, but equally successful[7] theory and instrumentation could have emerged. He does not

---

[6] Computer codes are a means of doing calculations on large amounts of data. They are used for almost everything in observational astronomy.

[7] Hacking notes that Imre Lakatos' (1970) idea of an empirically and conceptually pro-gressive research programme is one measure by which science can be judged as successful.

deny the existence of quarks, but says their existence is interdependent with the contingent scientific practice in which they appeared. The fact that there are different catalogues based on the same data within DES is certainly contingent (on the individual assumptions made by the people writing the codes), but that is not disputed by the scientists. Discussions are underway to set up a 'data vetting panel' to decide on the official catalogues to use, but choosing one set of data rather than another introduces a fundamental element of ambivalence into the final conclusions drawn from the catalogues, including the size of the error bars on the dark energy parameters. Applying Pickering's theory, the panel's decision is completely arbitrary, because whatever seems in the end to be the most robust choice is contingent on all the circumstances and could just as well have been otherwise.

While most scientists would not agree with Pickering, they would probably accept Quine's (1953) logical observation that many incompatible theories could be consistent with any one set of data. During an informal interview Professor M said:

> I don't think there's one answer to anything in this game. When we started there were only a few photo-z codes; now there's twelve. And by the way they're not all different from each other because there are similar elements. They make slightly different assumptions. So now what do you do with it? What I would like to see instead of someone telling me use only method X, I would like to see twelve ellipses – twelve different figures of merit for the dark energy parameters.

The amount of uncertainty in a measurement can be plotted as an ellipse around a central point that is deemed to be 'the truth', so that by combining different estimations of the truth from different cosmological probes, the measurement of the central point becomes more precise. The figure of merit is simply a number that is the inverse of the area of an ellipse. Professor M would like there to be transparency in the process, so that 'the outside world' can see that different assumptions lead to different conclusions. But it can be observed, in line with Quine, that in principle an infinite number of different codes could be written, producing an infinite number of marginally different catalogues of data.

When Postdoc 1 says having different codes will 'lead to trouble', he is not simply referring to epistemic confusion. If several people in the collaboration develop codes to do the same thing, a competition is set up between them, even if it remains unspoken. If DES results are going to be standardized (by the data vetting panel), then not every code can be used to generate results to go in papers.[8] Thus some people will do a lot of work and not end up gaining the credit that comes from having other people use their code. As Postdoc 5 put it: 'That's the whole politics of collaborations – you have endless and painful discussions'. To an extent, competition is encouraged because it pushes people to work harder and innovate, but it becomes apparent that in order to standardize and (ostensibly) reduce uncertainties within DES results, the uncertainty generated by competition is increased. With regard to his code, Postdoc 6 told me:

> I want it to be used widely and I have to push to make it used as widely as possible, so that's more of an unknown unknown. It's an effort to make that happen, it's a big effort. If I stopped doing [the code] now it would just disappear and no one would ever use it and I'd have wasted loads of time.

Another element of competition appears when scientists choose which projects to work on in the first place. There is a lot of opportunity for new science to be done in DES, but some projects are deemed to be more significant than others in the wider scientific community, which potentially means greater credit for (including more citations of) the resulting papers. Again, this competition results from the commitment of the collaboration to not producing too many conflicting results, which would reduce the credit of DES as a whole among the rest of the scientific community and the wider world. It follows that who gets to do what science, and thus what science emerges, is in part dependent on the effect of personality, which is inherently subjective. One senior scientist said firmly, 'You have to have visibility ... How successful you are depends on how you engage with other

---

[8] Having two or three codes for the same problem is thought to be a good, even essential, idea, if they give results within about the same error bars, because they act as a check on one another.

people', but some scientists find this randomness difficult to deal with. Postdoc 7 said, 'I think the theory is fun. It's really nice. I think the hard part of science is confrontation with others. When you have to prove that what you did is great, basically'. Professor M told me that when interviewing potential students and postdocs he gave 'a lot of weight to communication skills', but he worried that this was becoming a pre-requisite for a career in astronomy and wondered if there was enough space for more introverted people. 'There's an element of marketing and presenting your stuff, which is a skill by itself and some are very good at it.'

On the other hand, whatever the private fears about competition, the atmosphere when the DES UCL group gathered together was lively and occasionally somewhat tense, but not overtly competitive. People seemed willing to talk about difficulties they were having and to help one another. At the full collaboration meeting in Urbana-Champaign, within the smaller group sessions and during conversations over lunch, a sense of urgency and purpose prevailed. The uncertainty engendered by the pressure to be competitive was apparently counteracted in some measure by making connections with others. I witnessed a lot of new alliances being formed and when I spoke to the postdocs about it they told me:

> The best way to publish lots of papers is to have lots of good collaborators, I've found, and God knows I haven't published that many papers. And I've found my collaborators have got better and better as I've gone on. You'll find it very difficult to find good collaborators if you behave in a way that damages your reputation. [Postdoc 4]
>
> Definitely who I talk to has an effect on the science. There are some co-checks that I started because I was talking to people in the conference. [Postdoc 7]

While I remain neutral about the construction of facts, these statements reveal the extent to which social interactions and scientific knowledge are inextricably linked. You need good credit before you can attract collaborators and 'do science', thus accruing further credit. The desire to limit scientific uncertainty and establish a single 'truth' leads not only to uncertainty among individual scientists about how much credit they will get, but also motivates an increase in communication with others, which in turn directs the course of the science.

23.5 Credit and certainty: Truth and mythology

From the perspective of ANT, when a postdoc persuades people to use his code, he is in the process of reifying the parameters produced by that code into scientific fact. Social constructionists propose that there are no underlying Platonic truths in the universe, but the scientists I spoke to had no doubt about the reality of the physical world.

> It's as true as truth gets, what we're looking for right now. I've seen people devoting time, effort, careers, whatever, just because they want the truth. Veritas, right? [PhD 3]

Nevertheless, scientists recognize and identify three main types of uncertainty within scientific knowledge itself: systematic, statistical and theoretical. Systematic uncertainties, or 'systematics', are caused by something inherent to the system being observed, such as the distortion of the image by the telescope optics and the atmosphere (this quality is known as the point spread function or PSF). They cannot be reduced by simply gathering more data, only by trying to understand more deeply what is happening and trying to construct a model from the data available. But one can never be sure that everything is accounted for – there are always unknown unknowns. Statistical uncertainty is a random fluctuation in the data, such as that caused by different numbers of photons hitting a charge-coupled device (CCD) each time the telescope is pointed at a certain area of the sky. The image of a galaxy is different in every exposure, but this distribution can be modelled and the uncertainty will diminish as a greater number of images are taken. An example of theoretical uncertainty is that general relativity cannot be reconciled with quantum theory, and it is not certain whether it holds in every circumstance.

Some of these uncertainties can be quantified by the addition of error bars, which the scientists then try to shorten (equivalent to reducing the area of an ellipse), often by constructing a model. The aim is to reduce the systematics until they are significantly smaller than the statistical effects, which will be small when DES is complete because a very large volume of sky is being surveyed in depth. Ultimately, in order to plot a path through hundreds of uncertainties, scientists put their trust in the scientific method:

> Every problem is followed with the scientific method and it's science when you do it that way and it's belief when you don't. We try to avoid belief. Belief in *w* or belief in anything else. [Senior scientist]

I conjecture that the scientists' certainty in the division between belief and knowledge emerges or is reinforced in the course of their everyday work (whether or not this is officially classed as doing science), which is above all a process of decreasing systematic and statistical uncertainties in the data. The difficulty is finding out where the border between knowledge and belief actually lies. While people were keen to tell me how hard it is to separate the signal from the data, they refuted the constructionist idea that the signal emerges in tandem with the analysis, has no existence prior to this process and is no way pre-determined. Nevertheless, if they *are* getting closer to truth, it is only through a process of increasing complexity. At what point the error bars can be said to be small enough is an unresolvable question and the presence of even a tiny amount of uncertainty means that the theoretical model is potentially open to change. This makes DES scientists a lot less certain about any of their findings than the popular perception of science would allow. It is widely acknowledged that the current cosmological theory could switch at any time to something radically different, and the degree to which a model is trusted is thought to be a highly subjective judgement.

> People don't realise that blind faith is not something exclusive to religion. Interpretations are not possible from a purely objective point of view. As soon as a piece of information enters your brain you become subjective. [Senior scientist]
>
> The theory itself is a nice story that we tell ourselves, but in fifty years it's going to be completely different. You shouldn't take any theory seriously, that's the point. I think the point is that within the context of what we know today, what can we do to know more? [Postdoc 1]

I watched one postdoc discussing a science paper with his student, and the postdoc explained that 'science is a process of trying to convince yourself that you're doing the right thing'. For a scientist, reducing the uncertainties in one's scientific analysis is predominantly a process of convincing oneself – of reducing uncertainty in one's own mind. Constructionists would say that this process cannot be separated from the wider context of which the scientist is a part, but scientists would claim that persuading other people that the knowledge is 'robust', which is where credit comes in, is a separate matter: 'Other people believe you if you manage to sell it. That's a totally different issue' (Postdoc 1).

In order to do any work at all, scientists acknowledge they must make assumptions about the universe, which may or may not be correct. 'You can't analyse data or understand it without making some kind of assumptions about what you're doing and what all the things in it mean' (Postdoc 6). They may be realists (we are getting closer to the truth) but they have no doubt that everything is liable to change.

Collins (1985: 144) argues that the degree of certainty which is ascribed to scientific knowledge increases dramatically the further away one gets from its point of creation, both in space and time. He points to the way that science is taught in schools, where children learn there is a 'right' result for an experiment and stage-manage their methods to achieve it. Collins is a constructionist, but one does not have to be to agree with him.

I suggest that one explanation for this change in the conception of scientific knowledge as we go outwards from its origin, is that it is the result of pressure from critical points in the network[9] (to borrow the heuristic image from Science and Technology Studies). In order to amass credit with the public and with funding bodies, scientists must try to convince them to expect (and accept) some measure of certainty from the results, and this pressure increases as scientific projects get bigger and more conspicuous. The very title of the Dark Energy Survey indicates that it has been funded on the premise that it will be able to deliver some definitive statement about a fundamental property of the universe.

---

[9] Martin (1998), among others (e.g. Haraway, 1994), argues that a network is too definite and simple a metaphor for imagining the way in which science interacts with the rest of society. She suggests instead Deleuze and Guattari's (1980) image of the rhizome, a root-like underground plant stem that can be broken apart yet still continue to spread. It may more effectively depict the assemblage of fractured, discontinuous connections between scientists, scientific knowledge and the rest of the world.

The majority of the funding for DES came from the Department of Energy (DOE) in the United States, which only funds large projects focusing on single, holistic questions. Professor T, who moved to cosmology from particle physics, said:

> We actually managed, over the years, to change the DOE mission statement. Ten or fifteen years ago it was all about particle accelerators and particle physics, but they now include the fundamental science of the universe.

Historically, astronomy was a small, elite society and cosmological theory was highly speculative, but as the instrumentation improves the discipline is being restructured and becoming, at least outwardly, more certain of its scientific knowledge. As Postdoc 2 put it, 'No one will fund you if you say, "I have no idea what I'm going to find but give me millions to build a telescope and do these observations"'.

According to Schrempp (2012), the kind of tangibly credible statements about the cosmos that DES must deliver in order to receive public and political credit (in terms of credibility and reward), are most likely to transpose in ways that are reminiscent of the strategies characteristic of traditional mythologies. Myths tend to be unified cosmic visions with an anthropocentric slant and are often, outwardly, decried as anathema to science. But Schrempp suggests that when trying to parlay scientific findings into compelling visions 'there are some things that myth still does best and for which as yet we have found no alternative' (Schrempp, 2012: 232). The tendency was noticeable whenever people spoke in terms of 'the bigger picture', and my fieldwork suggested that the desire to synthesize a coherent vision of the universe with human (in this case Western scientist) values, will almost inevitably result in the creation of a contemporary mythology.

Senior scientists in particular, perhaps because they are generally less closely connected to the (uncertain) process of writing computer code, more often articulated a broader perspective. Professor M has been closely involved with DES for over ten years and it was noticeable how often he expressed a concern for 'what to tell the world'. He wanted to place work done in DES within the wider context of the progressive history of science.

> It would be nice if this project could make a difference, either in terms of much more strongly establishing that $w = -1$ or finding something

else about *w*, or estimating neutrino mass. It would be nice if this project could be important, from a perspective of twenty years from now. It would be nice if it makes a landmark discovery.

His social imaginary (Anderson, 1983) had a grander quality than that of the younger scientists and he was more likely to use poetic language when talking about the project. For example, he told me, 'Whether or not we'll find something new depends on whether nature is kind enough to unveil it'. The image of Nature as a goddess, hiding her secrets behind a veil, goes back at least to Renaissance Italy and has been a constantly recurring theme throughout the development of modern science. Unlike Francis Bacon (1561–1626), who thought the secrets of nature were 'better discovered under the torture of the [mechanical] arts' (Hadot, 2006:120), Professor M implies that DES is simply observing whatever happens to be detected by the telescope. He ignores the thousands of hours of data analysis required in favour of a poetic metaphor which suggests a respectful and admiring, or 'Orphic' (Hadot, 2006), curiosity about nature, rather than any admission of human and mechanical agency in the production of knowledge. In asking what message to give to the world, Professor M is concerned about the credit of the collaboration as a whole, with which his identity is firmly linked.

23.6 Conclusion

I have argued that scientists themselves establish a contentious boundary between what does and does not count as doing science. Since doing science is the principal means of amassing credit for a scientist, both in terms of wider recognition (publications) and of self-esteem, those who are not, or not obviously, working on the science side of the division feel additional uncertainty about their future within academia, or frustration with their current situation. Conversely, in order to try and make sure all scientists get some credit, the division is reinforced by defining the 'not science' as infrastructure work. The type of work scientists carry out, and how much credit they acquire, is central to constituting their self-identity as scientists.

Latour and Woolgar (1979) observed that a scientist's ability to produce scientific knowledge (information) can be thought of as a commodity and that, in order to have a successful career in academia, scientists must to some extent be strategists. I would agree with this and suggest that the concept of credit, if it is broadened to include reward, credibility and self-worth, can account for many of their decisions. However, Latour assumed that all kinds of associations and actors were part of the credit cycle, including money, data, prestige, credentials, equipment and papers (Latour and Woolgar, 1979: 200). I argue that some of these (related) features may not always make it onto the cycle in the first place, depending on whether or not scientists consider them to be science. In the wake of Latour's early study, a great deal of Science and Technology Studies scholarship (e.g. Callon, 1986; Derksen 2000), including ANT, has gone on to collapse what is internal and what is external to science, in order to explain how networks produce scientific facts. My point is that such boundaries are integral to the operation of 'science' as a whole, because they are the benchmark for credit, and credit is what makes science move.

The production of different astronomical catalogues and the use of different computer codes for the same problems, could lead to a lack of standardization across the collaboration, but all DES scientific papers need to be reproducible and not give contradictory results. My fieldwork indicates that the collaboration's concern with presenting one standard set of results to 'the world', generates competition between scientists as a consequence of an individual's need for credit. This leads to further uncertainty in the lives of individuals.

In order to maximize their chances of having a successful scientific career (which for postdocs means first of all finding a permanent job and for senior scientists is partly linked to winning grants), at least two considerations are important. First, scientists can try to ensure that at least some of the work they do is classed as science by other scientists and, second, they can aim to form good collaborations with others. In effect, this means that the emergence of scientific knowledge is dependent on the personality of individual scientists. Social constructionists might argue that this process is evidence that scientific facts come into existence contingent on circumstances and there is no objective reality 'waiting' to be discovered. The tendency to want to have a single 'truth', such as establishing only certain codes, could potentially undermine the vision of science as a progressive, universal venture.

However, I do not think this question is resolvable. Constructionists argue that knowledge is inseparable from the social structure and that 'a private language is no language at all' (Collins, 1985:147), and Latour and Woolgar (1979: 230) go as far as to suggest that it is the need to obtain credit that leads to the production of facts,[10] but I maintain that credit cannot explain what convinces a particular scientist that their conclusions (e.g. about DES data) are trustable and robust. Scientists clearly believe they are moving closer to fundamental truths, but they admit that ascertaining where the border between belief and knowledge lies is a subjective process. They have no expectation of ever attaining certain knowledge about anything, and indeed no desire to, since that would mean the 'end' of science. Every scientist I spoke to would be happy if DES data infer that the current best prediction for the dark energy parameter ($w = 1$) is incorrect. To a scientist, uncertainty is integral to knowledge.

I suggest, therefore, that the certainty scientists project to the wider world is primarily a consequence of the need to amass enough credit (in terms of credibility and reward) to keep doing science. The bigger, more expensive and more complicated astronomy surveys become, the more pressure there is on scientists to convince funding bodies and the general public that they will produce some guaranteed final result. When scientists spoke about the bigger picture, the language used seemed to convey a wish to regain, or impart, a coherent cosmic vision. Following Schrempp, the introduction of this mythic quality into scientific cosmology is prevalent whenever scientists communicate to a wider audience because it is enduringly effective in imparting the 'wonder' of science, even while what scientists are actually offering is an intrinsically uncertain conceptual framework. It was noticeable, too, how few DES scientists specified 'wonderment of the cosmos' as a motive for remaining in science, although it was often given as a reason for studying astronomy in the first place. It appears that the distribution of credit to some extent compensates for the diminishing role of wonderment, perhaps displaced by so much uncertainty.

---

[10] ANT is more circumspect and proposes only to offer a method to explore the relations within a network, yielding insights rather than explanations (Latour, 2005), although the distinction is not always obvious.

References

Anderson, B. (1983). *Imagined Communities*. London: Verso.

Bourdieu, P. (1972). *Outline of a Theory of Practice*. Cambridge: CUP.

Biagioli, M. (1998). The instability of authorship: Credit and responsibility in contemporary biomedicine. *The FASEB Journal* **12**: pp. 3–16.

Callon, M. (1986). Some Elements of a Sociology of Translation: The Domestication of the Scallops and the Fishermen of St. Brieuc Bay. In *Power, Action and Belief: A New Sociology of Knowledge?* Law, J. (ed.). London: Routledge and Kegan Paul.

Collins, H. (1985). *Changing Order: Replication and Induction in Scientific Practice*. Chicago: University of Chicago Press.

Deleuze, G. and Guattari, F. (1980). *A Thousand Plateaus*. London: Bloomsbury Academic.

Derksen, L. (2000). Towards a sociology of measurement: The meaning of measurement error in the case of DNA profiling. *Social Studies of Science*, **30**, (6): pp. 803–845.

Hacking, I. (1983). *Representing and Intervening*. Cambridge: CUP.

Hacking, I. (1999). *The Social Construction of What?* London: Harvard University Press.

Hadot, P. (2006). *The Veil of Isis: An Essay on the History of the Idea of Nature*. London: Harvard University Press. Translated by M. Chase.

Haraway, D. (1994). A game of cats cradle: Science studies, feminist theory, cultural studies. *Configurations* 2(1): pp. 59–71.

Lakatos, I. (1970). Falsification and the methodology of scientific research programmes. In *Criticism and the Growth of Knowledge*. Lakatos, I. and Musgrave, A. (eds), Cambridge: CUP.

Latour, B. (1987). *Science in Action*. Cambridge, Massachusetts: Harvard University Press.

Latour, B. (2005). *Reassembling the Social: An Introduction to Actor-Network-Theory*. Oxford: OUP.

Latour, B., and Woolgar, S. (1979). *Laboratory Life. The Construction of Scientific Facts*. Reprint, USA: Princeton University Press, 1986.

Martin, E. (1998). Anthropology and the cultural study of science. *Science, Technology and Human Values*, **23**: pp. 24–44.

Pickering, A. (1984). *Constructing Quarks: A Sociological History of Particle Physics*. Edinburgh: Edinburgh University Press.

Quine, W. O. (1953). *From a Logical Point of View*. Cambridge, Massachusetts: Harvard University Press.

Schrempp, G. (2012). *The Ancient Mythology of Modern Science*. Quebec: McGill-Queens University Press.

Traweek, S. (1988). *Beamtimes and Lifetimes: The World of High Energy Physicists*. London: Harvard University Press.

CHAPTER 24

A Philosopher's Look at the Dark Energy Survey: Reflections on the Use of the Bayes Factor in Cosmology

Michela Massimi

In this chapter, philosopher of science Michela Massimi reflects on the use of the 'Bayesian' method to select models and estimate parameters in cosmology. She highlights the advantages over the 'frequentist' approach, but cautions us to tread carefully if relying on Bayes for scientific 'evidence'.

24.1 Introduction

Cosmology is one of the most exciting contemporary areas of research for philosophers of science like myself. According to current estimates of the lambda cold dark matter (ΛCDM) model, the present-epoch universe consists of approximately 70% dark energy, 25% dark matter and 5% ordinary matter. Clarifying the nature of dark matter and dark energy remains however an open and pressing question for contemporary research both in particle physics and cosmology, and a treasure trove of ideas for philosophers. I started my first job at University College London (UCL) in 2005 as a Lecturer in Philosophy of Science, at the Department of Science and Technology Studies, within the Faculty of Mathematics and Physical Sciences. In that context, I met Ofer Lahav, who took an active interest in philosophy and over several working lunches at the Housman Room at UCL introduced me to DES project and some of the mind-boggling ideas behind dark energy. We published a short popular piece as a result of that dialogue across philosophy and cosmology (Lahav and Massimi, 2014).

More recently, for my European Research Council project *Perspectival Realism. Science, Knowledge and Truth from a Human Vantage Point*,[1] I have continued to look at the Dark Energy Survey (DES) and its fascinating upcoming results. This time, with a series of methodological questions in mind, I have been looking at probe integration in DES and in particular at the widespread use of Bayesian methods for parameter estimation and model selection.

Bayesianism has proved hugely successful as a statistical method in cosmology. Cosmological data are very complex to harvest. Cosmological models typically contain very many parameters that may prove difficult to estimate and may not be well constrained by the data. For these reasons, frequentist approaches (i.e. χ^2) looking for the best fit of models to data might not always be feasible; or, might end up unduly penalizing interesting models with not well constrained parameters (see Amendola and Tsujikawa, 2010, p. 363-4). Bayesianism offers statistical tools for parameter estimation and model selection more congenial to the specific challenges faced by cosmology: the Bayes theorem and the Bayes factor.

The Bayes theorem is defined as follows:

$$p\left(H|D\right) = \frac{p\left(D|H\right)p\left(H\right)}{p\left(D\right)} \tag{24.1}$$

where H is a hypothesis and D is the experimental data. The theorem says that the posterior probability of the hypothesis given data D is equal to the probability of D given H (the term $p\left(D|H\right)$ is called the 'likelihood') multiplied by the prior probability of the hypothesis H (the term $p(H)$ measures the prior probability of H before D is taken into account), divided by the 'evidence' – that is, the denominator $p(D)$.

The Bayes theorem allows us to constrain cosmological parameters (assuming say model M_1), by calculating the maximum posterior

---

[1]http://www.perspectivalrealism.org

probabilities for any parameter $\theta_i^{M_1}$ (which let us assume for simplicity can have $i = 1, 2$) as follows:

$$
\begin{aligned}
\frac{p(\theta_1^{M_1}|D, M_1)}{p(\theta_2^{M_1}|D, M_1)} &= \frac{p(D|\theta_1^{M_1}, M_1)p(\theta_1^{M_1}|M_1)}{p(D|M_1)} \times \frac{p(D|M_1)}{p(D|\theta_2^{M_1}, M_1)p(\theta_2^{M_1}|M_1)} \\
&= \frac{p(D|\theta_1^{M_1}, M_1)p(\theta_1^{M_1}|M_1)}{p(D|\theta_2^{M_1}, M_1)p(\theta_2^{M_1}|M_1)}
\end{aligned}
$$

$$(24.2)$$

where the $p(D|M_1)$ is called 'Bayesian evidence' (not to be confused with the term evidence more in general) and it cancels out. Thus, the posterior probability for each parameter value is given by the likelihood of the data sets D given the parameter value and the model — $p(D|\theta_1^{M_1}, M_1)$ — and the prior for the parameter value given model — $p(\theta_1^{M_1}|M_1)$ – and similarly for $\theta_2^{M_1}$.

The Bayes factor in turn is usually expressed in terms of the marginal likelihoods for two models, say M_1 and M_2:

$$
R = \frac{\int p(D|\theta_i^{M_1}, M_1) \; p(\theta_i^{M_1}|M_1) \; d^n\theta_i^{M_1}}{\int p(D|\theta_i^{M_2}, M_2) \; p(\theta_i^{M_2}|M_2) \; d^n\theta_i^{M_2}}
$$

$$(24.3)$$

where D are the data sets, $\theta_i^{M_1}$ are n theoretical parameters that enter in model M_1, $p(\theta_i^{M_1}|M_1)$ are the prior probabilities of the parameters, and $p(D|\theta_i^{M_1}, M_1)$ is the likelihood (i.e. the probability of the data sets given the values of the parameters). One advantage of the Bayes factor is that if there is a parameter among the n theoretical parameters $\theta_i^{M_2}$ that is not well constrained by cosmological data, the model selection is not affected by changing ever so slightly the value of the parameter at issue. Thus, model selection using the Bayes factor does not unduly penalize models that – albeit interesting to explore – might nonetheless have not very well constrained theoretical parameters (by contrast with frequentist best-fit approaches). However, the use of the Bayes factor in cosmological model selection is itself subject to specific epistemic limits. In this chapter, I raise some philosophical questions concerning the role of the Bayes factor in testing ΛCDM within DES.

24.2 DES Y1 results

In August 2017, the long-awaited DES Year 1 results were released. I attended DES collaboration meeting in Chicago in June 2017 and the sense of anticipation for the official release was clearly in the air. Several DES papers outlining the methods and the provisional results were officially released in August 2017. The main paper, entitled 'The Dark Energy Survey Year 1 Results: Cosmological Constraints from Galaxy Clustering and Weak Lensing' is the paper I concentrate on here (see Dark Energy Survey Collaboration, 2018). DES Year 1 results include a combination of two main probes for which DES has turned out to have a particularly good sensitivity: namely, galaxy clustering and gravitational lensing. The data were collected with the Dark Energy Camera (DECam) from August 31, 2013 to February 9, 2014 and analysed at DES Data Management (DESDM). By combining galaxy clustering with gravitational lensing, DES has an effective way of controlling galaxy bias and systematic errors. How can DES Y1 data be used to fix constraints on cosmological parameters? And how does model selection take place?

DES Year 1 compares the standard cosmological model ΛCDM with a possible rival wCDM. In ΛCDM, the relevant cosmological parameters are the matter energy density (Ω_m); the baryon density (Ω_b); the massive neutrinos density (Ω_ν, which is often set to zero in other cosmological surveys but in DES Year 1 takes various values by splitting the mass equally among the three eigenstates); the reduced Hubble parameter (h); the dark energy equation of state w, which is fixed to -1 to indicate that it is constant through cosmic time; the amplitude and the spectral index of the primordial scalar density perturbations, A_s and n_s. The universe is assumed to be spatially flat ($\Omega_\Lambda = 1 - \Omega_m$); other assumptions include the ratio of primordial tensor to scalar perturbations r is assumed to be zero, and a two-parameter primordial power spectrum of adiabatic and Gaussian fluctuations.

wCDM is not exactly a rival model, but more of a phenomenological proxy for a variety of physical models that treat the dark energy equation of state parameter w not as fixed at -1, but as a free parameter. Thus, while wCDM shares with ΛCDM, the aforementioned six parameters, it also has a seventh additional free parameter in w. In addition to these main cosmological parameters, there are 20 nuisance parameters including parameters for lens galaxy bias, photo-z shifts for both lens galaxies and source galaxies and shear calibration. These nuisance parameters are common to both ΛCDM and wCDM. Table 24.1 gives the priors for all these cosmological parameters and nuisance parameters. Priors are key in the model selection and parameter estimation procedure that follows.

DES made a choice for 'wide, flat priors that conservatively span the range of values well beyond the uncertainties reported by recent experiments' (DES 2018 043526-13). Posterior probabilities for the main seven

Table 24.1 List of priors for main theoretical and nuisance parameters. (DES 2018 043526 -7.)

Parameter	Prior
Cosmology	
Ω_m	flat $(0.1, 0.9)$
A_s	flat $(5 \times 10^{-10}, 5 \times 10^{-9})$
n_s	flat $(0.87, 1.07)$
Ω_b	flat $(0.03, 0.07)$
h	flat $(0.55, 0.91)$
$\Omega_\nu h^2$	flat$(5 \times 10^{-4}, 10^{-2})$
w	flat $(-2, -0.33)$
Lens Galaxy Bias	
$b_i (i = 1, 5)$	flat $(0.8, 3.0)$
Intrinsic Alignment	
$A_{IA}(z) = A_{IA}[(1 + z)/1.62]^{\eta_{IA}}$	
A_{IA}	flat $(-5, 5)$
η_{IA}	flat $(-5, 5)$
Lens photo-z shift (red sequence)	
Δz_l^1	Gauss $(0.001, 0.008)$
Δz_l^2	Gauss $(0.002, 0.007)$
Δz_l^3	Gauss $(0.001, 0.007)$
Δz_l^4	Gauss $(0.003, 0.01)$
Δz_l^5	Gauss $(0.0, 0.01)$
Source photo-z shift	
Δz_s^1	Gauss $(-0.001, 0.016)$
Δz_s^2	Gauss $(-0.019, 0.013)$
Δz_s^3	Gauss $(+0.009, 0.011)$
Δz_s^4	Gauss $(-0.018, 0.022)$
Shear calibration	
$m^i_{\text{METACALIBRATION}} (i = 1, 4)$	Gauss $(0.012, 0.023)$
$m^i_{\text{IM3SHAPE}} (i = 1, 4)$	Gauss $(0.0, 0.035)$

parameters in both ΛCDM and wCDM are then calculated by multiplying likelihoods (featuring evidence from different combined data sets) by the wide flat priors as per Table 24.2.

For the purpose of the present chapter, I am interested in looking at the role played by Bayesianism and in particular by the Bayes factor in this procedure. How to go about comparing and choosing between ΛCDM and its proxy wCDM? DES resorts to the Bayes factor that measures how much evidence D there is *in favour of* M_1 (in this case ΛCDM) over the rival M_2 (wCDM in this case). How to interpret the numerical values of the Bayes factor R? DES adopts the standard Jeffreys scale (Jeffreys 1939/1961), whereby $3.2 < R < 10$ is regarded as *substantial evidence* for M_1 over M_2, and $R > 10$ is regarded as *strong evidence* for M_1. Vice versa, M_2 is strongly favoured over M_1 if $R < 0.1$ and there is *substantial* evidence for M_2 if $0.1 < R < 0.31$ (see DES 2018 043526-15).

The Bayes factor is used for model selection and for ruling out the alternative wCDM phenomenological proxy. This is done by combining DES data sets with a number of other data sets about CMB, BAO and SNe from Planck, BOSS and JLA respectively, which overall give a Bayes factor

Table 24.2 Posterior probabilities (with margins of errors) for the main cosmological parameters in lambda cold dark matter (ΛCDM) and wCDM. (DES 2018 043526 -17.)

Model	Data Sets	Ω_m	S_8	n_s	Ω_b	h	$\sum m_\nu$ (eV) (95% CL)	w
ΛCDM	DES Y1 $\xi_\pm(\theta)$	$0.260^{+0.065}_{-0.037}$	$0.782^{+0.027}_{-0.027}$	...			...	...
ΛCDM	DES Y1 $w(\theta) + \gamma_t$	$0.288^{+0.045}_{-0.026}$	$0.760^{+0.033}_{-0.030}$	...	...	...	...	...
ΛCDM	DES Y1 3x2	$0.267^{+0.030}_{-0.017}$	$0.773^{+0.026}_{-0.020}$	...	...	...	...	...
ΛCDM	Planck (No Lensing)	$0.334^{+0.037}_{-0.026}$	$0.841^{+0.027}_{-0.025}$	$0.958^{+0.008}_{-0.005}$	$0.0503^{+0.0046}_{-0.0019}$	$0.658^{+0.019}_{-0.027}$	...	...
ΛCDM	DES Y1 + Planck (No Lensing)	$0.297^{+0.016}_{-0.012}$	$0.795^{+0.020}_{-0.013}$	$0.972^{+0.006}_{-0.004}$	$0.0477^{+0.0016}_{-0.0012}$	$0.686^{+0.009}_{-0.014}$	< 0.47	...
ΛCDM	DES Y1 + JLA + BAO	$0.295^{+0.018}_{-0.014}$	$0.768^{+0.018}_{-0.023}$	$1.044^{+0.019}_{-0.087}$	$0.0516^{+0.0050}_{-0.0080}$	$0.672^{+0.049}_{-0.034}$	...	...
ΛCDM	Planck + JLA + BAO	$0.306^{+0.007}_{-0.007}$	$0.815^{+0.015}_{-0.013}$	$0.969^{+0.005}_{-0.005}$	$0.0483^{+0.0008}_{-0.0006}$	$0.678^{+0.007}_{-0.005}$	< 0.22	...
ΛCDM	DES Y1 + Planck + JLA + BAO	$0.298^{+0.007}_{-0.007}$	$0.802^{+0.012}_{-0.012}$	$0.973^{+0.005}_{-0.004}$	$0.0479^{+0.0008}_{-0.0008}$	$0.685^{+0.009}_{-0.007}$	< 0.26	...
wCDM	DES Y1 $\xi_\pm(\theta)$	$0.274^{+0.073}_{-0.042}$	$0.777^{+0.036}_{-0.038}$	...	...	...	...	$-0.99^{+0.33}_{-0.39}$
wCDM	DES Y1 $w(\theta) + \gamma_t$	$0.310^{+0.049}_{-0.036}$	$0.785^{+0.040}_{-0.072}$	...	...	...	...	$-0.79^{+0.22}_{-0.39}$
wCDM	DES Y1 3x2	$0.284^{+0.033}_{-0.030}$	$0.782^{+0.036}_{-0.024}$	...	...	...	...	$-0.82^{+0.21}_{-0.20}$
wCDM	Planck (No Lensing)	$0.222^{+0.069}_{-0.024}$	$0.810^{+0.029}_{-0.036}$	$0.960^{+0.005}_{-0.007}$	$0.0334^{+0.0099}_{-0.0032}$	$0.801^{+0.045}_{-0.097}$	...	$-1.47^{+0.31}_{-0.22}$
wCDM	DES Y1 + Planck (No Lensing)	$0.233^{+0.025}_{-0.033}$	$0.775^{+0.021}_{-0.021}$	$0.971^{+0.006}_{-0.006}$	$0.0355^{+0.0050}_{-0.0039}$	$0.775^{+0.056}_{-0.040}$	< 0.65	$-1.35^{+0.16}_{-0.17}$
wCDM	Planck + JLA + BAO	$0.303^{+0.010}_{-0.008}$	$0.816^{+0.014}_{-0.013}$	$0.968^{+0.006}_{-0.006}$	$0.0479^{+0.0016}_{-0.0014}$	$0.679^{+0.013}_{-0.008}$	< 0.27	$-1.02^{+0.05}_{-0.05}$
wCDM	DES Y1 + Planck + JLA + BAO	$0.301^{+0.007}_{-0.010}$	$0.801^{+0.011}_{-0.012}$	$0.974^{+0.005}_{-0.005}$	$0.0483^{+0.0014}_{-0.0016}$	$0.680^{+0.013}_{-0.008}$	< 0.31	$-1.00^{+0.05}_{-0.04}$

$R = 0.1$, 'disfavouring the introduction of w as a free parameter' as in wCDM (see DES 2018 043526 -23).

Bayesian inferences are absolutely central both for model selection between ΛCDM and wCDM; for fixing new tight constraints on the dark energy equation of state $w = -1$ (by combining DES Y1 data sets with other surveys' data sets); and for improving on the estimates of relevant cosmological parameters. Yet in assessing results coming from a large survey such as DES (and what they might tell us about the correct model; or the 'true parameters'), a cautionary tale is in order. Such a cautionary tale has to do with the all-powerful, and ubiquitous use of the Bayes factor. I give some of the reasons for the cautionary tale in Section 24.3 where I review some well-known statistical phenomena and explain how they might affect cosmological results.

24.3 On the Bayes factor

24.3.1 Context-dependence of standards of evidence along the Jeffreys scale

How much evidence is evidence enough? What counts as strong, substantial or decisive evidence for some model? How to draw these lines of demarcation? One clear advantage of the Bayes factor along the Jeffreys scale is that it provides a convenient way of demarcating these standards of evidence in any given area of inquiry. The Jeffreys scale tells us that given two hypotheses (the null hypothesis H_0 and the rival hypothesis H_1), $3.2 < R < 10$ is *substantial evidence* for H_0 over H_1; $10 < R < 100$ is *strong evidence* for H_0 over H_1 and $R > 100$ is *decisive* evidence for H_0. Much as this way of carving different standards of evidence has proved convenient, there is a peculiar tyranny of the Jeffreys scale. Scales do matter in interpreting probability measures. And in the case of the Jeffreys scale, the tyranny comes from its sensitivity to the context of inquiry. As Kass and Raftery (1995, p. 777) have observed, the Jeffreys scale is only a 'rough descriptive statement about standards of evidence in scientific investigation'. And that it provides only rough descriptive statements (as opposed to clear cut-off points about what does or does not count as evidence in favour of H_0 over H_1) becomes clear when one considers the different contexts of use for the Bayes factor along the Jeffreys scale.

For example, in the context of forensic science, when evaluating evidence as 'decisive' in a criminal trial, a jury is likely to require posterior probabilities for the hypothesis H_1 (person a is guilty) over hypothesis H_0 (person a is innocent) on a much more expanded and fine-grained scale than a scale of up to or over 100. Unsurprisingly, the Swedish National Laboratory of Forensic Science uses a Bayes factor to assess the ratio between the likelihood that evidence e occurred given H_0 and the likelihood that evidence e occurred given H_1 on a more expanded Jeffreys scale of up to a million (see Kruse, 2012). Only when the likelihood ratio is above 1,000,000 is forensic evidence regarded as *extremely strongly supporting* the null hypothesis over the rival one. *Strong support* requires the likelihood ratio to be between 6000 and 1,000,000. Below 6000 there is just support, and below 100 there is only marginal support.

But there is more to this context-sensitivity of the Jeffreys scale. In the case of forensic science – when a jury has to assess the probability of evidence e given the hypothesis that say person a is guilty – a lot of circumstantial considerations tend to enter into the final verdict. It might happen, for example, that in cases of circumstantial evidence, one cannot claim just on the basis of the Bayes factor alone that there was indeed absence of scientific evidence connecting person a directly with the crime (see Evett, 2015 for a discussion of this example). Or, the other way around. It might be that there is an important difference (in cases of circumstantial evidence) between saying that for example 'a hammer could have made the mark' and concluding on that basis that 'there is moderate scientific support for the 'prosecution case'. Conflating the former with the latter would constitute 'prosecution bias' (see again Evett, 2015, p. 6). In general, reaching a verdict of 'not guilty' using the Bayes factor along the Jeffreys scale in forensic science is not equivalent to reaching a verdict of being 'innocent' precisely because of the context-sensitivity of the inquiry and possible presence of circumstantial evidence.

Switching from forensic science to cosmology, one is here reminded of the need to exercise caution and not conflate a sentence such as 'evidence e favours ΛCDM over wCDM' with the sentence 'ΛCDM is true'.[2] The sensitivity of the Bayes factor along the Jeffreys scale to the context of inquiry

---

[2] I am grateful to Ruth King for helpful comments on this point.

highlights two important features. First, depending on what the stakes are in a given context (very high in the case of forensic science, where error may result in either convicting an innocent person, or letting a criminal go free), the scale might need be adjusted to provide a proportionate way of evaluating evidence for or against rival hypotheses. Second, regardless of what the stakes are, standards of evidence need nonetheless be adjusted to evaluate in a proportionate way what counts as strong, substantial or decisive evidence in favour of or against a given hypothesis in any given context. This is a powerful reminder of how to take the Bayes factor along the Jeffreys scale with a pinch of salt, rather than as a gold standard for what counts as strong, substantial or decisive evidence in favour of a given hypothesis in absolute terms, which leads me to my next point.

24.3.2 Relative versus absolute evidential measures

The Bayes factor measures standards of evidence always *relative to* two (or more) rival hypotheses, as opposed to providing standards of evidence in *absolute terms*. It measures how likely the evidence e is given the null hypothesis or a rival hypothesis H_1. And whenever the Bayes factor suggests that there is substantial or strong evidence for the null hypothesis, this should not be understood as equivalent to saying that the null hypothesis is the correct true hypothesis. There might be other rival hypotheses that have not yet been considered, or whose evidence for (or against) has not yet been evaluated using the Bayes factor. Those rival hypotheses remain live candidates worth exploring and examining in future research.

24.3.3 Dimensionality and nested models

Calculating the Bayes factor in cosmological model selection is subject to specific implementation problems caused by the large number of parameters. With 26 free parameters and 431 degrees of freedom, ΛCDM is a good candidate for this kind of implementation problem, despite best-fit analysis in terms of standard deviation of individual data points. The risk of poor results affected by small likelihoods in the parameters is high when using the Bayes factor for hypotheses with high dimensionality (such as ΛCDM and wCDM). Moreover, if there is enough evidence in favour of ΛCDM, this is reason enough for not considering any further wCDM (even as a sheer phenomenological proxy, rather than a fully-fledged rival model). But, one would require much stronger evidence for wCDM, since ΛCDM might be regarded as a special case of wCDM when w is set equal to -1. Thus, more evidence would seem to be required to favour wCDM over ΛCDM than is required to favour ΛCDM over wCDM.

24.3.4 Simplifications and sensitivity to priors

The Bayes factor is also well known for being highly sensitive to the choice of the priors and simplifications concerning such a choice (see Kass and Raftery, 1995). Such simplifications must be fully justified. Genuine priors in Bayesianism are expected to be subjective, data-independent, personal degrees of beliefs. Looking at Table 24.1, which codifies those priors for both main cosmological parameters and nuisance parameters, it is clear that in the case of nuisance parameters such as photo-z for example (but also in parameters concerning the shear calibration such as m^i), the priors are obtained from previous analyses. The priors on the lens and source photo-z in Table 24.1 were obtained for example from selecting and sampling galaxies from an already existing database (COSMOS), which was taken as 'representative of DES sample with successful shape measurements based on their colour, magnitude and pre-seeing size. The mean redshift of this COSMOS sample is our estimate of the true mean redshift of DES source sample "(DES 2018, 043526 -8). In any Bayesian analysis, it is standard practice to conduct a prior sensitivity analysis, and check how much the posterior probability might change by a different specifications of priors. In the case of DES, the choice of priors and the prior sensitivity analysis relies on other galaxy catalogues. Choices are made every step of the way about which galaxy catalogue to use as a sample, which is the most representative for DES sample and the grounds on which the mean redshift of the COSMOS sample might count as the 'true' mean redshift for DES.

These remarks are not meant (and should not be read as) casting doubts on the validity and rigour that DES undoubtedly put into statistical and systematic-error analysis, when fixing priors in Table 24.1. But they are instead meant to highlight a distinctive way in which Bayesian analysis works in cosmology. Far from being data-independent prior degrees of belief, each and every one of these priors packs as much prior information as needed for the Bayes analysis to be computable. Some of this prior information comes from theoretical considerations (as is the case with the choice of priors for *w*). Others come from data analysis from similar or related surveys (such as COSMOS, but also the Sloan Digital Sky Survey, whose spectroscopic redshift is used for cross-correlation of DES RedMaGiC software used for lens photo-*z*). *Data-dependent* Bayesian priors are of course perfectly acceptable, but they also reinforce pre-existent cosmologists' choices about which data to count as relevant (and which one can be bracketed off as irrelevant). And it is worth keeping this aspect in mind whenever such choices have to do with ongoing debates among cosmologists about important cosmological parameters (such as the value of the Hubble constant) and how these choices in turn may feed into catalogues used to fix the data-dependent priors. This is another reminder of how the Bayes factor works best as 'guiding an evolutionary model-building process' (Kass and Raftery, 1995, p. 773) whereby there is a clear continuity and almost evolution in building and assessing model selection following (theoretically and experimentally) well-trodden paths.

24.3.5 The Bartlett paradox and wide priors

But there is an even more interesting feature of the Bayes factor that would seem to create a particular tension in delivering on two related tasks: model selection and parameter estimation. When it comes to parameter estimation, the Bayes theorem invites a choice of flat wide priors that are meant to be as uninformative as possible. Such choice is made so as to minimize the extent to which the posterior probabilities of parameters might be sensitive to the choice of priors. However, choosing flat wide priors is not a methodologically innocent move, and has side-effects in what statisticians call the Bartlett paradox (from Bartlett, 1957, who first raised it against Jeffreys). The problem concerns how prior choices impact on model selection in turn. It turns out that if priors on a given relevant parameter ψ under H_1 are wide, that is, have a large spread (precisely to make the prior choice as broad and as much uninformative as possible depending on the parametrization of a given model), the Bayes factor will tend to favour the null hypothesis (see Raftery, 1996). Jeffreys responded to the paradox by suggesting that priors with a large spread should be avoided for main theoretical parameters (although the ban on wide priors does not necessarily extend to nuisance parameters too).

Going back to DES: as already remarked above, the choice of priors is crucial in the process of both parameter estimation and model selection. On the one hand, it might be desirable to choose wide flat priors that assign uniform probabilities to a range of possible values within an interval in performing parameter estimation. On the other hand, the Bartlett paradox tells us that, when it comes to model selection, flat wide priors tend to force the choice on the null hypothesis (ΛCDM, in this case). Thus, it seems that choosing the right priors for model selection is somehow in tension with the desideratum of having broad and more uninformative priors for parameter estimation. DES collaboration (2018 043526 -13) is clearly aware of the problem of using wide priors: 'overly broad parameter ranges would distort the computation of the Bayesian evidence, which would be problematic as we will use Bayes factors to assess the consistency of the different two-point function measurements, consistency with external data sets, and the need to introduce additional parameters (such as w) into the analysis. We have verified that our results below are insensitive to the prior ranges chosen'. Cosmologists with a pragmatic attitude may at this point reply that a difference is a difference only if it does make a difference. If the results prove insensitive to the range of priors chosen and prior sensitivity analysis has been done correctly, no one should be worried about the Bartlett paradox.

In response, there are reasons, I think, why the Bartlett paradox needs be taken into account when dealing with model selection in cosmology. Jeffreys argued that while the Bayes factor is sensitive to the choice of wide priors for theoretical parameters, nuisance parameters (and their priors) are less relevant to the final outcome of model selection. In principle, this division is correct whenever theoretical and nuisance parameters are sufficiently insulated and independent of one another. But this is often not the case in cosmology. For example, the nuisance parameters about lens photo-z shifts feed directly into the gravitational lensing data that are used to estimate the value of the main parameter Ω_m (among others). And Ω_m enters in turn in model selection, whereby (as per Figure 24.1) constraints are fixed on the value of the dark energy equation of state w that ultimately disfavours wCDM over ΛCDM. Thus, in practice, theoretical parameters and nuisance parameters are never sufficiently insulated from each other. More to the point, the Bartlett paradox is an important reminder that in model selection using Bayesianism one should avoid the 'fallacy of acceptance': interpreting 'accept model M_1' (because there is no inconsistent evidence against it) as evidence for model M_1.

24.3.6 The Bayes factor as Occam's razor?

But there is more. Further worries might accompany the use of the Bayes factor in cosmological model selection. This time because of another feature of the Bayes factor well known to statisticians, namely that the Bayes factor can be misleading whenever the models have the same number of parameters (see Atkinson, 1978). It has been shown that if two models are nested and both models are true, the Bayes factor tends to favour the

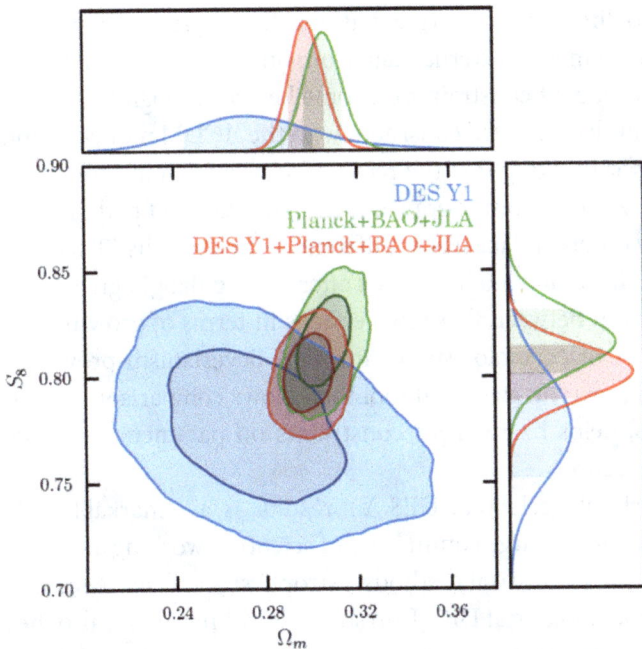

Fig. 24.1 Fixing constraints within lambda cold dark matter (ΛCDM) by combining on the $\Omega_m - S_8$ plane the Dark Energy Survey (DES) Y1 data (blue) with Planck + baryon acoustic oscillations (BAO) + joint lightcurve analysis (JLA) data (in green) and overlap region in red. Images from DES 2018 043526 -21, Fig. 13.

simpler model (i.e. the one with the smaller number of parameters, going back to the issue about dimensionality). In other words, the odds in favour of the simpler model increase as the number of parameters n increases. In defence of the Bayes factor, Smith and Spiegelhalter (1980) have argued that it acts as Occam's razor in model selection: choosing the simpler model is more expedient for predictive purposes. And so once again, one might wonder whether in the case of cosmology wCDM (or similar rival models) might get penalized by the use of the Bayes factor because they are less 'simple' than ΛCDM (by having w as an extra free parameter).

24.4 Philosophical coda

In an era characterized by large cosmological surveys searching for dark energy, DES Y1 results have proved a remarkable confirmation of the validity of the current standard model in cosmology. These results integrate data from two different probes (gravitational lensing and galaxy clustering) and compare them with external data from other cosmological surveys so as to fix very stringent constraints on some key cosmological parameters. Such constraints have proved consistent with the ΛCDM model and offered new reasons for thinking that the received view, which interprets dark energy as a non-zero vacuum energy, is indeed on the right path. Explaining how and why different data sets at different cosmic epochs (380,000 years after the Big Bang *and* ten billion years after the Big Bang) agree – and what has happened in between these two epochs in terms of growth of structure – is the task of cosmologists to find out. Bayesianism provides a ubiquitous and powerful statistical tool that allows comparison among different data sets, helps fix rigorous constraints on parameters, and delivers on model selection.

We should celebrate DES Y1 results as a remarkable collaborative achievement of a large community of scientists working at the intersection of cosmology, statistical methods, astrophysics and space technology. As a philosopher looking at DES, I am struck by the power but also the epistemic limits of using Bayesian inferences in cosmology. These epistemic limits, well known to statisticians and cosmologists alike, should entice the scientific community to continue and further expand the search for alternative models to the ΛCDM model (rather than rest on ΛCDM's laurels), even if as of today ΛCDM remains on the right path, for all we know.

Acknowledgements

I am very grateful to Ofer Lahav for inviting me to contribute a chapter to this book, and for the many fruitful conversations over the years about DES. I thank Ruth King, Niall Jeffrey, Pablo Lemos, Julian Mayers and Joe Zuntz for helpful comments on earlier versions of this paper. This article is part of a project that has received funding from the European Research Council (ERC) under the European Union's Horizon 2020 research and innovation programme (grant agreement European Consolidator Grant H2020-ERC-2014-CoG 647272 *Perspectival Realism. Science, Knowledge, and Truth from a Human Vantage Point*).

References

Amendola, L., and Tsujikawa, S. (2010). *Dark Energy. Theory and Observations.* Cambridge: Cambridge University Press.

Atkinson, A. C. (1978). 'Posterior probabilities for choosing a regression model', *Biometrika* **65**, 39–48.

Bartlett, M. S. (1957). 'Comment on a statistical paradox' by D. V. Lindley', *Biometrika* **44**, 533–534.

Dark Energy Survey Collaboration (2018). Dark energy survey year 1 results: Cosmological constraints from galaxy clustering and weak lensing, Physical Review **D 98**, p. 043526, doi:10.1103/PhysRevD.98.043526.

Evett, I. W. (2015). 'The logical foundations of forensic science: Towards reliable knowledge', *Philosophical Transactions of the Royal Society* **B370**, 20140263.

Jeffreys, H. (1939/1961). *Theory of Probability*, 3rd ed., Oxford: Oxford University Press.

Kass, R. E., and Raftery, A. E. (1995). 'Bayes factors', *Journal of the American Statistical Association* **90**, No. 430, 773795.

Kruse, C. (2012). 'The Bayesian approach to forensic evidence: Evaluating, communicating, and distributing responsibilities', *Social Studies of Science* **43**(5), 657680.

Lahav, O., and Massimi, M. (2014). 'Dark energy, paradigm shifts, and the role of evidence', *Astronomy & Geophysics* **55**, 3.133.15.

Raftery, A. E. (1996). 'Approximate Bayes factors and accounting for model uncertainty in generalised linear models', *Biometrika* **83**(2), 251266.

Smith, A. F. M., and Spiegelhalter, D. J. (1980). 'Bayes factor and choice criteria for linear models', *Journal of the Royal Statistical Society* **B42**, 213–220.

CHAPTER 25

Artists' Reflections

Judy Goldhill & Jane Grisewood

As artists-in-residence in the Physics and Astronomy Department at University College London (UCL), Judy Goldhill and Jane Grisewood continue their passion for the interaction between art and science. They are compelled by questions surrounding who we are and why we are here, and how these concerns are expressed through art practice. They are captivated by dark energy, blackness and darkness, quasars and black holes, the visible and the invisible.

Through the National Optical Astronomy Observatory (NOAO) in Arizona, we were invited to spend a month as the first artists-in-residence at the NOAO. Six months later we spent a month in the southern hemisphere in Chile at Cerro Tololo Inter-American Observatory (CTIO). Cerro Tololo was awe-inspiring. It was a thrilling experience to be on the magnificent 2700m summit, with light from thousands of stars sparkling through the dark universe. Even to naked eye observers, the immense southern skies revealed the Magellanic Clouds, the Milky Way and the dramatic Leonid Meteor Showers that split the sky with lines of light. We experienced first-hand the perpetual motion of remote galaxies and nearby planets.

Our time in Chile coincided with the observatory's fiftieth anniversary and the launch of the Dark Energy Camera (DECam), where we met University College London (UCL) Professor Ofer Lahav. Witnessing DECam's 'first light' in the 4m Victor M. Blanco Telescope left an intense impression on us and inspired a compulsion to inquire further. Observing the vast clear skies through the powerful telescope, we saw images of incredible objects billions of light years distant. These phenomenal encounters

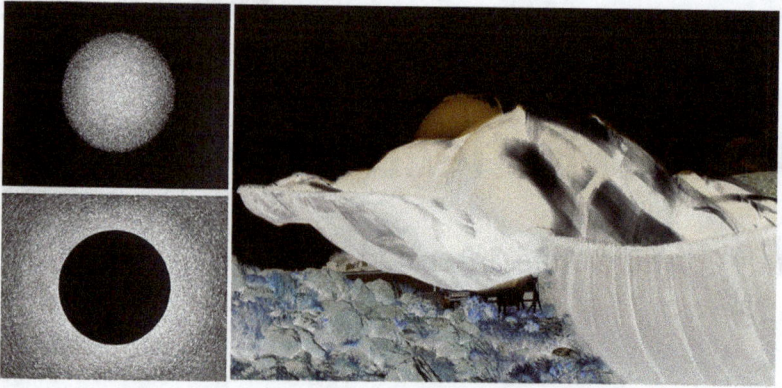

Fig. 25.1 Left: **Black Hole And Quasar, 2016.** Jane Grisewood. The cosmological ink drawing (85 x 52 cm) reflects the positive and negative, the dark and light, probing the mystery of the universe through the Dark Energy Survey (DES). The diptych points up the connection between monster black holes and quasars. The black hole is so powerful in the gravitational field that it absorbs all light and matter, whereas the quasar itself is a supermassive black hole with intense light circling the accretion disc of luminous cloud and gas. Image and text appear in the Phaidon book *Universe*. Right: **Dark Energy Fills My Sails.** Judy Goldhill. This image is from a series of photographs made in response to the inauguration of DES in 2012, to chart the origins and nature of 'dark energy'. The presence of the colour-inverted images is an attempt to refigure the unknown nature of this concept, presenting it as a structured, time-based phenomenon. The Dark Energy Camera, situated inside the dome of the Victor Blanco Telescope, Chile, appears as a point of reference in several photographs.

were intensified further by 'observing the observers' in the control rooms through the night. The spectra of distant stars appearing on monitors could be translated into a wealth of information, making the invisible visible. Observations and discussions with scientists and engineers on the summits and in the downtown headquarters had a radical impact on our art practice and our way of 'seeing' the world. The experience at CTIO further focused our interests in the dark universe and led to us working as artists-in-residence in the UCL Physics and Astronomy Department. Two examples of our work, inspired by the Dark Energy Survey, are shown in Figure 25.1.

Jane Grisewood recalls the experience which led to the work *Dancing with Sirius: Lines of Light* (see Figure 25.2): 'In the high altitude of the Chilean Andes, the cloudless sky revealed a dazzling canopy of stars as night

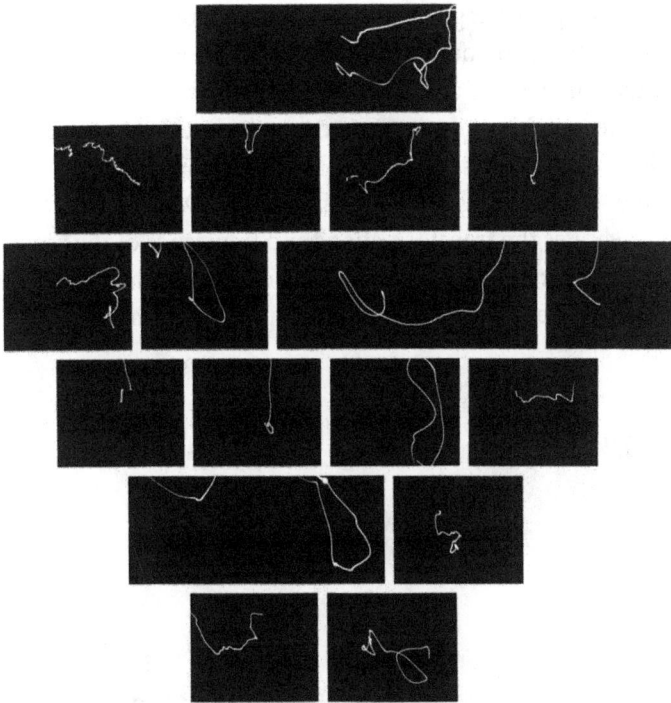

Fig. 25.2 *Dancing with Sirius: Lines of Light.* Jane Grisewood. Invisible lines made visible through photographs of Sirius, the brightest star, taken on the summit of Cerro Tololo in the Chilean Andes during the artist residency. Following Sirius with a handheld camera, random choreographic gestures were drawn in the air, creating ephemeral ghost-like threads in the night sky. *Photographic installation: seventeen mounted matt black prints with the white lines of Sirius; Hahnemuhle photo rag: size 150 x 150 cm.*

fell over the Atacama Desert. Standing on the summit of Cerro Tololo, 'the mountain in front of the abyss' and surrounded by the imposing white observatories in the cold dark atmosphere, there was one star that captivated me: Sirius, the brightest star in the night sky, visible in both southern and northern hemispheres. Hypnotized by its brightness and movement, and aware that it is through darkness that the light of stars is made visible, I spent most of the night, until sunrise, photographing it with a handheld camera. I was intent on following this one dominant star. The random choreographic gestures of my arms, like drawing in the air, dissolved as the next movement occurred, and in turn gave rise to a visible trace. Lines of light emerged, ephemeral ghost-like threads dancing across the sky that, while dynamic and temporal, became suspended in time in the camera. Each line was made possible by my actions, but I had no control over the outcome. My enduring interest in the line draws inspiration from Gilles Deleuze's notion that lines are the basic components of things and events: lines of becoming, always in movement. *Dancing with Sirius: Lines of Light* celebrates the transformations and transience of the night sky, the dark and the light, recording the fleeting moment of invisible lines made visible through black-and-white photographs of light trails.

CHAPTER 26

At the Edge of the Abyss: A Poem for the Dark Energy Survey

Amy Catanzano

Amy Catanzano is Associate Professor of English and poet-in-residence at Wake Forest University in North Carolina. Her work as a writer explores the boundaries between literature, science and art and focuses on quantum physics and cosmology. From December 8 to 14, 2018, with a W.C. Archie Endowed Fund for Faculty Excellence grant from Wake Forest, she conducted research on the Dark Energy Survey at the Cerro Tololo Inter-American Observatory in Chile. As a result of these creative and scholarly investigations, Amy wrote a book-length poem entitled, At the Edge of the Abyss: A Poem for the Dark Energy Survey *(2019). We include the opening lines of the poem here, which reference the phrase, 'mountain at the edge of the abyss', one English translation of 'Cerro Tololo'.*

> Here at Cerro Tololo,
> at the dark
>
> edge of the abyss,
> the abyss is outer space
>
> and the edge is a mirror
> catching light

in a camera pointing
to it from Earth.

Here at Cerro Tololo,
a furnace of galaxies
spirals the heart,
weaving sparks
and thick
mists –

Each galaxy is a performance

hypnotizing space

with whorled motions
that reach the eyes,

which telescope
outward to see the universe
and inward to see
themselves.

The eyes that see themselves
have supermassive
black holes
in their centers,
dark pupils

surrounded by
galactic structures
moved away
from the other galaxies
beyond them,

and those galaxies
are moved away, too,
resulting in

more space

between galaxies
and at an accelerating rate.

CHAPTER 27

The Dark Energy Survey and the Future of Dark Energy

David Weinberg

With its multi-probe approach, the Dark Energy Survey (DES) has done much to shape our conception of dark energy as an experimental field. Over the next two to three years, DES and other large surveys will, for the first time, sharpen measurements of the amplitude of dark matter clustering to the ~ 1% level already achieved for measurements of the expansion history, allowing stringent new tests of modified gravity alternatives to dark energy models. The results of DES and its 'Stage III' brethren will define the agenda for the still more ambitious 'Stage IV' experiments of the 2020s. With rapid growth of observational and modelling capabilities, dramatic changes in our understanding of cosmic acceleration could come quickly, but we are dependent on the kindness of nature as well as our own ingenuity.

27.1 Past

Authors of these chapters are invited to explain how they became involved with the Dark Energy Survey (DES) and what role they play in the survey. Identifying a beginning is difficult even in hindsight, but if I trace my involvement with DES back to a starting point then the best date I can pick is ... January 1992. As a freshly arrived postdoc at the Institute for Advanced Study, I joined the Sloan Digital Sky Survey (SDSS, though in 1992 it was still just the DSS), at a point where we were beginning to define the project's technical requirements and target selection strategies. Over the ensuing decade, the SDSS pioneered many critical aspects of what we now recognize as survey cosmology: construction of extremely powerful instruments for a dedicated multi-year experiment; creation of public data sets with broad scientific applications; collaboration of DOE national laboratories with traditional astronomy institutions and development of a scientific culture that could take full advantage of the collective expertise of hundreds of scientists and engineers, in hardware, software, experimental design, data analysis and theoretical interpretation. Building on the advances made by the SDSS, DES exemplifies all of these traits.

For my seminar talks in the early 1990s, I usually opened with an overhead transparency listing the big questions (as I saw them, anyway) in the field of large-scale structure:

1. Did galaxies and large-scale structure form by gravitational instability?
2. If so, what were the properties of the primordial density fluctuations, and how did they arise?
3. What is the dark matter?
4. Is $\Omega = 1$?
5. What is the relation between the distribution of galaxies and the distribution of mass?

These were all open questions in 1990 and they have all been largely (though not entirely) answered today. This remarkable progress has emerged thanks to the combination of data from the cosmic microwave background (CMB); large galaxy redshift surveys; supernova surveys; primordial deuterium abundance measurements and weak lensing, X-ray and Sunyaev-Zeldovich probes of matter clustering. We now know that structure formed by gravitational instability, from primordial fluctuations with properties very close to those predicted by straightforward versions of inflationary cosmology – that is, adiabatic, Gaussian and nearly but not perfectly scale-invariant. We know that dark matter is non-baryonic, and indeed that it is not composed of *any* of the fundamental particles discovered to date. We know that a model of cold, weakly interacting particles allows a good account of a wide range of observations, with some anomalies that could be a sign of novel dark matter properties or may be explainable with baryonic physics. We know that observed galaxy clustering, across a wide range of redshift and galaxy types, can be explained by populating cold dark matter halos with galaxies in ways that are mathematically straightforward and physically sensible.

Question 4 turned out to have the most startling answer: No *and* Yes. When I wrote this question on my transparencies, I certainly had in mind the density parameter of matter, which we now know to be ≈ 0.3. Although the cosmological constant was a subject of polite conversation, and the occasional journal article, in 1992, I considered it an extreme hypothesis, and an improbable way to reconcile observational evidence for a low-matter density with the theoretical preference for a flat universe. But within a decade we learned that space *is* flat (or nearly so) and that it expands at an accelerating rate. Wow.

Cosmic acceleration is the most surprising cosmological discovery of my scientific lifetime, a demonstration that gravity on cosmic scales repels rather than attracts. If I look for other cosmological discoveries of the last two centuries that have a comparable combination of surprise and importance, I come up with four: the anomalous precession of the orbit of Mercury, the expansion of the universe, dark matter and the CMB. Cosmic acceleration joins an august list of transformational scientific advances and we hope that our efforts to understand it will lead to yet another startling breakthrough, one that will tell us new things about the nature of gravity, or the nature of the quantum vacuum, or exotic new forms of energy, or the properties of the universe over ultra-long time and length scales, or about several of these subjects at once. For now, the field is framed by two broad questions:

1. Does acceleration arise from a breakdown of General Relativity (GR) on cosmological scales or from a new energy component that exerts repulsive gravity within GR?
2. If acceleration is caused by a new energy component, is its energy density constant in space and time?

From its initial conception through to its current analyses, DES has emphasized the ability to pursue four complementary methods to investigate cosmic acceleration: weak lensing, clusters of galaxies, baryon acoustic oscillations (BAO) and supernovae. This way of framing the field's experimental opportunities proved highly influential, in part because it was adopted by the Dark Energy Task Force (DETF; Albrecht et al., 2006), which in turn shaped the discussion and implementation of dark energy experiments by Department of Energy (DOE) prioritization panels and by the Astro2010 decadal survey. It also shaped one of my own most significant contributions to the field, the Weinberg et al. (2013b) review of 'Observational Probes of Cosmic Acceleration', which I advocate unabashedly as an ideal introduction to these methods, to other techniques being used to measure cosmic expansion and growth of structure, and to the integration of these methods into comprehensive constraints on dark energy and modified gravity. While DES is advancing the state of the art on all four of these cosmic acceleration probes, it is weak lensing and cluster cosmology where its potential is truly revolutionary relative to existing data sets.

My direct role in DES has been much smaller than I anticipated in 2006, when Ohio State joined the project, mainly because I became Project Scientist of SDSS-III (Eisenstein et al., 2011) the following year. The Baryon Oscillation Spectroscopic Survey (BOSS) became, in effect, DES's Stage III sister project, leaping forward in BAO measurements from redshift-space galaxy and Lyman-α forest clustering while DES leaped forward in deep, wide-area imaging for weak lensing measurements. Although my observational energies have gone mainly to the SDSS, I have learned a great deal from the talented DES postdocs who have passed through Ohio State over the past decade, and DES has been the prime motivation for much of the theoretical work I have done with students and other collaborators, on combinations of galaxy clustering, galaxy–galaxy lensing and cluster–galaxy lensing (Yoo et al., 2006; Weinberg et al., 2013b; Zu et al., 2014; Wibking et al., 2017; McEwen and Weinberg, 2018). By the time DES is analyzing its final data set, I hope that the non-linear modelling methods we are developing will become practical tools for extracting sharper cosmological inferences from the best weak lensing data set the world has yet seen.

27.2 Present

Over the past decade, improvements in supernova surveys and BAO surveys have sharpened measurements of the cosmic distance scale and expansion rate from the 5% to 10% level down to 1% to 2%, with correspondingly tighter constraints on the equation of state of dark energy. Most current supernova, BAO and CMB measurements are in excellent agreement with the expansion history predicted by ΛCDM, a cosmological model based on cold dark matter, flat space, inflationary initial conditions and a cosmological constant. DES weak lensing data is, for the first time, sharpening measurements of the growth of matter clustering to this 1% to 2% level. The Kilo-Degree Survey (KiDS) and Hyper-Suprime Camera (HSC) weak lensing surveys will achieve similar advances, though the larger survey area of DES should ultimately yield the highest precision. Matched precision in expansion history and growth of structure measurements is critical for distinguishing dark energy explanations of cosmic acceleration from modified gravity theories, since the latter generically alter matter clustering in addition to the global expansion rate.

As I write, the cosmological analyses of the Year 3 DES data set are far advanced, but still hidden inside their blinding box. Like many cosmologists inside and outside of DES, I am eager to see what these analyses have to say about the tension between the predicted and observed amplitudes of matter clustering. The prediction comes from extrapolating the CMB anisotropies observed by the *Planck* mission forward in time to low redshift, assuming ΛCDM and GR-based structure growth. Many but not all weak lensing and cluster studies from the past decade infer an amplitude of matter clustering that is 5% to 10% lower than predicted by *Planck*+ ΛCDM. The significance of this tension has been difficult to quantify because the discrepancy with any individual experiment is only $\sim 2\sigma$, because some of the analyses *are* consistent with the *Planck*+ ΛCDM prediction, because the statistical uncertainty in the *Planck* measurement is itself not negligible, and because any or all of the analyses could be affected by systematic errors larger than those reflected in the quoted error bars. Recent action on this front comes from cosmic shear, galaxy–galaxy lensing and galaxy clustering analyses that draw on SDSS, CFHT, KiDS-450 or DES Year 1 weak lensing measurements and the BOSS, 2dFLens or GAMA redshift surveys.

For KiDS-450, Hildebrandt et al. (2017) and Joudaki et al. (2018) infer values of $S_8 \equiv \sigma_8 \Omega_m^{0.5}$ that are significantly below the *Planck* prediction, but van Uitert et al. (2018) find good agreement with *Planck*+ ΛCDM. Leauthaud et al. (2017) measure a galaxy–galaxy lensing signal for BOSS galaxies that is 10% to 30% lower than predicted by *Planck*-normalized mock catalogues, but the strongest discrepancy is on non-linear scales where the predictions may be sensitive to the assumed relation between galaxies and dark matter halos. DES Year 1 analysis Dark Energy Survey Collaboration (2018) finds an amplitude of matter clustering that is fully consistent with *Planck*, but it is also consistent with the earlier analyses that were below *Planck*.

DES Year 3 analyses should yield substantially tighter statistical errors *and* better control and understanding of systematic uncertainties. They will also yield the first strong cosmological results from DES cluster weak lensing, which should have power comparable to cosmic shear and galaxy–galaxy lensing and systematics that are at least partly independent. Analyses of the expanding KiDS and HSC data sets will independently probe low redshift matter clustering, again with precision better than that of any measurements that exist today. The final cosmological analyses from the *Planck* team may slightly shrink the error bars on predicted matter clustering; more importantly they may, or may not, shift best-fit parameter values as a consequence of more complete characterization of instrumental effects. Continuing analyses of ongoing CMB surveys from the South Pole Telescope and the Atacama Cosmology Telescope will provide partial checks on the *Planck* results. Within a year, therefore, the mild but long-simmering tension over matter clustering could dissipate, or it could solidify into the central issue of contemporary cosmology.

The other prominent tension in cosmology today is between the Hubble constant $H_0 = 67.8 \pm 0.9$ km s^{-1} Mpc^{-1} inferred from *Planck*, assuming ΛCDM, and many of the direct H_0 determinations built on the local distance ladder, for example, $H_0 = 73.24 \pm 1.74$ km s^{-1} Mpc^{-1} by Riess et al. (2016), where the quoted error bar includes a detailed accounting of the identified systematic uncertainties. While the CMB prediction depends on the specific assumptions of flatness and a cosmological constant, the combination of CMB, BAO and supernova data leads to a 'cosmological' value of H_0 that is less model dependent. For example, Aubourg et al. (2015) combine these data sets into an 'inverse distance ladder:' the CMB calibrates the BAO sound horizon in absolute units, BOSS measures the BAO distance scale to ∼ 1% precision at $z = 0.32$ and $z = 0.57$, and supernovae provide a high precision *relative* distance scale to transfer this measurement to $z = 0$. Their analysis yields $H_0 = 67.3 \pm 1.1$ km s^{-1} Mpc^{-1} even with very flexible assumptions about dark energy. And other CMB+BAO+supernova analyses obtain similar central values and error bars for loosely parameterized dark energy models. If the measurements driving this conflict are themselves correct, then reconciling the inverse and direct distance ladder values requires changing *pre-recombination* physics in a way that alters the sound horizon and thus rescales the BAO distances, for example, by adding a neutrino species or dark energy that is dynamically significant at early times. However, these possibilities are increasingly well constrained by the CMB power spectrum itself, which is well reproduced assuming a standard recombination history.

Will either of these tensions prove to be a harbinger of new physics? A conservative guess is that both will go away as the data improve and systematic uncertainties are more fully understood. Still, it is interesting to look at three historical analogies. In the late 1980s and early 1990s, there were numerous indications of 'excess large-scale power' relative to the predictions of $\Omega = 1$ CDM: the cluster correlation function, peculiar velocity flows, the angular correlation function of the APM galaxy survey and the galaxy power spectrum of the CfA2 redshift survey. There were (reasonable) concerns about observational systematics and speculative explanations based on galaxy formation physics, but here the eventual resolution was a low-matter density (implying either dark energy or an open universe). The mid-1990s gave rise to the 'age crisis', globular clusters older than the inferred age of the universe. Systematic errors in many estimates of H_0 and cluster ages exacerbated the apparent severity of this conflict, but at least in hindsight we can see that the resolution lay partly in cosmic acceleration, which gives $t_0 \approx 1/H_0$ instead of the smaller age of a matter-dominated low-Ω universe. In 2003 to 2005, halo-based clustering models with WMAP1 cosmological parameters ($\sigma_8 \approx 0.9$, $\Omega_m \approx 0.3$) over-predicted the mass-to-light ratios of galaxy clusters (van den Bosch et al., 2003; Tinker et al., 2005; Vale and Ostriker, 2006). Even as one of those who was highlighting this conflict, I thought the most likely explanation was that our model of non-linear galaxy bias was insufficiently accurate. However, the source of the discrepancy turned out to be systematic errors in CMB polarization foreground corrections, which affected the WMAP1 parameters and went away in WMAP3. The galaxy clustering analyses got it right.

There is no consistent message in these analogies, but they show that observational discrepancies sometimes do point to new physics even when the systematics might be large enough to explain them, and that when systematics *are* driving the tension it can be difficult to guess which ones are at fault.

27.3 Future

Even as Year 3 analyses are underway, the final six-year DES data set is nearly in hand. This timing seems consistent with that of other recent experiments: it typically takes two to three years to go from a completed data set to a high-quality cosmological analysis. In terms of data volume, the growth from Year 3 to Year 6 is a smaller factor than that from Year 1 to Year 3. However, there are many ideas in the pipeline for ways to mitigate limiting systematics, and to more fully exploit DES data by modelling non-linear scales and by optimally combining information from complementary probes. It is reasonable to expect, therefore, that the final DES analyses will yield a factor of two to three better precision on most of the fundamental measurements *and* much better control and tests of systematic effects, relative to Year 3 results. As theorists are always keen to point out, modelling improvements that shrink error bars by a factor of two have the same impact as extending a six-year survey to 20 years!

If there is indeed a several percent discrepancy between ΛCDM predictions and the true level of matter clustering in the low redshift universe, then the final results from DES and corroborating analysis from KiDS and HSC will establish the conflict definitively. Alternatively, these analyses could remove the current tension and reveal a new one at the $\sim 1\%$ level, again with $2 - 3\sigma$ significance, or they could demonstrate excellent agreement between the observed expansion history and the GR-predicted history of matter clustering. Improvements in expansion history measurements from DES supernovae, BAO and weak lensing could reveal percent-level discrepancies with ΛCDM, or they could yield tighter constraints around $w = -1$.

The path to establishing or removing the H_0 tension is less clear. *Hubble* and eventually *James Webb Space Telescope* measurements of Cepheid distances to more SNIa host galaxies will tighten what is currently the loosest rung of the distance ladder (see the elegant Figure 10 of Riess et al., 2016). *Gaia* parallaxes of Galactic Cepheids will improve the statistics and mitigate the systematics of the lowest rung, the calibration of the Cepheid period-luminosity relation. Larger samples of gravitational lens time delays, which currently yield $H_0 = 71.9^{+2.4}_{-3.1}$ km s^{-1} Mpc^{-1} (Bonvin et al., 2017), should achieve enough statistical precision to provide an independent check of the distance ladder results. In the long term, standard siren distances to merging neutron stars could provide an alternative route to H_0, but reaching percent-level accuracy will require thorough understanding of selection biases in the systems for which electromagnetic counterparts enable redshift measurements.

Table 27.1 summarizes a selection of current and planned dark energy experiments, taken from the Snowmass 2013 Dark Energy Facilities review, Weinberg et al. (2013a). The 2020s will see the advent of five extremely powerful dark energy facilities ('Stage IV' in the parlance of the Dark Energy Task Force, see Table 27.1): the Dark Energy Spectro-scopic Instrument (DESI), the Subaru Prime Focus Spectrograph (PFS), the Large Synoptic Survey Telescope (LSST, recently renamed the 'Vera Rubin Observatory's Legacy Survey of Space and Time'), the European Space Agency's *Euclid* mission and NASA's Wide Field Infrared Survey Telescope (WFIRST, recently renamed 'Nancy Grace Roman Space Tele-scope'). Roughly speaking, Stage III experiments such as DES and BOSS can measure observables such as the distance scale, expansion rate and amplitude of matter fluctuations with aggregate precision of about 1%. Stage IV experiments aim to do an order-of-magnitude better, with larger data volumes, higher angular resolution and wider redshift range. Deliv-ering on this goal will require corresponding improvements in control of systematics.

DESI is an ambitious successor to BOSS (with SDSS-IV's eBOSS bridging the two). Instead of a 1000-fibre spectrograph on a 2.5-metre telescope, DESI employs 5000 robotically positioned fibers on a 4-metre telescope, enabling a spectroscopic survey of thirty million galaxies and the Lyman-α forest absorption towards one million high redshift quasars, over a footprint of 14,000 deg^2. DESI will achieve near cosmic-variance limited BAO measurements out to $z \approx 1$, plus high precision BAO measurements at $z = 1 - 1.5$ from emission-line galaxies and at $z = 2 - 3.5$ from the Lyman-α forest. For the first time, DESI will allow redshift-space distortion measurements of structure growth at a level competitive with current and future weak lensing measurements. This matched capability is a major asset for testing modified gravity theories, as these generically predict differences between the gravitational potentials that govern light-bending and those that govern the acceleration of non-relativistic tracers such as galaxies. More generally, the combination of weak lensing and redshift-space galaxy clustering allows a much richer set of tests for the behaviour of gravity on cosmological scales. The Subaru PFS, with its 2400 fibers, has similar goals for cosmological studies over 1500 deg^2 using galaxies and Lyman-α forest over the redshift ranges specified in Table 27.1.

Table 27.1 A selection of major dark energy experiments.

Project	Date	Area/deg^2	Data	Spec-z Range	Method
BOSS	2008–2014	10,000	Opt-S	0.3–0.7 (gals)	BAO/RSD
DES	2013–2018	5000	Opt-I		WL/CL SN/BAO
eBOSS	2014–2020	7500	Opt-S	0.6–2.9 (gal/QSO)	BAO/RSD
SuMIRE	2014–2024	1500	Opt-I		WL/CL
HETDEX	2014–2019	300	Opt-S	$1.9 < z < 3.5$ (gals)	BAO/RSD
DESI	2019–2024	14,000	Opt-S	0–0.1.7 (gals)	WL/CL
				2–3.5 (LyαF)	SN/BAO
PFS	2019–2025	1500	Opt-S	0.7–1.7 (gals)	BAO/ RSD+WL
				2.5–3.5 (LyαF)	
				2.7–7.0 (gals)	
LSST	2020–2030	20,000	Opt-I		WL/CL SN/BAO
Euclid	2020–2026	15,000	Opt-I		WL/CL
			NIR-S	0.7–2.2 (gals)	BAO/RSD

Table 27.1 (*Continued*)

Project	Date	Area/deg²	Data	Spec-z Range	Method
WFIRST	2024–2030	2,200	NIR-I		WL/CL/SN
			NIR-S	1.0–3.0 (gals)	BAO/RSD

Notes: Abbreviations in the 'Data' column refer to optical (Opt) or near-infrared (NIR) imaging (I) or spectroscopy (S). For spectroscopic experiments, the 'Spec-z' column lists the primary redshift rangefor galaxies (gals), quasars (QSOs) or the Lyman-αforest (LyαF). Abbreviations inthe 'Methods' column are weak lensing (WL), clusters (CL), supernovae (SN), baryonacoustic oscillations (BAO), and redshift-space distortions (RSD).
Source: Table and caption have been reprinted from Weinberg and White (2018) based on data from Weinberg et al. (2013a).

Although they were conceived independently, as a dark energy experiment, LSST is the natural successor to DES. (In GRE-speak, LSST:DES :: DESI:BOSS.) With an 8.4-m telescope, 3.2-gigapixel camera and ten-year survey duration (vs. 4-m, 0.57-gigapixel, six-year for DES), LSST aims to measure $\sim$ 2 billion weak lensing shapes over 20,000 deg^2 (vs. $\sim$ 0.2 billion over 5000 deg^2 for DES). LSST will also measure multi-band light curves of many thousands of Type Ia supernovae per year, and the limits of its supernova cosmology will be set by how many of these can be followed spectroscopically and by how much can be achieved with photometry alone. Many of the lessons being learned now from DES will be directly applicable to LSST and many of the challenges will be similar. There is already substantial overlap between DES and LSST dark energy science teams, and many of those now sweating the details of DES Year 3 cosmology will be sweating the details of mid-survey LSST cosmology a decade down the road.

Euclid and *WFIRST* both seek to exploit the higher angular resolution and greater photometric and astrometric stability achievable from space. Each mission plans a large weak lensing survey, *Euclid* over $\sim$ 15,000 deg^2 in the optical and *WFIRST* over $\sim$ 2000 deg^2 in the infrared. I would characterize *Euclid*'s approach as 'optimized for statistics' and *WFIRST*'s as 'optimized for systematics control,' and there are good reasons to hope that LSST, *Euclid* and *WFIRST* can accomplish much more in combination than any one of them could on its own. *Euclid* and *WFIRST* will also carry out infrared slit-less spectroscopic surveys to measure BAO and redshift-space distortions from tens of millions of Hα emission-line galaxies at

$z = 1 - 2$. The relative power of DESI, *Euclid* and *WFIRST* surveys in this redshift range depends on still-uncertain performance forecasts and on the ability of non-linear modelling to exploit the higher density (but factor ~ 7 smaller sky area) of the *WFIRST* sample. The third pillar of *WFIRST*'s dark energy programme is a supernova survey; relative to LSST, the number of supernovae will be smaller, but infrared sensitivity and space-based angular resolution and calibration stability will allow *WFIRST* to probe to higher redshift and achieve tighter systematics control.

All four of these Stage IV facilities will produce spectacular data sets that enable a wide range of science, from denizens of the outer solar system to galaxies and quasars at the epoch of reionization. Their surveys will continue the centuries-long human quest to map the universe in ever growing detail over an ever larger fraction of its observable volume. What will they teach us about cosmic acceleration? The answer depends on our ingenuity in exploiting them and, above all, on what nature has hidden behind the veil of 1% precision.

Dark energy research in the 2020s will be shaped heavily by what DES and its Stage III brethren discover over the next couple of years. If they confirm one or both of the current conflicts with ΛCDM at high significance, then Stage IV efforts will focus on mapping out these discrepancies, quantifying their phenomenology with multiple probes over a range of redshift to get better constraints on their physical origin. If Stage III experiments instead suggest mild ($2-3\sigma$) discrepancies with ΛCDM, then Stage IV experiments can confirm or refute them fairly quickly because they acquire data so rapidly. Percent-level deviations are still large enough to be mapped in detail by the collection of Stage IV experiments. Finally, if we emerge from Stage III in excellent agreement with ΛCDM, then there is still room for the Stage IV experiments to uncover a deviation and confirm it at high significance, but they will have to work well, and we will have to be lucky.

What if we get to the 2030s with no clear evidence against a cosmological constant? Some physicists consider a cosmological constant to be natural and the value of it to be not especially puzzling, just a quantity one has to measure. They may consider the problem solved. However, I regard current explanations for the tiny magnitude of Λ relative to the obvious particle physics scales to be sufficiently unconvincing that I assign Λ only a moderate bump in the Bayesian prior of possible dark energy models. In this view, unfortunately, measuring $w = -1.00 \pm 0.01$ tells us that $w = -1.00 \pm 0.01$ and nothing more, since deviations could lurk anywhere in $\log|1 + w|$. While there are numerous ideas about how to improve on the planned Stage IV experiments, there may not be strategies that can achieve order-of-magnitude further improvements at reasonable cost. In Weinberg et al. (2013b), the most promising ideas we could identify were radio weak lensing surveys that use HI (neutral atomic hydrogen) velocity fields to drastically reduce shape noise or a futuristic space-based gravitational wave mission that could measure hundreds of thousands of standard siren distances to white dwarf binaries in other galaxies (Cutler and Holz, 2009).

It is worth remembering that hard cosmological problems can take many decades to solve, and that the resolution can come from unexpected directions. It was six decades from the first evidence for dark matter to convincing demonstrations that it is non-baryonic, and we now search for it using that methods that were not conceived until the 1970s to 1990s. Understanding the anomalous precession of Mercury also took sixty years, and there the crucial development was a theoretical breakthrough driven by entirely different considerations about the speed of light and its implications for the measurement of space and time. We can only hope that the explanation of cosmic acceleration, when it comes, is similarly convincing, and similarly profound.

References

Albrecht, A., Bernstein, G., Cahn, R., Freedman, W.L., Hewitt, J. et al. (2006). Report of the Dark Energy Task Force. arXiv:astro-ph/0609591.

Aubourg, É., Bailey, S., Bautista, J. E., Beutler, F., Bhardwaj, V., et al. (2015). Cosmological implications of baryon acoustic oscillation measurements, *Physical Review D* **92**, 12, p. 123516, doi:10.1103/PhysRevD.92.123516.

Bonvin, V., Courbin, F., Suyu, S. H., Marshall, P. J., Rusu, C. E., et al. (2017). H0LiCOW - V. New COSMOGRAIL time delays of HE 0435-1223: H_0 to 3.8 per cent precision from strong lensing in a at ΛCDM model, *Monthly Notices of the Royal Astronomical Society* **465**, pp. 4914–4930, doi:10.1093/mnras/stw3006.

Cutler, C. and Holz, D. E. (2009). Ultrahigh precision cosmology from gravitationalwaves, *Physical Review D* **80**, 10, pp. 104009–+, doi:10.1103/PhysRevD.80.104009.

Dark Energy Survey Collaboration (2018). Dark Energy Survey Year 1 results: Cosmological constraints from galaxy clustering and weak lensing, Physical Review **D 98**, p. 043526.

Eisenstein, D. J., Weinberg, D. H., Agol, E., Aihara, H., Allende Prieto, C., et al. (2011). SDSS-III: Massive spectroscopic surveys of the distant universe, the Milky Way, and Extra-Solar Planetary Systems, *The Astronomical Journal* **142**, pp. 72, doi:10.1088/0004-6256/142/3/72.

Hildebrandt, H., Viola, M., Heymans, C., Joudaki, S., Kuijken, K., et al. (2017). KiDS-450: Cosmological parameter constraints from tomographic weak gravitational lensing, *Monthly Notices of the Royal Astronomical Society* **465**, pp. 1454–1498, doi:10.1093/mnras/stw2805.

Joudaki, S., Blake, C., Johnson, A., Amon, A., Asgari, M., et al. (2018). KiDS-450 + 2dFLenS: Cosmological parameter constraints from weak gravitational lensing tomography and overlapping redshift-space galaxy clustering, *Monthly Notices of the Royal Astronomical Society* **474**, pp. 4894–4924, doi:10.1093/mnras/stx2820.

Leauthaud, A., Saito, S., Hilbert, S., Barreira, A., More, S., et al. (2017). Lensing is low: cosmology, galaxy formation or new physics? *Monthly Notices of the Royal Astronomical Society* **467**, pp. 3024–3047, doi:10.1093/mnras/stx258.

McEwen, J. E., and Weinberg, D. H.(2018). The effects of assembly bias on the inference of matter clustering from galaxy-galaxy lensing and galaxy clustering, *Monthly Notices of the Royal Astronomical Society* **477**, pp. 4348–4361, doi:10.1093/mnras/sty882.

Riess, A. G.,Macri, L. M., Hoffmann, S. L., Scolnic, D., Casertano, S., et al. (2016). A 2.4% determination of the local value of the hubble constant, *The Astrophysical Journal* **826**, 56, doi:10.3847/0004-637X/826/1/56.

Tinker, J. L., Weinberg, D. H., Zheng, Z., and Zehavi, I. (2005). On the mass-to-light ratio of large-scale structure, *The Astrophysical Journal* **631**, pp. 41–58, doi:10.1086/432084.

Vale, A., and Ostriker, J. P. (2006). The non-parametric model for linking galaxy-luminosity with halo/subhalo mass, *Monthly Notices of the Royal Astronomical Society* **371**, pp. 1173–1187, doi:10.1111/j.1365-2966.2006.10605.x.

van den Bosch, F. C., Mo, H. J., and Yang, X. (2003). Towards cosmological concordance on galactic scales, *Monthly Notices of the Royal Astronomical Society* **345**, pp. 923–938, doi:10.1046/j.1365-8711.2003.07012.x.

van Uitert, E., Joachimi, B., Joudaki, S., Amon, A., Heymans, C., et al. (2018). KiDS+GAMA: Cosmology constraints from a joint analysis of cosmic shear, galaxy-galaxy lensing, and angular clustering, *Monthly Notices of the Royal Astronomical Society* **476**, pp. 4662–4689, doi:10.1093/mnras/sty551.

Weinberg, D.,Bard, D., Dawson, K., Dore, O., Frieman, J., et al. (2013a). Facilities for Dark Energy Investigations, *ArXiv e-prints, 1309.5380.*

Weinberg, D. H., Mortonson, M. J., Eisenstein, D. J., Hirata, C., Riess, A. G., et al. (2013b). Observational probes of cosmic acceleration, *Physics Reports* **530**, pp. 87–255, doi:10.1016/j.physrep.2013.05.001.

Weinberg, D., and White, M. (2018). Dark Energy, *Review of Particle Physics.*

Wibking, B. D.,Salcedo, A. N., Weinberg, D. H., Garrison, L. H., Ferrer, D., et al. (2019). Emulating galaxy clustering and galaxy-galaxy lensing into the deeply nonlinear regime: Methodology, information, and forecasts, *Monthly Notices of the Royal Astronomical Society,* **484**, pp. 989–1006 doi:10.1093/mnras/sty2258.

Yoo, J., Tinker, J. L., Weinberg, D. H., Zheng, Z., Katz, N., et al. (2006). From galaxy-galaxy lensing to cosmological parameters, *The Astrophysical Journal* **652**, pp. 26–42, doi:10.1086/507591.

Zu, Y., Weinberg, D. H., Rozo, E., Sheldon, E. S., Tinker, J. L., et al. (2014). Cosmological constraints from the large-scale weak lensing of SDSS MaxBCGclusters, *Monthly Notices of the Royal Astronomical Society* **439**, pp. 1628–1647, doi:10.1093/mnras/stu033.

The US Department of Energy Approval Process

The process begins with Critical Decision 0, CD-0, approved in November 2005. This established, within the Department of Energy (DOE) Office of High Energy Physics, the mission need for a Ground-Based Dark Energy Experiment. DOE was considering three concepts for a Ground-Based Dark Energy Experiment. One was the Dark Energy Survey (DES), another was the Large Synoptic Survey Telescope (LSST) and the third was an early concept for the Dark Energy Spectroscopic Instrument (DESI). This was important because it was the first time that DOE determined that dark energy was part of the DOE mission and that DES and the two other alternatives were appropriate for the DOE Mission of High Energy Physics. The three alternatives required a partnership with NSF, and each alternative would either use an existing NSF telescope or NSF would build a new telescope. This was the first time that DOE partnered with NSF Division of Astronomy and it required a lot of work. There was a fourth alternative, which was to do nothing. CD-0 Fermilab was authorized to spend some R&D money on DECam, which did not count towards the Total Project Cost (TPC).

CD-1: Approve Alternative Selection and Cost Range. DOE selected the Dark Energy Survey Experiment and CD-1 approval was granted on 10 October 2007. DECam was allowed to spend R&D money that was appropriate for DECam. This was included in other project costs, which were part of the Total Project Cost (TPC). The amount of money was small but critical. The Head of DOE High Energy Physics and the Fermilab Director negotiated this amount. The 2006 P5 Report, which was based on the spring and summer meetings of P5 in 2005, and a subsequent HEPAP meeting in late 2005 or early 2006, determined that DES was ready for Major Item of Equipment (MIE) funding in FY2006. It determined that LSST and DESI were not mature enough for MIE funding. LSST did not have a mature cost estimate and the DOE-NSF partnership had not been worked out. Note: October 2007 is the first month in FY2008. At this point there was a preliminary cost estimate for DES of $25 million and a spending plan. This was submitted as part of the FY2008 Congressional Budget Request (DOE). LSST and DESI were not rejected, they were ultimately put in the DOE long range plan and the NSF long range plan. LSST construction began in 2014 and full operations are estimated to begin in 2022. DESI received approval and construction began in 2015. The R&D line for DES in CD-1 shows the expected R&D funding for 2007 and 2008. These are preliminary cost estimates and the final cost estimate is not set until the completion of CD-2 and CD-3a reviews, which are largely cost and organization reviews. This decision was reported to the US Government's Office of Management and Budget (OMB) and the top level of DOE. The LSST consortium continued to work on LSST and developed a complete proposal, with the roles of the partners clearly defined. At this point, DESI was still just a concept.

CD-2: Approve the Performance Baseline of the Dark Energy Camera Project (for the Dark Energy Survey Experiment) at the Fermi National Accelerator Laboratory. This document is a consequence of the intensive CD-2 and CD-3a reviews, which were held on 29–31 January 2008 by the DOE Office of Science. This has a final Total Estimated Cost (TEC), which is the MIE cost, and a final TPC (includes the MIE and R&D costs) and a final schedule, subject to Congressional Approval. The figure $35.1 million appears for the first time. CD-2 was approved on 29 April 2008.

CD-3a: Approve limited construction for the Dark Energy Camera Project (for the Dark Energy Survey Experiment) at the Fermi National Accelerator Laboratory. This is for long-range procurements that need to be made quickly in order to meet the approved schedule. This was approved on 20 May 2008.

CD-3b: Approve full construction of the Dark Energy Camera Project (for the Dark Energy Survey Experiment) at the Fermi National Accelerator Laboratory. This approved the full MIE cost and authorized Fermilab to spend the money. The total amount is $35.1 M. CD-3b approval also required a signed Memorandum of Understanding that defined the responsibilities of Fermilab, NCSA and NOAO. The Directors of Fermilab, NCSA and NOAO signed the Understanding. The Understanding provides the details of how the Dark Energy Experiment will be executed. The legal authorities of Fermilab, NOAO and NCSA signed the Memorandum of Understanding and delegated the actual execution of DES to the three Directors. CD-3b was approved on 24 October 2008.

There is even a CD-4 that approves the start of operations and allows Fermilab to spend operations money on the execution of DES. This was approved by DOE in November 2012.

The Dark Energy Survey Collaboration Meeting Group Photos

Since 2004, the Dark Energy Survey (DES) collaboration has met twice a year in various venues around the world, including Barcelona, Berkeley, Cambridge, Chicago, Philadelphia, Portsmouth, La Serena (Chile), London, Madrid, Michigan, Munich, Ohio, Rio de Janeiro, Stanford and Sussex. We include here group photos from some of those meetings.

Fig. B.1 The inaugural ceremony of DES at the Blanco Telescope, Chile, November 2012. Photo credit: Judy Goldhill.

Fig. B.2 Barcelona, Spain, September 2013.

Fig. B.3 University of Michigan, US, May 2015.

Fig. B.4 University of Chicago, US, June 2017.

Fig. B.5 University of Queensland, Brisbane, Australia, November 2017.

Fig. B.6 Rio, Brazil, December 2018.

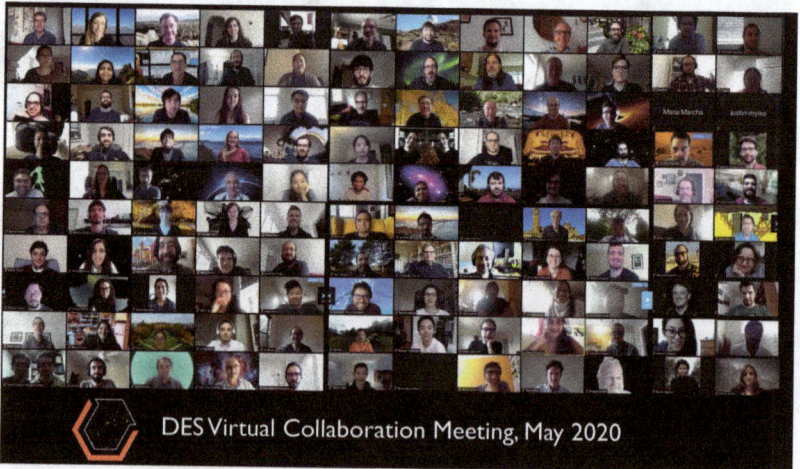

Fig. B.7 Virtual DES collaboration meeting, May 2020.

Index